RADIO AND LINE TRANSMISSION

VOLUME 3

TELECOMMUNICATION TECHNICIANS' SERIES

RADIO AND LINE TRANSMISSION

VOLUME 3
Radio Communication

G. L. DANIELSON
M.SC. TECH., B.SC., C.ENG., M.I.E.E.
Formerly Head of Department of Telecommunication and Electronics, Norwood Technical College

R. S. WALKER
C.ENG., M.I.E.R.E.
Senior Lecturer, South London College, formerly Norwood Technical College

NEWNES–BUTTERWORTHS
LONDON BOSTON
Sydney · Wellington · Durban · Toronto

The Butterworth Group

United Kingdom	**Butterworth & Co (Publishers) Ltd** London: 88 Kingsway, WC2B 6AB
Australia	**Butterworths Pty Ltd** Sydney: 586 Pacific Highway, Chatswood, NSW 2067 Also at Melbourne, Brisbane, Adelaide and Perth
Canada	**Butterworth & Co (Canada) Ltd** Toronto: 2265 Midland Avenue, Scarborough, Ontario M1P 4S1
New Zealand	**Butterworths of New Zealand Ltd** Wellington: T & W Young Building 77–85 Customhouse Quay, 1, CPO Box 472
South Africa	**Butterworth & Co (South Africa) (Pty) Ltd** Durban: 152–154 Gale Street
USA	**Butterworth (Publishers) Inc** Boston: 19 Cummings Park, Woburn, Mass. 01801

First published 1969 by Iliffe Books Ltd
Revised reprint 1972
Reprinted 1975 by Newnes-Butterworths
Reprinted 1977, 1978

ISBN 0 592 05969 3

Reproduced and printed by photolithography and bound in Great Britain by J. W. Arrowsmith Ltd, Bristol BS3 2NT

PREFACE

Following RADIO AND LINE TRANSMISSION VOLUMES 1 AND 2, by the same authors, the present book concentrates on radio communication. For the most part a knowledge of the work covered in the earlier volumes is assumed. A brief recapitulation of previous matter is included, however, where it is thought advantageous.

This volume covers the syllabus of the City and Guilds of London Technicians' Certificate examination in 'Radio Communication C' and should be of interest and help to students who have completed a one or two year full-time course, or a three or four year part-time course in telecommunications and radio.

Symbols and nomenclature are in accordance with the recommendations of the British Standards Institution. The S.I. system of units is used throughout.

There are worked examples in the text and questions are provided at the end of each chapter. It is suggested that students should use these to test the efficiency of their study. Questions should be returned to at intervals to check the amount of knowledge which has been retained.

A number of questions are taken by kind permission from past examination papers of the City and Guilds of London Institute. The Institute, of course, bears no responsibility for any worked answers.

The serious student is recommended to maintain an adequate notebook. Such notes are essential for revision and the effort demanded by their making maintains concentration and increases learning efficiency.

Acknowledgement is gratefully made to several companies who generously provided information about their products: Eddystone Radio, The General Electric Co., The Marconi Co., Pye Telecommunications, and Standard Telephones and Cables.

G.L.D.
R.S.W.

CONTENTS

1

Propagation

1.1. INTRODUCTION

Radio waves may travel from a transmitter aerial to places where they induce e.m.f.s in receiving aerials by one or more of several methods. It is assumed that the reader has already studied at an elementary level most of the methods which are listed below. This chapter will add some detail to methods of propagation applicable to frequencies up to 300 MHz.

1.1.1. Methods of propagation

Ground wave. The signal travels round the surface of the earth by a process of diffraction.

Ionosphere. The signal reaches the receiving point after being returned to earth by an ionised layer between 45 km and 400 km high.

Reflection or re-radiation from man-made satellites. These would be moving through space along with the earth but fixed in position relative to the earth.

Direct wave propagation. The radio wave travels directly from the transmitting to the receiving aerial without earth loss and without the need of reflection or refraction by a transmitting medium.

Earth reflection. At short ranges, v.h.f. waves are reflected by a point on the earth between the transmitter and the receiver.

Troposphere refraction and scattering. Waves are returned to earth by refraction due to variation in the refractive index of the earth's atmosphere between sea level and a height of approximately 12 km.

1.2. GROUND WAVE

Were it possible to radiate a ground or surface wave over a terrain which absorbed no power from the wave, then the r.m.s. electric field strength of a vertically polarised wave would be given by:

$$E = \frac{\sqrt{(30\,PG)}}{d} \qquad \text{V/m} \qquad (1.1)$$

where P is the radiated power in watts, G is the gain of the aerial as compared with an isotropic radiator, and d is the distance in metres from the aerial of the place for which the field strength is calculated.

The field strength falls off inversely as the distance which the wave has travelled, because the power of the wavefront is being dispersed over a wider area. However, earth losses make it necessary to modify Equation 1.1 by multiplying it by a factor known as the *Sommerfeld reduction factor.* The factor depends in a complex manner on frequency, earth conductivity and earth permittivity.

The surface wave travels both over and within the earth. Each unit cube of earth through which the alternating fields of the radio wave pass can be represented as a resistance shunted by a capacitance. At v.l.f. and l.f. the reactance of the capacitance is very large compared with the resistance so that at these frequencies the relative permittivity

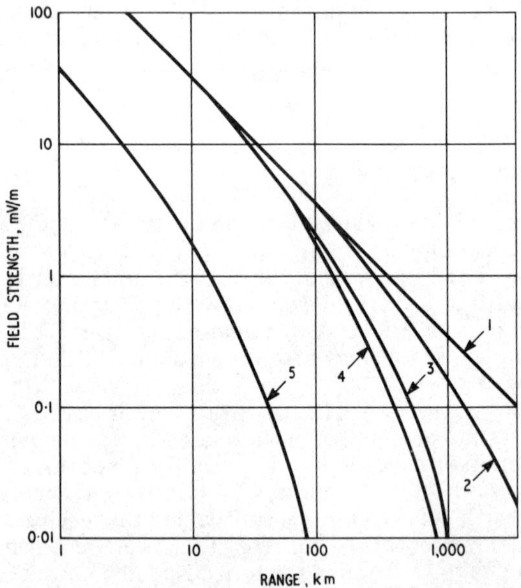

Fig. 1.1 Graphs of field strength against range

of the earth is of little importance. Effective ranges of transmitters radiating frequencies below about 1 500 kHz depend mainly on transmitter power, frequency and earth conductance. Higher frequencies are partially decoupled from the earth resistance by the earth capacitance, the reactance of which falls as the frequency is increased. At a given frequency, dielectric hysteresis loss is less for an earth with a large relative permittivity than for an earth of smaller relative permittivity.

Fig. 1.1 shows how the field strength measured in millivolts per metre decreases with distance from a transmitter radiating a power of

1 kW from a vertical dipole aerial. The numbered curves have the following applications:

1. Applies to propagation over an earth of zero loss.
2. Applies to l.f. propagation over the sea.
3. Applies to l.f. passing over an earth of poor conductivity.
4. Applies to m.f. propagation over the sea.
5. Applies to m.f. passing over an earth of poor conductivity.

Two important features are revealed by the curves. The first is that a transmitter operating over the sea has a longer range for the same power than one whose transmission path is overland, and the second is the even greater difference of range over the two types of terrain which obtains for m.f.

A conductance figure of between 4 and 5 S/m is typical for sea water. This compares with figure of between 1 and 2 mS/m for land of poor conductivity.

In the v.l.f. band the attenuation of the surface wave is even smaller than in the l.f. band. The frequency band 10–14 kHz is, therefore, reserved for long distance radio navigation and radio location purposes, and the band 14–19·95 kHz for long range maritime telegraphy transmissions. A transmitting station typical of those operating in the v.l.f. bands is the long range telegraphy station completed at Anthorn on Solway Firth in 1964, which is capable of radiating with a power of 50 kW on a frequency of 16 kHz or with a power of 100 kW on a frequency of 20 kHz. The purpose of this station is to communicate with naval vessels at sea and at long distances from their bases. The size of the station is evidenced by the circular area of radius 1 340 m covered by its six multi-cable rhombic aerials and the 13 600 kg of copper wire used in its buried earth system. The power fed into the aerial, as distinct from the radiated power, is 550 kW. Such immense projects are considered necessary to provide unbroken communication facilities which are not subject to the fade-out conditions which are experienced at times on long distance h.f. sky-wave circuits.

1.3. IONOSPHERE

The behaviour of sky-waves depends on the heights and densities of the layers of ionised gases which comprise the ionosphere. The gases which form the earth's atmosphere are in the main ionised by ultra-violet radiation reaching the earth from the sun. The energy from this extremely high-frequency radiation is able to release electrons from gas molecules so that they become positive gas ions while the released electrons are free negative charges in space. The number of electrons and ions in a unit volume, that is the density of ionisation, depends directly on the intensity of the ultra-violet radiation, the number of gas molecules per unit volume available for ionisation, and inversely on the rate at which electrons and ions come into collision and thus recombine.

Energy is absorbed from the ultra-violet radiation as it moves further into the gases of the atmosphere and releases electrons from their parent molecules. This factor considered alone suggests a decreasing density of ions towards the earth's surface. However, the gas density is greater nearer to the earth's surface so that there is an optimum height for ionisation. The rate at which electrons and ions recombine increases towards the earth's surface, since with increasing gas pressure the ions and electrons are closer together and more liable to come into collision. For this reason the ionisation density falls off more rapidly below the maximum density height than it does above it.

The formation of two or more layers may be accounted for by the presence of ultra-violet radiation from the sun on two discrete frequency bands. A different optimum height for ionisation obtains for each of these bands. Variation of the mixture of gases in the earth's atmosphere is a contributory factor. At great heights, where no wind blows to stir the gases into uniform mixture, the lighter gases may float outwards leaving heavier molecules in greater concentration lower down. Because

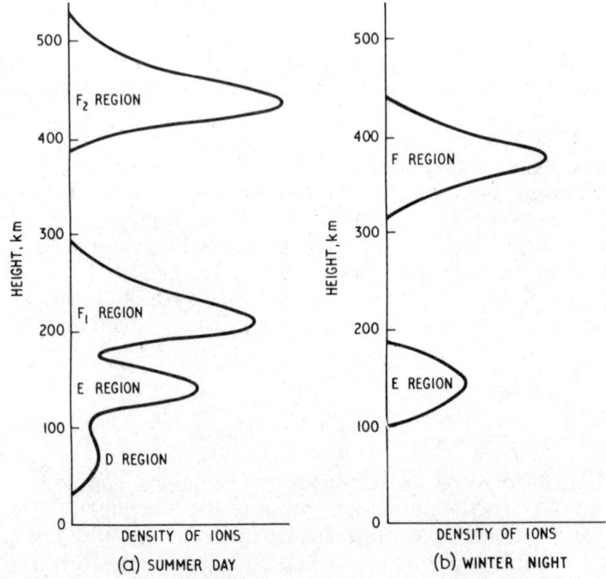

Fig. 1.2 Distribution of layers in ionosphere

different gases require different field energies for ionisation, this could lead to the formation of distinct layers of ionisation.

Over 100 research stations distributed over the world take radio soundings of the ionosphere. Collection and comparison of their findings gives information about layer heights and probable best frequencies for long distance h.f. communication in the month ahead. Fig. 1.2 (a) and (b) indicate a possible pattern of variation of ionisation density with height for two widely different circumstances: a summer day when the effect of the sun is at its greatest, and a winter night when the effect of the sun is most withdrawn. The D layer is not very marked and exists only during the day. The chief effect of this layer is to attenuate waves which pass through it. The degree of refraction is very small. The E layer, more strongly ionised during daylight than darkness and of greater average density in summer than in winter, is able to reflect l.f. and m.f. signals to earth. This gives increased range at night for transmitters using such frequencies. For broadcast stations this is usually a disadvantage as it results in one station interferring within the service area of another.

In daylight two distinct F layers, F_1 and F_2 form. These merge at night to form an F layer between 50 and 100 km above the height of the F_1 daylight layer. The F_2 layer does not follow very closely the changes in intensity of sunlight. It has been suggested that particle radiation from outer stars may contribute to its ionisation. A further factor at this high altitude is the expansion of gases with a rise of temperature and their contraction on cooling. Expansion in summer can result in a reduced density of ionisation even though the total number of molecules ionised increases. Contraction at nightfall may increase the density of the gas ions despite the reduced rate at which ions are formed. Variation of ionisation density in the F layers is of commercial importance as it governs the frequencies available for long distance h.f. radio traffic.

1.4. ATTENUATION OF SKY-WAVES

The alternating electric field of a radio wave passing through the ionosphere gives up energy in setting the free electrons in the ionised gases in a state of oscillation. Providing the oscillating electrons do not come into collision with positive gas ions, they re-radiate their energy so that the wave does not sustain a loss. Some of the oscillating electrons collide with positive gas ions. The energy of these electrons is lost as heat, and attenuation of the radio wave by the extent of this loss results. The loss of energy suffered by a radio wave in the ionosphere increases if the probability of such electron–ion collisions increases. The loss becomes greater if the density of ions increases or if the path length of oscillation of the free electrons becomes longer. The latter factor is inversely proportional to frequency. A lower frequency of electric field exerts its displacing force on the free electrons for a longer period and the oscillating electrons achieve higher velocities and greater displacements with lower frequencies. A higher radiation frequency thus means reduced ionosphere loss.

The earth's magnetic field is a further factor in determining the attenuation of a radio wave in the ionosphere. If the free electrons set in oscillation by the radio wave have some component of velocity at right angles to the earth's magnetic field, they constitute a current alternating in a direction at right angles to the earth's field. A force is thus exerted on them at right angles to their direction of motion and at each instant proportional to their velocity. If the frequency of the wave is high, the oscillating electrons do not attain a very high velocity so that the transverse force due to the earth's field is small. The path of oscillation is a narrow ellipse. A lower frequency of field accelerates the free electrons to a higher velocity and their path of oscillation becomes a longer and wider ellipse.

There is one frequency at which the alternating transverse force due to the earth's magnetic field is equal to the alternating force exerted on the electrons by the electric field of the radio wave. As these two forces are spatially at right angles, the path of the electrons tends to be circular though a forward motion of the electrons in the direction of travel of the wave results in a spiral path. The frequency which produces this condition is called the *gyro frequency*. The value of the gyro frequency varies a little from place to place according to the strength of the earth's magnetic field. An average value is about 1·4 MHz. Frequencies of this order suffer extreme attenuation in the ionosphere and are, therefore, of little use for sky-wave communication.

1.5. REFRACTION OF SKY-WAVES

As a wave progresses along a transmission path its phase at any point along the path relative to its phase at the beginning of the path, lags by an amount dependent upon the velocity of travel, distance travelled and the frequency of the wave. The phase change in one metre of path length is the *phase constant* of the medium through which the wave is travelling. The frequency of the wave is constant and fixed by the frequency of current in the transmitting aerial.

If ω is the angular frequency of the wave in radians per second and β the phase constant in radians per metre, then

$$\omega/\beta = \frac{\text{rad/s}}{\text{rad/m}} = \text{m/s}$$

This ratio of distance to time is called the *phase velocity*. A decrease in the phase constant β of the transmission medium causes an increase in the phase velocity. Neglecting the losses of a transmission path, the phase constant is equal to:

$$\beta = \omega\sqrt{(LC)} \text{ rad/m}$$

where ω = angular frequency in rad/s
L = inductance of the transmission path in H/m
C = capacitance of the transmission path in F/m.

In free space the alternating electric field of a radio wave may be represented by an alternating electric flux. The rate of change of this flux in coulombs per second is called a displacement current. The current can be calculated from the expression:

$$I = 2\,\pi\,f\,V\epsilon_0 \text{ A},$$

where f = wave frequency in Hz.
V = electric field strength in V/m
ϵ_0 = the permittivity of free space is the capacitance between two parallel and opposite surfaces each 1 m^2 in area and separate by a distance of 1 m in *vacuo.*

If electrons are present in the path of the wave and are set oscillating in a path parallel to the electric flux, these oscillating charges constitute a current of negative charge which renders the resultant current less than that of the displacement current alone. The effect is as though the permittivity of free space or the capacitance per unit volume of free

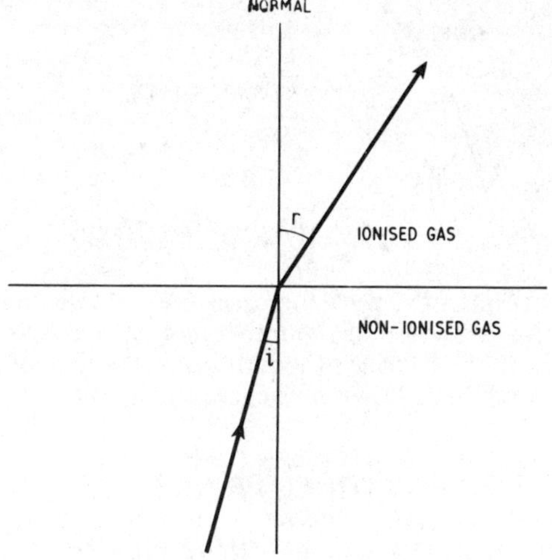

Fig. 1.3 Angles of incidence and refraction

space had been reduced. The phase constant of the transmission path is thereby reduced by the presence of free electrons and the phase velocity is increased.

Whenever an electromagnetic wave passes into a medium and suffers a change of phase velocity refraction may occur. Fig. 1.3 simplifies the situation by assuming that a sharp division exists between a layer of gas which is uniformly ionised and a layer of non-ionised gas beneath it. The arrowed line is the direction of progression of the radio wave. Angle *i* is the angle of incidence. Angle *r* is the angle of refraction.

The ratio $\sin i/\sin r$, is called the refractive index of the two media of transmission. The refractive index can be calculated from:

$$\mu = \sqrt{(1 - 81\, N/f^2)}$$

where N is the electron density in electrons/m^3 and f is the wave frequency in hertz.

This formula indicates that the index is smaller if the density of ionisation increases or if the frequency is reduced. A small refractive index which leads to a sharp refraction, or bending of the wave path,

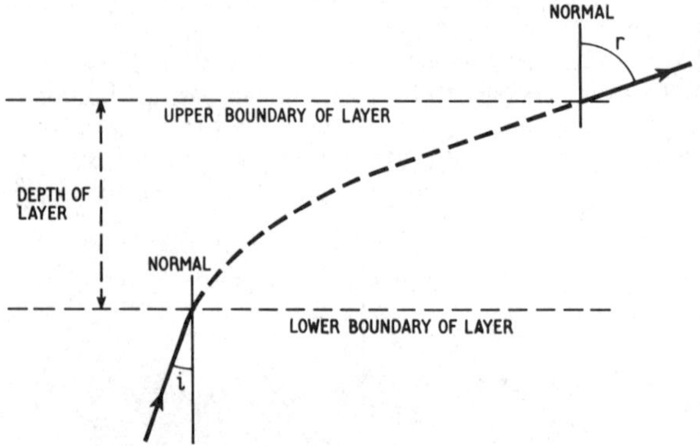

Fig. 1.4 Refraction within an ionised layer

hence the refraction of the wave increases for an increase in ionosphere density, or with a reduction of the frequency radiated. Assuming a gradual increase in the density of ionisation of the ionosphere with height, the path of the radio wave is curved as in Fig. 1.4.

1.6. CRITICAL FREQUENCY

If a radio wave is to be returned to earth by a process of refraction, the density of ionisation must be sufficient to make the angle r in Fig. 1.4 equal to 90°. Sin r is then equal to unity, and

$$\sin i = \sqrt{(1 - 81\, N/f^2)}$$

This gives the sine of the minimum angle to incidence for which the wave is refracted sufficiently to return it to earth from the ionosphere. This minimum angle is thus a function of the electron density and the wave frequency.

The highest frequency which at a given time and place is able to be reflected with a vertical angle of incidence is called the *critical frequency*. The value of the critical frequency can be evaluated if the electron density and the frequency are known by equating sin i to zero. This shows the critical frequency to be:

$$f_c = 9\sqrt{N_{max}}$$

where N_{max} is the maximum density of electrons in the layer.

1.7. MAXIMUM USABLE FREQUENCY

As attenuation is less at higher rather than lower frequenies, the maximum possible usable frequency is generally used for communication. The value of this frequency depends on the height of

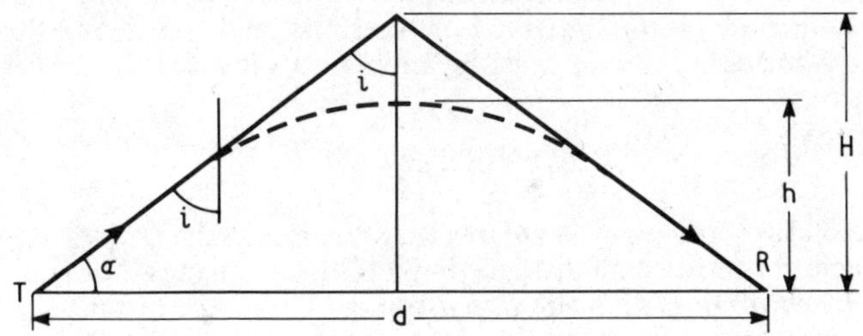

Fig. 1.5 Actual and virtual reflection heights

the ionosphere, the density of ionisation, and the distance between the transmitter and the receiver.

In Fig. 1.5 the dotted line shows the path followed by a radio wave through the ionised gas. The actual height from which the wave returns to earth is h. Yet the effect is as if the reflecting height were H. H is called the *virtual reflection height* and is the height which is measured by radio soundings of the ionosphere using vertically propagated pulses.

The distance d which separates the transmitting and receiving stations is known. Angle α can be found since the tangent of this angle is $2H/d$. The highest frequency for which the geometry of Fig. 1.5 can apply is that for which the equation

$$\sin i = \sqrt{(1 - 81\, N_{max}/f^2)}$$

is true.

Squaring both sides of this equation gives

$$\sin^2 i = 1 - 81\, N_{max}/f^2$$

Or substituting

$$f_c^2 = 81\, N_{\max} \text{ gives } \sin^2 i = 1 - f_c^2/f^2$$

Therefore,

$$\begin{aligned} f_c^2/f^2 &= 1 - \sin^2 i \\ &= \cos^2 i \end{aligned}$$

Rearranging this shows,

$$\begin{aligned} f &= f_c/\cos i \\ &= f_c \sec i \\ &= f_c \sec (90 - \alpha) \end{aligned}$$

This result is referred to as the *secant law,* and sec i as the maximum usable frequency *(m.u.f.) factor.* For transmission distance of 1 000 km or more a correction factor must be applied to allow for the curvature of the earth.

1.8. SKIP DISTANCE

If the radiation frequency is greater than the m.u.f. the sky-waves will first reach the earth some distance beyond the receiving station and the receiver will lie within the *skip distance.* This is the distance between the transmitter and the nearest point to which a sky-wave returns. Skip distances increase if the transmission frequency is raised or if the density ionisation of the reflecting layer decreases. Thus although considerations of minimum attenuation suggest a high frequency for the most efficient sky-wave propagation, the frequency chosen must nonetheless be chosen so that it is low enough for the receiver to lie outside the skip area. To allow for the continuous fluctuation of skip distances which occur while traffic is being sent, the frequency selected for transmission is usually a little lower than the m.u.f. This *optimum traffic frequency* is approximately 85% of the m.u.f. The *lowest usable frequency* for communication between two points depends on the transmitter power available and the prevalent density of ionisation between the transmitter and receiver.

1.9. MULTIPLE HOP TRANSMISSIONS

When distances of the order of 4 000 km or more separate the transmitter and receiver, communication is likely to be by multiple hop propagation in which the signal travels to and from the ionosphere two or more times before reaching the receiver. If the number of hops for a particular transmission path is increased, the angle of elevation at which radiation must leave the transmitter in order to reach the receiving point must be greater. Table 1.1 shows the approximate values of the best angles of elevation at which radiation must leave the transmitter aerial to reach a receiver at a distance of 6 000 km, using different numbers of hops. For the purpose of this table the height of the F layer is assumed to be 300 km.

Table 1.1

Number of hops	2	3	4	5	6
Best elevation	5°	13°	19°	25°	30°

The higher the required angle of elevation the more sharply is the wave refracted at the ionosphere and the lower is the m.u.f. Minimum attentuation, therefore, requires a minimum number of hops and the highest usable frequency. The practice of using a working frequency which is only about 85% of the maximum possible may result in the signal travelling to the receiver by two different paths, e.g. by a two-hop path as well as a three-hop path. Under these conditions severe fading can occur, especially if the signals reaching the receiver by diverse paths arrive with similar amplitudes. Fluctuation of the ionosphere density causes two such signals to drift in and out of phase as the difference in length of the two signal paths varies.

Fading is not the only difficulty. Distortion results too, since with double sideband radio-telephony one sideband may fade at a time when the other does not. Alternatively, the carrier frequency may fade and leave the side frequencies disproportionately large in amplitude. The resultant signal is effectively over modulated. This selective fading gives rise to both phase and amplitude distortion in the detected signal. Interference between signals reaching the receiver after different numbers of reflections between earth and the ionosphere can be reduced by using directional aerials at the transmitter which concentrate radiation at the correct angle of elevation for the number of hops in the intended path of propagation.

1.10. ABNORMAL IONOSPHERE CONDITIONS

Isolated patches of well above average densities of ionisation have been found to exist at E layer height. These are called sporadic-E layers. In temperate latitudes sporadic-E patches have been noted which extend for a 100 km or more. Frequencies of 60 MHz or over may be reflected by sporadic-E layers but, as these layers are unpredictable, they are not very useful. Between the latitudes of 15° N and 15° S, sporadic-E ionisation is less irregular and its presence is taken into account when assessing the values of m.u.f. for transmissions which cross the equator. The breaking up of this sporadic-E layer around sunset may cause fluttering and fading of signals on communication routes between the U.K. and the southern hemisphere.

Solar flares, which occur unpredictably, cause radiation of unusual intensity to reach the earth, and increase ionisation densities in the lower levels of the ionosphere. Such conditions cause very heavy attenuation of h.f. signals, so that even the highest available traffic frequencies fade out. These fade-outs may last several minutes or even for hours. They are worse at the equator than at higher latitudes. The name of the American J. H. Dellinger is assoicated with these fade-outs, hence the term *Dellinger fade-out.*

Solar flares not only cause ultra-violet radiation of exceptional intensity, they also emit charged particles. Some of these enter the earth's atmosphere about 30 hr after the ultra-violet radiation has arrived. The charged particles are deflected by the earth's magnetic field towards the polar regions where they ionise atoms in the upper atmosphere with the release of light frequency energy. This is visible as the Northern Lights or *Aurora Borealis* in the Arctic circle and their counterpart the *Aurora Australis* in the Antarctic regions.

The streams of charged particles swirling in the upper atmosphere constitute electric currents with magnetic fields which violently disturb the usual stability of the earth's field. These conditions are described as a *magnetic storm.* Such storms tend to recur at intervals of 27 days which is the period of rotation of the sun about its axis. Ionosphere storms obliterate communications with paths in higher latitudes perhaps for hours or even days. The ill effects continue by day and night until the shower of particles from the sun abates. Lost communication time may be partly recovered by re-routing traffic. For example, if telephone circuits by radio between New York and London are unworkable, the traffic may be relayed via the West Indies which is a more southerly route.

1.11. THE V.H.F. BAND

The band of frequencies between 30 MHz and 300 MHz is classified as the very high frequency (v.h.f.) band. Within this range are the television bands 1 and 3, and the f.m. broadcast band 2. Police and fire services also have both fixed and mobile stations operating at v.h.f. Generally v.h.f. services are not intended to provide long distance communication and the expected limits of the service area probably lie about 60–80 km from the transmitter.

As wavelengths range between 1 m at minimum and only 10 m at maximum, directional aerials are possible and these may be mounted many wavelengths above the earth over which they radiate. It is, therefore, possible for waves to travel directly through the space between the sending and receiving aerials without being influenced by the ground beneath.

1.12. DIRECT WAVE

The direct wave which passes between the transmitter and the receiver without reflection by the earth has a range which is chiefly limited by

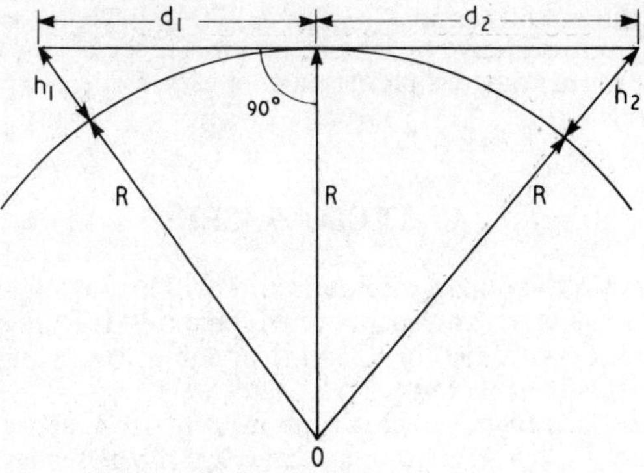

Fig. 1.6 Optical range of transmission

the heights of the terminal aerials. Neglecting atmospheric refraction, the range would be the sum of the distance d_1 and d_2 in Fig. 1.6. The heights of the two aerials above the earth's surface are shown in the same diagram as h_1 and h_2. R is the radius of the earth compared with which h_1 and h_2 are much smaller than they appear on the diagram. The range can be calculated by application of the result of Pythagoras' theorem, as follows:

$$d_1^2 + R^2 = (R + h_1)^2$$

or $$d_1^2 + R^2 = R^2 + 2Rh_1 + h_1^2$$

Subtracting R^2 from both sides of the equation

$$d_1^2 = 2Rh_1 + h_1^2$$

But since h_1 is exceedingly small compared with R, h_1^2 is negligible compared with $2Rh_1$ therefore

$$d_1^2 = 2Rh_1$$

and taking the square root of each side:

$$d_1 = \sqrt{(2Rh_1)}$$

Similarly it may be shown that

$$d_2 = \sqrt{(2Rh_2)}$$

so that the range is given by:

$$d_1 + d_2 = \sqrt{(2R)}\ [\sqrt{(h_1)} + \sqrt{(h_2)}]$$

The radius of the earth is approximately 6 370 km. The above reasoning determines the optical range of which the limit is set by the earth's curvature. The actual range is increased to a varying degree by tropospheric refraction.

1.13. TROPOSPHERE

The troposphere is the region of clouds and of the changing weather. It is the part of the earth's atmosphere which extends from sea level to an altitude of approximately 12 km. It is subject to change and turbulence especially at its lower levels.

Temperature falls from values of the order of 15°C at sea level to as low as –50°C at 12 km. Pressure may be 1 000 mb at sea level but decreases to about a third of that value at maximum altitude. A further variable physical factor of the troposphere is water vapour pressure, a nominal value for which is 10 mb at sea level but decreasing in direct proportion to altitude. The refractive index of the troposphere and the bending of the paths of light or radio waves depends on all the three variable factors of temperature, pressure and water vapour pressure. It is the rate of change of refractive index with height which determines the refraction of the wave path.

In a *standard atmosphere* the refractive index is approximately 1·0003 at sea level and decreases at the rate of $0{\cdot}039 \times 10^{-6}$ per metre of altitude.

The refraction of v.h.f. waves by the troposphere increases their range to beyond the optical limit evaluated in Section 1.12. The increased range available due to refraction by a standard atmosphere may be calculated by assuming the earth's radius to be 4/3 times its physical value. The factor 4/3 is called the *earth radius factor.* This factor changes if the rate of change of the refractive index with height alters. It is in the general case given by:

$$k = (1 + R.\ \mathrm{d}n/\mathrm{d}h)^{-1}$$

where R is the earth's radius, and $\mathrm{d}n/\mathrm{d}h$ is the rate of change in refractive index with height, a gradient which is usually negative. The following example illustrates how this formula is applied:

Example 1.1

A transmitter uses an aerial with a height of 100 m above sea level. Find the maximum distance at which a receiving aerial placed at the same height could receive signals from this transmitter if the refractive index of the troposphere decreases at $0{\cdot}05 \times 10^{-6}$/m of altitude. Take the earth's physical radius to be 6 370 km. The effective earth radius factor is first calculated from:

$$k = (1 + R.\ \mathrm{d}n/\mathrm{d}h)^{-1}$$

Inserting values given

$$k = (1 + 6\,370 \times 10^3 \times -0{\cdot}05 \times 10^{-6})^{-1}$$

$$= (1 - 0{\cdot}3185)^{-1}$$

$$= (0{\cdot}6815)^{-1}$$

$$= 1{\cdot}47$$

The effective earth's radius is thus

$$R' = 6\,370 \times 1{\cdot}47$$

$$= 9\,360 \text{ km}$$

The horizon distance is given by $\surd(2R'\ h)$ or

$$\surd(2 \times 9\,360 \times 10^3 \times 100) \text{ m}$$

$$= 10^3 \times 43{\cdot}2 \text{ m}$$

$$= 43{\cdot}2 \text{ km}$$

As the height of the receiving aerial is also 100 m, the total distance between the stations is limited to 2 × 43·2 km, or 86·4 km.

The calculation of the extended range of a v.h.f. transmitter in the above example assumes a uniform gradient of refractive index. This method is not sufficiently accurate for distances exceeding about 160 km, since it assumes that all the effective radiation leaves the transmitting aerial in a horizontal direction and is refracted in the lower regions of the troposphere.

In times of fine calm weather unusual atmospheric conditions sometimes follow a hot day when a clear night sky allows the earth's surface temperature to fall more rapidly than that of the atmosphere above it. This causes temperature inversion with temperature increasing rather than falling with altitude up to a limited altitude. Under these conditions a radio wave may reflect back and forth between the layer of warm air and the earth, and so be led or ducted to extreme ranges of several thousand kilometres. *Ducting* is, however, more frequent at frequencies greater than 300 MHz.

1.14. REFLECTED WAVE

A wave which is reflected from the ground or from some other object may reach the receiving aerial in addition to the direct wave (Fig. 1.7).

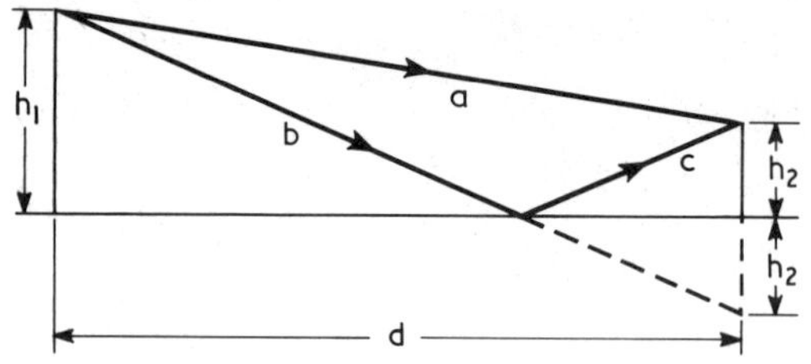

Fig. 1.7 Direct and reflected waves

The total space wave reaching the receiving point is the combination of the direct and reflected waves. As in other examples of multipath transmission, destructive interference between the two components of signal can occur. If the difference between the two paths of travel is any odd number of half wavelengths and the phase change on reflection is negligible, the two waves arrive with opposing phase and the resultant signal may be small. Referring to Fig. 1.7 it is seen that destructive interference between the two waves occurs if

$$(b + c) - a = n\,\lambda/2$$

and n is an odd integer. If n is an even integer, the two waves reach the receiving aerial in phase and the resultant signal is strong. For a

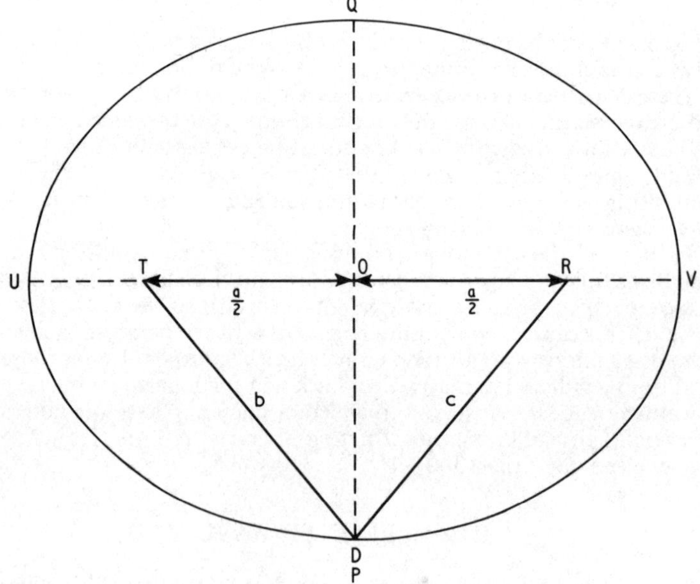

Fig. 1.8 A Fresnel zone

particular value of n, since the distance a is fixed, $(b + c)$ is also constant and equal to $(a + n\,\lambda/2)$.

The locus of points in one plane which represent the changing position of point D (Fig. 1.8) as the ratio of b/c alters while $(b + c)$ is constant, is an ellipse. The focal points of the ellipse are the positions of the transmitting and receiving aerials. The solid figure which would be described by rotating this ellipse about its major axis is an ellipsoid. The volume contained within such an ellipsoid pertaining to n equal to unity, is called the first *Fresnel zone.*

In Fig. 1.8 the major axis of the ellipsoid UV is equal to the distance $(b + c)$ or to $(a + n\,\lambda/2)$ m. By application of the theorem of Pythagoras, the minor axis QP can be shown to be equal to $\sqrt{(b^2 - a^2/4)}$. If the distance a is many wavelengths this minor axis approximates to $\sqrt{(n\,a\lambda)}$. The deduction of this is as follows:

In Fig. 1.8, $b = c$

$$= (b + c)/2$$

$$= \frac{(a + n\,\lambda/2)}{2}$$

By the theorem of Pythagoras:

$$OP = \sqrt{[b^2 - (a/2)^2]}$$

$$= \sqrt{[(a/2 + n\,\lambda/4)^2 - a^2/4]}$$

$$= \sqrt{(n\,a\,\lambda/4 + n^2\,\lambda^2/16)}$$

$$PQ = 2OP$$

Thus, $$PQ = \sqrt{(n\,a\,\lambda + n^2\,\lambda^2/4)}$$

$$= \sqrt{(n\,a\,\lambda)}\sqrt{(1 + n^2\,\lambda^2/4\,na\lambda)}$$

But if $a \gg n\lambda$

$$PQ \simeq \sqrt{(n\,a\,\lambda)}$$

The longer the transmission path in terms of a wavelength, the thinner is the shape of the ellipsoid. As the value of n becomes larger, the boundary of the equivalent Fresnel zone becomes wider and reflected signals reaching the receiving aerial are weaker. If reflections can be avoided within the first Fresnel zone, conditions of operation will probably be satisfactory as other reflections are then much weaker than the direct-wave signal. An example to show the relevance of the first Fresnel zone to the siting of v.h.f. aerials follows.

Example 1.2

Transmission takes place between two similar vertical aerials spaced in the same horizontal plane by 20 km. The frequency used is 300 MHz. Assuming a level ground over the distance of transmission, find the minimum height for the aerials above the ground in order that ground reflection within the first Fresnel zone may be avoided. Show that destructive interference could occur from an object situated 0·25 m behind one of the aerials in a direction in line with the direct transmission path.

The wavelength corresponding to 300 MHz is equal to: wave velocity/frequency, or

$$\frac{3 \times 10^8 \,\text{m/s}}{3 \times 10^8 \,\text{Hz}} = 1 \text{ m}$$

The minor axis of the Fresnel zone is given by $\sqrt{(n\,a\,\lambda)}$ or, if n is unity, by $\sqrt{(a\,\lambda)}$. Substituting the given values, the minor axis is found to be $\sqrt{(2 \times 10^4)}$ or 141·4 m. To avoid ground reflection within the first Fresnel zone the horizontal transmission path must be at least half this distance above the ground. Thus the aerials should be at a minimum height of 70·7 m.

If in Fig. 1.8 it is imagined that D is moved along the ellipse to point U, then $UR = c$ and $UT = b$. Also $RV = UT = b$. The major axis of the ellipse, $UV = UR + RV$ and thus, $UV = c + b$. From previous working, $c + b = a + n\lambda/2\alpha$ or, substituting values given, $UV = (2 \times 10^4 + 0{\cdot}5)$ m.

$$RV = UT = \frac{UV - TR}{2}$$

$$= \frac{(2 \times 10^4 + 0{\cdot}5) - 2 \times 10^4}{2}$$

$$= 0{\cdot}25 \text{ m}.$$

Thus reflection from an object 0·25 m behind either aerial is from the boundary of the first Fresnel zone and would be in antiphase to the direct wave, assuming that no phase change took place on reflection.

Although the boundary of the Fresnel zone takes no account of any phase change occurring on reflection, any calculation of the phase

difference between the direct and reflected waves must allow for a phase change at the point of reflection. Especially is this so for horizontally polarised waves which are reflected with a phase change of almost 180°. Referring again to Fig. 1.7, the following reasoning applies:
By the theorem of Pythagoras

$$(b+c)^2 = d^2 + (h_1 + h_2)^2$$

$$= d^2 \left[1 + \left(\frac{h_1 + h_2}{d}\right)^2\right]$$

Taking square roots of both sides of the equation

$$(b+c) = d\left[1 + \left(\frac{h_1 + h_2}{d}\right)^2\right]^{1/2}$$

Expanding this by the binomial expansion but neglecting the third and higher terms of the result gives

$$b + c = d\left(1 + \frac{h_1^2 + 2\,h_1\,h_2 + h_2^2}{2d^2}\right)$$

$$= d + \frac{h_1^2 + 2\,h_1\,h_2 + h_2^2}{2d}$$

Also

$$a^2 = d^2 + (h_1 - h_2)^2$$

which by similar working becomes

$$a = d + \frac{h_1^2 - 2\,h_1\,h_2 + h_2^2}{2d}$$

Therefore, the difference of the path lengths followed by the direct and reflected waves is

$$(b+c) - a = \frac{2\,h_1\,h_2}{d}$$

This difference in path lengths causes a phase difference between the direct and reflected signals of

$$\frac{2 h_1 h_2}{d} \cdot \frac{2\pi}{\lambda} \text{ rad}$$

or

$$\frac{4\pi h_1 h_2}{d\lambda} \text{ rad}$$

If the reflection process results in further phase change in the reflected wave of α rad, the total phase difference between the two signals is:

$$\phi = \left(\alpha + \frac{4\pi h_1 h_2}{d \lambda}\right) \text{rad}$$

Whenever ϕ is an odd integral multiple of π, the two signals are out of phase by 180° and the resultant amplitude is the difference of their individual amplitudes.

To avoid weakening the signal by a reflected wave, careful siting of the aerials of a v.h.f. system is essential. Aerials having vertical as well as horizontal directional properties will help to prevent reception of an unwanted reflected wave. Reflected signals from hills, buildings etc. are a cause of *ghosting* in television pictures. A weaker image appears on the screen to the right of the wanted picture. This is caused by the reception of the reflected picture signal delayed by a few micro-seconds on the direct wave.

1.15. ALLOCATION OF FREQUENCY BANDS

Frequencies allocated to specific services are published by the International Radio Consultative Committee (C.C.I.R.). This is one of the permanent committees of the International Telecommunications Union at Geneva. In allocating frequencies to particular services the following points are relevant:

1. The type of service to be provided
2. The bandwidth necessary for the type of communication intended.
3. The distance of the transmitter from expected reception points.
4. The most suitable mode of propagation to be adopted.
5. The number of other similar services required and their geographic location.

Here are six illustrative examples:

Low speed telegraphy using keyed C.W.

Bandwidth–less than 100 Hz.

Range–very long distance, if possible a world-wide range for communication with naval vessels in all oceans.

Propagation mode–by ground wave and by multiple reflections between the lower levels of the E layer and the earth and using high power to ensure adequate signal strength at extremity of range.

Number of similar stations–because of their limited application, high cost and the small number of channels available, relatively few.

Band allocated to fixed and mobile maritime services: 14–19·95 kHz.

Low power telegraphy service for ships communicating with other ships and with local shore stations using C.W. or M.C.W. type emission with modulation at approximately 1 kHz.

Bandwidth per channel–less than 3 kHz.

Range–usually less than 300 km although greater ranges are often available at night via E layer reflections.

Propagation mode–by ground wave over the sea by day; sky-wave propagation at night is accepted as a welcome bonus.

Number of stations required–several thousand, according to the number of ships in use by all maritime nations, but frequencies are shared since not all ships require to transmit at the same time and they are distributed over the world sea routes.

Band allocated to mobile maritime use: 405–525 kHz. (This band includes the international calling and distress frequency of 500 kHz.)

Low power radiotelephony for communication between fixed and mobile stations which may be ships or aircraft.

Bandwidth per channel–less than 7 kHz.

Range–usually less than 150 km.

Propagation mode–by ground wave, sky-waves if present are subject to distortion due to selective fading.

Number of similar services–numerous but limited at any one time in a given area.

Bands allocated–2·045–2·498 MHz.

Broadcast service of news and entertainment.

Bandwidth per channel–9 kHz.

Range–maximum ranges of the order of 70 km.

Propagation mode–by ground wave; E layer reflections at night cause interference with services in other areas and in other countries.

Number of stations–in Europe there is congestion of the m.f. broadcast channels; the B.B.C: have 28 transmitters radiating regional programmes alone for the U.K.

Band allocated to m.f. broadcasting: 525–1 605 kHz.

Broadcast service of high quality for news and entertainment, free from interference from noise sources and other services even in densely populated areas. An f.m. system is required.

Bandwidth per channel– 200 kHz.

Range–maximum range usually less than 50 km.

Propagation mode–direct wave but with numerous reflections from buildings, hills etc.

Number of stations required–approximately 20 v.h.f. transmitters are distributed over the U.K.; local broadcast stations on v.h.f. are growing in number.

Frequency band allocated: 87·5–100 MHz.

Black and white television service having 625 lines with picture frequency 25 Hz using interlaced scanning and a sound channel with f.m. Vestigial sideband may be used on vision signal.

Bandwidth per channel–8 MHz.

Range–of the order of 40 km, but mainly dependent on the transmitter aerial height and the absence of intervening hills etc.

Propagation mode–direct wave only required; reflections cause picture distortion.

Number of stations required–relatively few programme services will be economically possible but the number of transmitting and relay centres will be more numerous than required for v.h.f. sound broadcasting.

Frequency band: television band 4: 470–582 MHz.

It will be noted that the larger the bandwidth required for a given service, the higher the frequency level at which the carrier must be fixed.

Table 1.2 sets out some frequency allocations.

Table 1.2

Frequency Band	*Service*
10–14 kHz	Radio-navigation and radio-location
115–126 kHz	Fixed stations and maritime mobile stations, radio-navigation and location
160–255 kHz	Long-wave broadcast
1 800–2 000 kHz	Frequencies for use by amateurs
2·85–3·155 MHz	Aeronautical mobile stations
6·2–6·525 MHz	Ship station h.f. working
7·0–7·1 MHz	Frequencies for amateur use
7·1–7·3 MHz	H.f. broadcasting
20 MHz	Fixed space and earth to space communication
27·5–28 MHz	Meteorological aids
100–108 MHz	Mobile v.h.f. stations
174–216 MHz	Television band 3
322–329	Radio astronomy

Many other frequencies are available for each of the services listed in Table 1.2. The few examples are selected to show the variety of applications provided for.

QUESTIONS

1. What do you understand by the term *ground wave?* State briefly the factors which determine the field strength provided by the ground wave of a transmitter at a given point.

2. Explain why radio waves passing through the ionosphere are attenuated and state briefly the factors which influence the degree of attenuation.

3. Explain briefly the factors which affect the ionisation density of the ionosphere. If the maximum density of free electrons in the F_2 layer $1{\cdot}69 \times 10^{12}$ electrons/m^3, find the critical frequency for this layer density.

4. Assuming a critical frequency for an F layer to be 10 MHz and a virtual reflection height of 400 km, find the m.u.f. for a transmission path 800 km in length.

5. Explain the meaning of the following terms and briefly indicate the factors on which they depend:

(a) maximum usable frequency,
(b) optimum traffic frequency,
(c) lowest useful frequency,
(d) critical frequency of vertical incidence.

Explain why it is desirable to use as high a frequency as possible for a long distance shortwave radio link. *(C & G).*

6. Explain briefly how the following affect h.f. radio communication:

(a) the sunspot cycle,
(b) Dellinger fade-outs,
(c) sporadic–E layers,
(d) ionosphere storms,
(e) variation of season.

7. A v.h.f. transmitting aerial is situated 100m above sea level. Assuming an effective earth radius of 9 000 km, find the height at which a receiving aerial must be placed in order that the maximum range shall be 80 km.

8. What are the disadvantages which arise from the presence of a reflected wave at a v.h.f. receiving aerial and how can the transmission path be arranged so as to reduce the risk of a reflected wave being received?

9. How can the range of a v.h.f. wave be extended by (a) ducting, and (b) tropospheric scattering?

10. Discuss the factors which influence the allocation of a particular frequency band to specific types of radio communication service.

2

Superheterodyne Receiver—1

2.1 RADIO RECEIVERS–GENERAL

A small number of radio receivers are required to operate on a number of spot frequencies only. Most receivers, however, must tune over a range of frequencies, such as the medium wave band of sound broadcast transmissions, and are often required to cover each of several, switched ranges.

The input signal may be very small–perhaps only a few microvolts: a considerable reserve of amplification may thus be necessary in the receiver.

2.1.1. Distribution of amplification between different frequencies

For conciseness of expression we often talk of amplification at a frequency; what we really mean is amplification of a band of frequencies; the band may be relatively wide or relatively narrow compared with the absolute value of the centre frequency.

Amplification may be applied at the signal frequency, at some frequency to which the signal frequency may be converted (Section 2.2), or at the output frequency–for example, the audio frequency for a sound receiver or the video frequency for a television receiver.

Some, possibly a great deal of amplification is required prior to the detector which needs upwards of one volt at its input for good results. The smaller the input level which may be encountered the greater the amplification resources which must be available.

There are difficulties in providing a high degree of amplification at one frequency, particularly if the frequency is high and if it is required to be variable. A receiver in which these difficulties are surmounted and which provides all the predetector amplification at one frequency is called a *tuned radio frequency (t.r.f.)* receiver

2.1.2. Disadvantages of t.r.f. type of receiver

The difficulties in designing and operating a t.r.f. type of receiver are as follows:

The greater the amount of amplification at a given frequency, the smaller is the fraction of the output power which may feed back to the input without introducing instability (that is the tendency of the amplifier to generate oscillations). If for instance an amplifier unit provides a gain of 10^6, it is only necessary to feed back one millionth part of the output, in the correct phase, for the amplifier to be completely unstable. If the amplifier gain is reduced to 10^4 the feedback factor may increase 100 times before the same conditions of instability arise.

To reduce feedback the various stages of an amplifier must be well screened from one another. It is difficult to screen a number of stages if each must be provided with all the mechanical paraphernalia of variable tuning.

For a multi-stage amplifier to be tunable in a number of stages it is necessary for either the operator to adjust a number of separate tuning controls, or for the various controls to be mechanically *ganged* together so that one adjustment varies all the controls in the correct manner. It is not easy to arrange that the tuning of each circuit is nearly exact over the whole of even one tuning range, the difficulties multiply when a number of ranges is involved.

Again, the discrimination exerted by a tuned circuit in favour of a signal of one frequency with respect to a signal of a different frequency depends not on the absolute frequency separation, but on the separation expressed as a fraction of the frequency to which the circuit is tuned. Thus a circuit which accepts a 10-kHz band of frequencies and substantially rejects frequencies outside this band when tuned to a frequency of 1 MHz will, when tuned to 2 MHz, accept a 20-kHz band of frequencies if the change in tuning has not altered the selectivity of the circuit. Thus the width of the accepted band changes with frequency. It is normally preferable to have a uniform acceptance bandwidth at all frequencies of tuning within a given frequency band.

Similarly, while it may not be difficult to separate transmissions with a given frequency spacing in a relatively low-frequency band, it may be quite difficult, or impossible, to effect the same separation in a high-frequency band.

2.1.3. Amplification at a fixed intermediate frequency

The provision of a considerable amount of amplification at a frequency, the *intermediate frequency,* which is fixed and different from, and usually lower than, the signal frequency reduces the difficulties outlined in the last section considerably because:

1. The high-frequency gain is divided between two (or more than two) frequencies, and feedback problems are much reduced.
2. There is a further easing of feedback difficulties because the intermediate-frequency amplifier circuits are fixed tuned which means that the circuits may be compact and, therefore, more easily screened than those for variable frequency tuning.

3. The several tuned circuits of the intermediate-frequency amplifier, operating at a fixed frequency, may be pre-set so that the number of variable frequency tuning controls in the equipment is much reduced: thus ganging problems are diminished.
4. With a large proportion of the high-frequency amplification taking place at one frequency, the selectivity (in this part of the receiver) is uniform whatever the signal frequency. There is no variation between different parts of one frequency band, nor between one frequency band and another.

Other advantages accrue from the incorporation of some fixed frequency amplification in that special circuits for very selective tuning (i.e. very narrow bandpass circuits) may readily be included and variable bandwidth facilities can be provided, thus: (a) enabling a powerful interference-free signal to be accepted in wide bandwidth circuits with corresponding high quality, while weak, interference-ridden signals may be received with narrow bandwidth (and, therefore, poor quality but relatively free of interference), and (b) the bandwidth of the receiver may be adjusted to the channel bandwidth of the received signal.

2.1.4. Noise. Noise factor. Signal-to-noise ratio

Electrical *noise* is any voltage variation not created by man-made signals. The presence of noise in some degree is inevitable. It may interfere with the enjoyment of a radio programme; in bad cases it may interfere with the interpretation of signals. The fact of its existence sets a limit to the amount of amplification which may be employed usefully. Unlike an interfering signal, noise usually covers a wide range of frequencies—the influence of some types of noise may be experienced over all the ranges of a receiver.

Some noise derives from sources external to the receiver—for example, lightning and electrical machinery.

Some noise is developed in the receiver, in particular in resistors, transistors and valves. Noise created in resistance, *thermal noise,* is due to the random movement of electrons in a conductor and covers an infinite frequency range. It increases with the temperature T in Kelvins (that is degrees Celsius plus 273), with the value of the resistance R in ohms, and with the bandwidth B in hertz of the circuits following the resistance. The effect, for a given resistance, is the same as would result from the application of a voltage (the *noise voltage*) E_n volts in series with a noise-free resistance of the same value R where

$$E_n^2 = 4kTBR$$

k is Boltzmann's constant ($= 1{\cdot}372 \times 10^{-23}$ J/K)

R is the effective resistance of the component or circuit considered—thus for a parallel low loss circuit (of inductance L henries, series loss resistance r ohms and capacitance C farads) at resonance $R = L/Cr$ the dynamic resistance of the circuit.

Some resistors, for example carbon ones, introduce additional noise due to their composition.

Transistors and valves cause noise because the random movements of holes and electrons create noise voltages at the collector and anode. There is also some thermal noise set up in the resistance of the emitter, base and collector material of transistors.

Valve noise increases with the number of electrodes, hence a pentode, or a frequency changer like a triode-hexode, introduces more noise than a triode. In addition, the relative noise level at the output of a given transistor or valve is greater when used as a frequency changer than when used as an amplifier.

Valve and transistor noise increases with operating frequency. The increase begins to assume importance with most valves at about 50 MHz (for transistors the frequency depends on the type of transistor).

In general the relative importance of internal noise increases with frequency.

The effect of a given noise level is greatest when it exists at the receiver input—the introduction of 20 μV of noise into an i.f. stage where the signal level is, say, 100 mV, is negligible. The same amount of noise at the input, where the signal level may be only a few microvolts, has a serious effect. The noise level at the input consists of:

(1) *External noise.* The magnitude of this depends on the frequency —the noise level decreases as the frequency rises; on the type of location—whether in open terrain or in close proximity to noise producing elements like motor cars and electrical equipment; and on the type and situation of the aerial—a well sited directional aerial can do a lot to mitigate the effects of a bad general location.

(2) *Internal noise.* This is developed for the most part in the first tuned circuit and the first active device (transistor or valve).

Because frequency changers introduce more noise than amplifiers, it is necessary to have a reasonably large signal at the input to the frequency changer. This is one of the reasons for the inclusion of one or two stages of signal frequency amplification in communication receivers (where the input voltage may be only a few microvolts).

The absolute level of noise is less important than its relative value in comparison with that of the signal. This relationship is expressed in the *signal-to-noise ratio.*

Example 2.1

At one point in a receiver the signal level is 150 μV and the noise voltage is 10 μV.

The signal-to-noise ratio is 15.

Expressed in decibels this is 20 log 15 = 23·5 dB.

The signal-to-noise ratio may be expressed in terms of power. At the input it is P_{1s}/P_{1n}, at the output it is P_{2s}/P_{2n}. The subscripts 1 and 2 are used to denote input and output respectively, the subscripts s and n refer to signal and noise respectively.

The receiver is in no way responsible for external noise. If this is present at the receiver input the receiver can not eliminate it–although the effects may be modified by changing the receiver bandwidth (the wider the bandwidth the greater the noise which is permitted to pass to the receiver output). It is true, of course, that by the use of a suitable aerial system in front of the receiver the amount of external noise which is brought to the receiver may be reduced.

Internal noise is another matter–this does depend on the properties of the receiver. It is convenient to have a means of expressing the quality of the receiver in terms of gain per unit level of noise introduced.

The *noise factor* is a commonly used measure of the extent to which a receiver, or amplifier, introduces noise in relation to the amount of gain which it provides. The noise factor is the ratio:

$$\frac{\text{TOTAL NOISE POWER AT THE OUTPUT}}{\text{NOISE POWER OUTPUT DUE TO NOISE INPUT}} = \frac{P_{2n}}{P_{1n} \times G}$$

Where G is the power gain P_{2s}/P_{1s}

Rearranging,

$$\text{Noise factor} = \frac{P_{2n}}{P_{1n} \times P_{2s}/P_{1s}}$$

$$= \frac{P_{1s}/P_{1n}}{P_{2s}/P_{2n}}$$

$$= \frac{\text{INPUT SIGNAL-TO-NOISE POWER RATIO}}{\text{OUTPUT SIGNAL-TO-NOISE POWER RATIO}}$$

The greater the noise factor the greater the amount of noise power introduced for each unit of gain, and, therefore, the less good the performance from this point of view, of the amplifier or receiver.

For an approximate idea of the orders of value which may be achieved in practice, we can say that for audio-frequency amplifiers the noise factor may be a little in excess of unity, for h.f. amplifiers using thermionic valves up to 10 or 20 MHz the value may be a little over 2, while further increase in frequency results in a progressive increase in noise factor, the figure being about 4 for a frequency of 50 MHz.

Provided that appropriate types are used, the values obtained for transistorised units are about the same as those for valved units.

2.2. SUPERHETERODYNE RECEIVERS

In the superheterodyne receiver much of the high-frequency amplification is effected at one fixed frequency so that the advantages outlined in Section 2.1 may be obtained. In some receivers more than one fixed frequency is employed–in the double superheterodyne receiver (Section 2.2.6.) two fixed frequencies are used.

2.2.1. Outline

A block diagram illustrating the superheterodyne receiver principle is given in Fig. 2.1.

The signal from the receiver aerial is amplified at signal frequency in the *signal-frequency amplifier.* (In some receivers, particularly broadcast

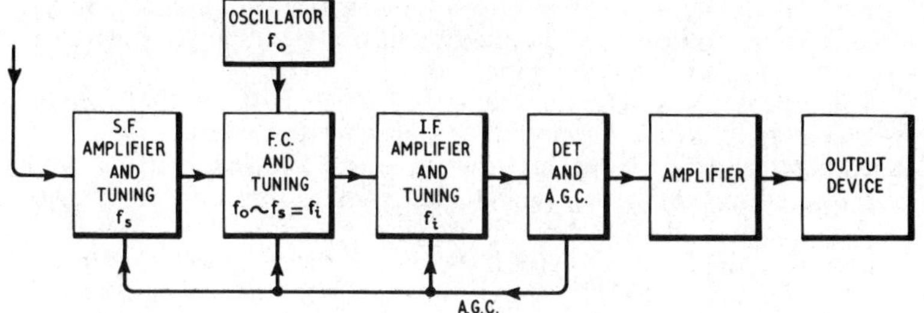

Fig. 2.1 Block diagram of basic superheterodyne receiver

receivers, signal-frequency amplification stages are sometimes omitted.) Some frequency selection is effected in the signal frequency tuned circuits.

Frequency changing is carried out in the *frequency changer.* Here any signal to which the receiver is tuned is converted to the intermediate frequency. We call the desired signal the *wanted signal,* or the *wanted frequency* f_W.

The absolute frequency separation of signals is not altered in the frequency changer, so that if, for instance, the intermediate frequency is 470 kHz, a wanted signal of frequency 20 MHz is changed in the frequency changer to 470 kHz, while an unwanted signal of 20·005 MHz, i.e. one separated from the wanted signal by 0·005 MHz, retains the same frequency separation at the output of the frequency changer; its percentage frequency difference from the wanted signal is thus much increased and its elimination correspondingly more simple.

Example 2.2

Using the values just quoted, the fractional spacing of the unwanted from the wanted signal at the signal frequency is 5/20 005 or 0·025%.

The fractional spacing at the intermediate frequency is 5/470 = 0·0106 or 1·06%.

After the frequency changer is the intermediate-frequency amplifier which provides most of the gain. Detection and further amplification (at the output frequency) follows in the usual way.

2.2.2. Frequency changing

Frequency changing is carried out in the frequency changer or *mixer* by suitably mixing the signal with a locally generated oscillation of the proper frequency. If the frequency of the signal is f_s and of the local oscillation is f_0 the output of the frequency changer contains the original frequencies f_s and f_0 together with frequencies equal to the sum of and the difference between f_s and f_0: that is, if f_0 is greater than f_s, $f_0 + f_s$ and $f_0 - f_s$. In fact, f_s represents not one frequency but the range of frequencies which constitutes the signal. The similarity to the modulation process and the production of upper and lower sidebands will be recognised (see also Chapter 6).

The process by which one frequency f_a beats with another signal of frequency f_b to create, on modulation, new frequencies $f_a \pm f_b$ is called *heterodyning.* From this derives the name supersonic heterodyne (abbreviated to superheterodyne)–a heterodyne creating a frequency too high for audibility.

Generally the local oscillator frequency is above that of the signal and the tuned circuits following the mixer eliminate all frequencies except $f_0 - f_s$.

Example 2.3

A receiver has an intermediate frequency f_i = 470 kHz. The receiver is tuned to a transmission employing a carrier frequency f_w = 1·5 MHz and having a bandwidth extending 5 kHz each side of the carrier. For the local oscillator frequency to be above the signal frequency, its value needs to be f_0 = 1 970 kHz when the receiver is tuned to this transmission.

Fig. 2.2 (a) displays the frequencies, f_0 and f_w beat together to yield $f_0 - f_w$ (the intermediate frequency f_i = 470 kHz) and $f_0 + f_w$ (3 470 kHz).

Note that because the intermediate frequency is derived as the difference frequency $f_0 - f_w$, the upper sideband (shown shaded in Fig. 2.2 (b)) appears in the intermediate frequency amplifier as the

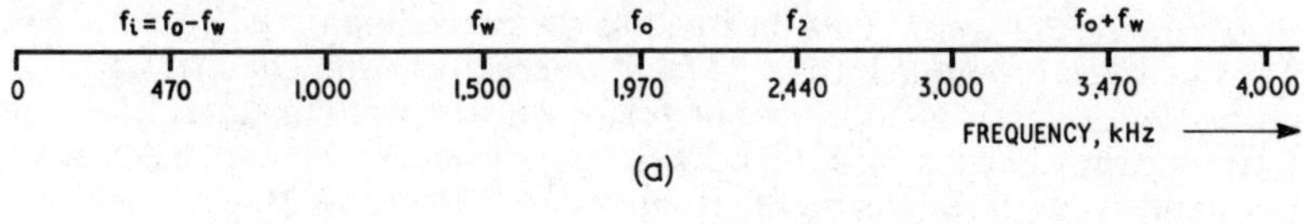

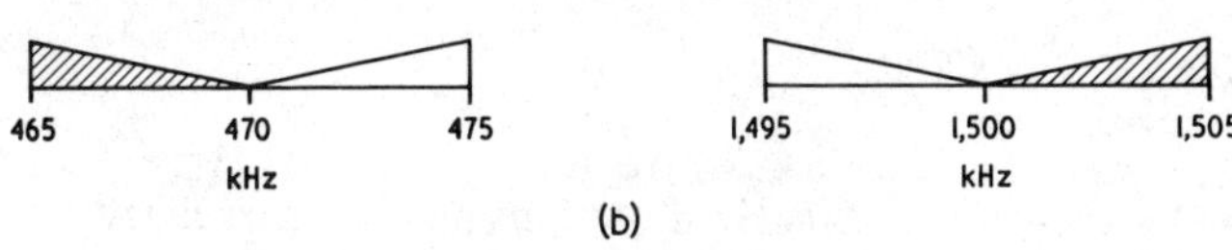

Fig. 2.2 Example 2.2

lower sideband, while the lower sideband (unshaded) appears in the intermediate-frequency amplifier as the upper sideband.

$$(1\,970 - 1\,505)\text{ kHz} = 465\text{ kHz}$$
$$(1\,970 - 1\,500)\text{ kHz} = 470\text{ kHz}$$
$$(1\,970 - 1\,495)\text{ kHz} = 475\text{ kHz}$$

This in no way affects the ultimate output signal.

Of those so far discussed the only signal to be accepted by the i.f. amplifier tuned circuits is the one comprising the (translated) carrier and sidebands centred on the intermediate frequency (that is the one centred on 470 kHz in the example above). The other frequencies in the frequency changer output f_w (= 1 500 kHz in the example), f_o (= 1 970 kHz in the example) and $f_o + f_w$ (= 3 470 kHz in the example) are rejected.

Observe, however, that frequency $f_2 = f_o + f_i$ (1 970 + 470 = 2 440 kHz in the example), if present at the input to the frequency changer beats with the local oscillator frequency of f_o to yield the intermediate frequency f_i (because $f_o + f_i - f_o = f_i$). Should this happen there is nothing which can prevent this (unwanted) signal from receiving full amplification in the i.f. amplifier and appearing in the output with the wanted signal. The signal frequency which causes this type of interference is that which is as much above the oscillator frequency as the wanted frequency is below it—that is, it is $f_2 = f_o + f_i = f_w + 2f_i$ (and vice versa if the oscillator frequency is below the signal frequency). The interfering frequency is called the *image channel,* or *second channel.*

To create second channel interference, of course, it is not necessary for f_2 to be of exactly the value quoted: any frequency within a few kilohertz of this value beats with the local oscillator signal to produce an interfering signal within the pass band of the i.f. amplifier.

To prevent second channel interference it is necessary to prevent any transmissions on or about the second channel frequency from reaching the frequency changer. This means that the preceding tuned circuits must be selective enough to remove all traces of second channel signal no matter how powerful it may be relative to the wanted signal. How difficult it is to accomplish this depends on how low the intermediate frequency is with respect to the frequency of the wanted signal.

Image channel rejection ratio is dealt with in Section 12.1.4.

Example 2.4

For a receiver tuned to a wanted frequency of 1 000 kHz, the second channel frequency is 1 220 kHz if the intermediate frequency is 110 kHz. The second channel is thus spaced 220 kHz from the wanted signal. This is only 22% and the second channel may well be very difficult to eliminate.

For an intermediate frequency of 470 kHz the second channel to the same wanted frequency is 1 940 kHz. This is spaced 94% from the wanted signal and, other things being the same, it is much easier to eliminate than the interfering signal in the first part of the example.

An objection to the introduction of a frequency changing stage into a receiver is the proneness to the generation of *interference whistles.* The frequency changer is a device which is specifically designed to facilitate heterodyning processes, so that in it not only the local oscillator and wanted signal interact to produce sum and difference frequencies, but also any other frequencies and their harmonics beat with one another. Any resultant frequencies which lie within the acceptance band of the intermediate frequency amplifier cause either direct interference or an interference whistle of audible frequency. Careful design reduces the effect of interference whistles to what is nearly always an acceptable minimum.

2.2.3. Choice of intermediate frequency

There are a number of considerations affecting the choice of intermediate frequency:

1. The tuning of the intermediate-frequency stages is fixed. Despite screening and the rejection of intermediate-frequency signals in the tuned circuits preceding the i.f. amplifier, some pick-up is likely to take place of any nearby transmission on the intermediate frequency. The intermediate frequency is, therefore, chosen to be outside the more widely used communication bands.
2. The intermediate frequency is normally selected to be outside the tuning range of the receiver, otherwise the signal-frequency amplifier and the intermediate-frequency amplifier operate on the same frequency with the likelihood of the instability already mentioned. Also, the i.f. circuits cannot separate the signal and the

intermediate frequency, with the result that an unwanted beat frequency is set up in the detector.

3. To remove the possibility of second channel interference, the intermediate frequency must be reasonably high. We remember that the second channel frequency is $f_2 = f_w + 2f_i$, and that second channel signals can only be removed in the tuned circuits of the s.f. stages. The greater f_i, the greater the spacing between the wanted and second channel frequencies and the better the prospect of the elimination of the second channel transmission.
4. Signals employing frequencies close to the wanted transmission are called *adjacent channel signals.* In the frequency changing process the same absolute spacing between signals is maintained. The lower the intermediate frequency, the greater is the percentage difference between the wanted and an adjacent channel transmission, and the greater is the reduction in the relative amplitude of the adjacent channel signal.
5. The wider the bandwidth of the wanted signal, the higher the intermediate frequency needs to be. The bandwidth of possibly 200 kHz needed in a frequency modulation receiver can not be sustained on the 470 kHz intermediate frequency commonly employed in amplitude-modulation receivers. An intermediate frequency of 10.7 MHz is commonly used in f.m. receivers (which of course operate in the v.h.f. band).
6. The lower the intermediate frequency, the more stable the intermediate-frequency amplifier.
7. The lower the intermediate frequency, the greater is the tendency for the oscillator to be *pulled* into the signal frequency (because the lower the intermediate frequency, the closer is the spacing between the oscillator and signal frequencies). If the oscillator does pull in to the signal frequency there is, of course, no ouput at the intermediate frequency–and hence no input to the detector.
8. The higher the intermediate frequency, the fewer are the interference whistles likely to be produced.

These various considerations are mutually conflicting–for a receiver covering a very wide range of frequencies it may be necessary to employ more than one intermediate-frequency amplifier. The first of these would be necessary at high frequency to enable second channel transmissions to be eliminated, and one would be required at a fairly low frequency for the removal of adjacent channel interference. The double superheterodyne, with two intermediate frequencies, is dealt with in Section 2.2.6. Otherwise a satisfactory compromise is sought using one intermediate frequency.

2.2.4. Choice of oscillator frequency and amplitude

For a given intermediate frequency in a receiver tunable over one or more frequency ranges, the oscillator frequency must be variable so that the difference between the signal and oscillator frequencies shall always be equal to the intermediate frequency. The oscillator frequency range may thus be above or below the signal frequency range. If the oscillator frequency is chosen to be below the signal frequency the ratio maximum/minimum frequency which must be covered by the oscillator is greater than that needed if the oscillator frequency is higher than the signal frequency.

Example 2.5

For a sound radio receiver covering the medium wave range of approximately 550–1 500 kHz the signal frequency range is about 3:1. With an intermediate frequency of 470 kHz, the oscillator range, if above the signal frequency, is 1 020–1 970 kHz a ratio of less than 2:1. The oscillator range, if below the signal, is 80–1 030 kHz, a ratio of about 13:1.

For a high range of signal frequencies, the difference between the necessary oscillator frequency ranges is not very marked, but the general principle remains and, since it is clearly easier to design and manufacture an oscillator with a small frequency range than one with a large range, the oscillator frequency is usually chosen to be above the signal frequency.

A further consideration is that the oscillator tuning control has to be ganged with the signal-frequency tuning control. It is possible to arrange this reasonably satisfactorily for a small frequency range but it is more difficult, or even impossible, for a large range.

The higher the operating frequency, the less stable is the oscillator. Hence for high signal frequencies the mixer may be designed to operate on a harmonic of the oscillator frequency. This has the advantage that the oscillator is basically more stable, but there is also the disadvantage that there is much more likelihood of the introduction of interference whistles.

The level of signal transferred from frequency changer to intermediate-frequency amplifier increases with the level of the local oscillator signal. However, too high a level of local oscillator voltage leads to an increase in interference whistles, and for this reason the oscillator signal level is kept as small as possible consistent with the efficient functioning of the frequency changer.

If the circumstances are such that this should happen, it will be radiated and may cause interference in other receivers. The inclusion of a signal-frequency amplifier before the frequency changer virtually eliminates the possibility of the coupling of the oscillator output to the aerial except at very high frequencies. With no s.f. amplifier, care must be taken to prevent stray coupling between oscillator and aerial circuit.

2.2.5. Signal-frequency amplification

We have seen that frequency changers are more noisy than amplifiers. For this reason it is necessary to provide a reasonably large signal at the frequency changer input. This calls for the provision of signal-frequency amplification except where the signal is always large and/or relatively free from interference.

It is also a fact that the amount of noise developed in a frequency changer, for a given gain, increases with the bandwidth of the preceding (s.f.) circuits. To decrease the bandwidth, tuned circuits are needed. For this reason also, therefore, it is beneficial to include one or two stages of s.f. amplification before the frequency changer.

To prevent image channel interference, a certain amount of pre-frequency changer selectivity is needed. For the medium wave broadcast band, the wanted frequency may be about 1 MHz and the image channel is, therefore, about 2 MHz (for an i.f. of 470 kHz). The separation of these is not much of a problem unless the wanted station is very weak and the image channel transmission is very powerful. Broadcast receivers, therefore, often omit s.f. amplifying stages. At high frequencies affairs are very different. With the same intermediate-frequency value, the image channel to a transmission of 10 MHz is at about 11 MHz–only 10% removed. The tuned circuits associated with one or two s.f. amplifiers may thus be needed to provide sufficient selectivity to eliminate a powerful second channel transmission.

In fact the inclusion of one or two stages of s.f. amplification not only guards against second channel interference but also substantially reduces the possibility of interaction between unwanted signals and the oscillator with a consequent beneficial reduction in the number and level of spurious signals in the i.f. amplifier.

2.2.6. Double superheterodyne receiver

When the signal frequency is high it is difficult to eliminate second channel stations unless there is a corresponding increase in the intermediate frequency value. With a high intermediate frequency, however, it is difficult or impossible to reject unwanted adjacent channel transmissions. For communication receivers, the dilemma is often resolved by using a double superheterodyne receiver. The term *communication receiver* is one which is broadly applied to receivers designed to cover a wide frequency range (including several high-frequency ranges) and to provide amplification, suitably low noise factor, adequate selectivity, and appropriate refinements to enable low-level signals of various types to be received satisfactorily.

The term double superheterodyne implies no more than the employment of two intermediate-frequency amplifiers, the first at

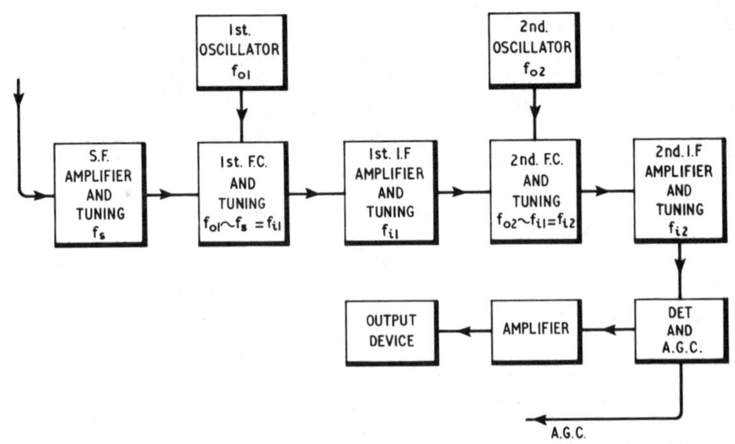

Fig. 2.3 Block diagram of basic double superheterodyne receiver

relatively high frequency, the second at relatively low frequency. Usually, however, a receiver of the double superheterodyne type (Fig. 2.3) includes:

1. One or two stages of s.f. amplification.
2. The (first) frequency changer which translates the frequency to that of the first i.f. amplifier.
3. The first i.f. amplifier. This often functions at a frequency of about 1·6 MHz so that the problem of removing any second channel interference (located at a frequency 3·2 MHz away from the wanted frequency) in the preceding s.f. stages is not unduly difficult. The respective selectivities of the s.f. amplifier and the first i.f. amplifier are not adequate to remove all possible adjacent channel interference. Accordingly the first i.f. amplifier is followed by the second frequency changer.
4. The second frequency changer (which unlike the first, usually operates with a fixed frequency input and, therefore, does not need a tunable oscillator) translates the frequency to that of the second i.f. amplifier.
5. The second i.f. amplifier operates at a low frequency–about 100 kHz. Here there is no second channel interference to be concerned with and at this low frequency any adjacent channel interference remaining is readily eliminated.

Following the second i.f. amplifier are the usual detector and following amplifier stages.

2.2.7. Superheterodyne reception–summary

With the major part of the high-frequency amplification concentrated in the (fixed) intermediate-frequency amplifier the gain is substantially constant over all the signal frequency bands. With a t.r.f. receiver the gain normally rises with frequency in a given band and falls between one band and another higher frequency band.

The selectivity is substantially constant for all signal frequencies. For a t.r.f. receiver the selectivity falls with increase of frequency. However, the rejection of adjacent and second channel transmission is less effective at the higher frequencies.

The receiver is more stable than a t.r.f. receiver because the high-frequency amplification is effected at more than one frequency. Or, for a given stability, more gain is possible.

Ganging difficulties are less than for a t.r.f. receiver, but the problem of tracking is introduced.

On the other hand, the superheterodyne tends to generate interference whistles and there is the possibility of coupling the oscillator to the aerial circuit with the consequent danger of radiating signals at the oscillator frequency.

QUESTIONS

1. Why is it advantageous in a radio receiver to apply amplification at a number of different frequencies rather than at a single frequency?

2. Comment succinctly on ways in which the existence of electrical noise influences radio receiver design.

3. Describe how the following effects occur when using superheterodyne receivers and explain how they can be reduced:

(a) second channel interference,
(b) adjacent channel interference,
(c) local oscillator radiation,
(d) interference caused by i.f. and local oscillator harmonics. *(C & G)*

4. Explain the meaning of the term *noise factor* and describe a method of measuring the noise factor of a receiver.

Why is it desirable for receivers operated at frequencies above about 5 MHz to employ radio-frequency amplification? *(C & G)*

5. Upon what factors does the magnitude of noise voltage in a circuit depend?

Other things being the same, compare the noise voltages due to (a) a parallel, and (b) a series arrangement of a loss-free capacitor of 200 pF and an inductor of 200 μH and effective series loss resistance of 10 Ω.

The voltage signal-to-noise ratio at the input to a receiver is 20 dB, at the output it is 15 dB. What is the noise factor?

6. Discuss the various factors which influence the choice of intermediate frequency in a superheterodyne receiver.

7. Discuss the factors which influence the choice of oscillator frequency and amplitude in a superheterodyne receiver.

A carrier of 16 MHz conveys one intelligence signal in the upper sideband and a different intelligence signal in the lower sideband. The signal is to be translated to an i.f. of 1·6 MHz. What local oscillator frequency must be employed to reverse the positions of the sidebands relative to the carrier?

What is the image frequency?

8. With the aid of a block schematic diagram describe the principle of operation of a superheterodyne receiver suitable for amplitude-modulated sound broadcast reception. What are the reasons for the use of a tuned radio-frequency amplifying stage?

Define the terms:

(a) sensitivity,
(b) image channel response ratio
in relation to the performance of a superheterodyne receiver.

Why is the image channel response ratio generally lowest when the receiver is tuned to the highest frequency in its range? *(C & G)*

9. Compare the relative advantages and disadvantages of superheterodyne in comparison with tuned radio-frequency receivers.

10. Define the terms *second channel, image channel, adjacent channel, frequency pulling* and *heterodyne.*

Sketch a block diagram of a double superheterodyne receiver. Explain the points of difference as compared with a single superheterodyne receiver.

3

Superheterodyne Receiver—2

3.1. SIGNAL-FREQUENCY AMPLIFICATION

The signal-frequency amplifier is required to amplify the incoming transmission, and in so doing to cause the minimum amount of noise. The amplification is included so that the magnitude of the signal presented at the input to the frequency changer will be great enough to swamp frequency changer noise. The overall frequency response curve should permit the acceptance, with little relative loss, of all the frequencies in the sidebands of wanted transmissions while still providing:

1. adequate discrimination against interfering signals (including the prevention of cross modulation, see below), and
2. complete elimination of image channel signals.

Generally the signal-frequency circuits must be tunable over one or several frequency ranges each of which usually embraces a maximum/minimum frequency ratio of about 2 : 1. Tuning is usually by variable (ganged) capacitors.

The s.f. amplifier stage(s) must introduce minimum distortion. Undue curvature of the valve or transistor characteristic causes the modulation of an unwanted transmission to become associated with the carrier of the wanted transmission. Such cross-modulation is to be avoided, because once it occurs the cross-modulation signal cannot be separated from the wanted signal.

To prevent cross-modulation there should be adequate selectivity before the first active device (transistor or valve) and the characteristic should have (ideally) no curvature components of a higher order than two. Also, excessive signal level is to be avoided: for this reason an attenuator is often included in front of the s.f. amplifier stage(s).

For v.h.f. one or two stages of s.f. amplification are nearly always employed. The inclusion of one stage is necessary to eliminate (or adequately reduce) oscillator radiation from the aerial. High quality tuned circuits are essential to keep the noise level to a minimum. They may be of inductance and capacitance in parallel, often with slug tuning—for very high Q and good selectivity, parallel or coaxial $\lambda/4$ or $\lambda/2$ transmission lines are used. The tuned circuits are tapped in order:

1. to give the required (possibly fairly low) level of gain, and

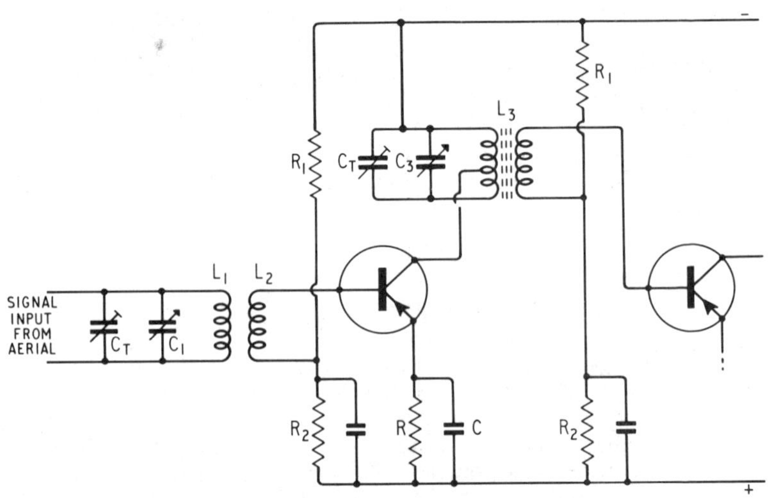

Fig. 3.1 Transistor signal-frequency amplifier stage

2. to reduce the damping imposed by the following circuit the input impedance of which is necessarily rather low.

The active device may be transistor, pentode or earthed grid triode (Section 3.1.2). Cascode circuits may be used (Section 3.1.1) as may push-pull circuits.

Particularly at v.h.f., considerable care must be taken in design and construction: very small stray capacitance values provide considerable amounts of coupling or shunting. Short lengths of wire have a relatively large reactance. Input leads must be screened; all leads must be short and well spaced, and they should cross at right angles. There should be a common connection point for each stage in order to avoid stray coupling. Separate stages should be screened from one another by earthed screens. Transformers should be completely screened. Decoupling circuit capacitors must be carefully chosen so that the impedance due to the self inductance is very low.

3.1.1. Signal frequency amplification using transistors

A typical transistor s.f. amplifying stage is shown in Fig. 3.1. The transistor must have a cut-off frequency which is higher than the highest frequency at which the amplifier is to operate. The use of field-effect transistors can give improved signal handling with lower noise figures and improved cross-modulation performance compared with the so-called unipolar (i.e. junction) transistors.

Coupling between aerial and transistor is via the input transformer L_1L_2. R_1R_2 sets the base potential of the transistor and renders it independent of the base current (provided that the current drawn by R_1R_2 from the supply is about ten or more times greater that the base current). R is the stabilising resistor which tends to prevent changes in collector and emitter current which are not due to the signal. C and R provide a time constant large enough to prevent changes in the voltage across R because of short-term signal variations. Thus R prevents an undue rise in collector current caused by a rise in transistor temperature, a rise which, in the absence of the stabilising influence of R, could lead to thermal runaway. R prevents this happening because a rise in current causes an increased voltage across R (positive to the base, negative to the emitter), and this automatically reduces the current. In the same way the presence of R provides an allowance for the spread of characteristics between transistors of the same type and nominally the same specification, and also provides a correction for the change in transistor characteristics with age. A transistor with higher than usual current values automatically creates a higher bias than transistors of lower current values, and correction is thus automatically developed.

The tuned circuit L_3C_3 in the collector output presents too high an impedance for good matching to the transistor output impedance. The collector is, therefore, taken to a tap on the coil. The position of the tap for maximum gain may be calculated or it may be determined experimentally—the selectivity, however, increases steadily as the tap is moved towards the earthy end of the coil (i.e. upwards in the Fig. 3.1.).

The output capacitance of the transistor is part of the output tuned circuit and so has no by-passing action as exists with a resistive load. The pre-set capacitors C_T are *trimmers,* they have a maximum value of about 50 pF (for a signal frequency of about 1 MHz) and are used to equalise variations in stray capacitance between circuits and possibly in the minimum value of ganged tuning capacitors (see Section 4.3). The output may be taken to a following similar stage.

There is a limit to the frequency at which a given transistor may be operated, but types are available which operate satisfactorily up to about 1 GHz. The reason for the deterioration in performance with increase of frequency is the finite time needed for the current carriers to move from emitter to collector across the base region. The current amplification obtained at low frequencies is taken as the criterion and the frequency at which the current gain falls to $1/\sqrt{2}$ of this value (i.e. falls by 3 dB) in the common base configuration is called the *cut-off frequency* (f_α). In the common-emitter configuration, the fall in gain with frequency rise is greater and the fall of 3 dB from the low frequency value occurs at a frequency, often called f_β, equal to $f_\alpha(1 - \alpha)$. This is about 2% of the value of f_α since α, the current gain in the common-base configuration, is usually equal to about 0·98.

Thus for a given transistor there is some frequency above which it is better to operate in the common-base configuration rather than common-emitter. However, because the low-frequency gain with the latter connection is so much greater than with the common-base connection, a considerable loss due to the high frequency (something like 34 dB) can be accepted before the current amplification in the common-emitter mode falls to as low a level as for the same transistor in the common-base configuration.

For frequencies in the 100–300 MHz region a typical type of circuit is that shown in basic form in Fig. 3.2.

Depending on the type of transistor and on the circuit configuration, coupling between output and input circuit may result in instability or in reduced gain. With some transistor types the effect is serious at quite low frequencies, with others there is no trouble even at high frequencies. The tendency to instability may be adequately resisted by simply keeping the stage gain down to a level below that at which instability is experienced. Otherwise, and in all instances in which feedback causes degeneration of the signal, the effect of feedback must be neutralised by providing a similar feedback path from equal voltage of opposite phase. A simple way of doing this is to centre tap the collector tuned circuit and connect series capacitance and resistance of the right value between the input (base) and the end of the tuned circuit remote from the collector. This is shown in Fig. 3.3(a) for a common-emitter circuit. The feedback through *CR* neutralises the internal feedback between collector and base. This is called *unilateralisation.* Correct choice of *CR* values provides approximate nullification of the feedback over the operational range of frequencies.

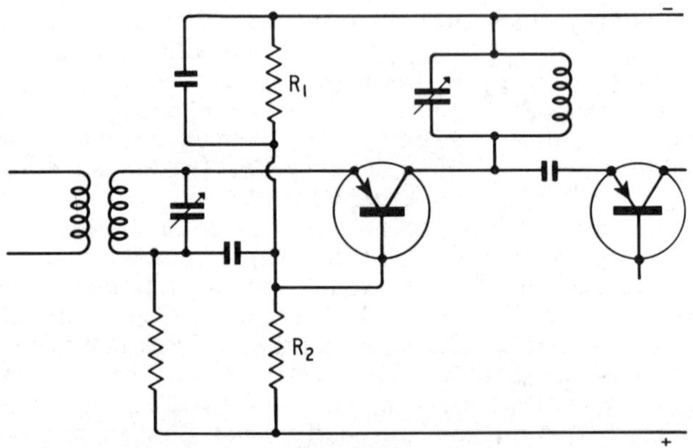

Fig. 3.2 Transistor v.h.f. amplifier stage (common base configuration)

When the collector circuit load consists of a transformer with untuned secondary winding, the feedback of the correct phase may be obtained from the secondary (Fig. 3.3(b)). Often a capacitor alone, instead of a capacitor and resistor, provides adequate neutralisation.

Different transistors of the same type show a spread of characteristics, so that circuit design can be for approximate neutralisation only.

Stage gain is kept at a level low enough to prevent excessive feedback even with the most disadvantageous combination of characteristics and values.

Good performance up to 200 or 300 MHz, with a pass band of several megahertz, may be achieved with two transistors in a *cascode*

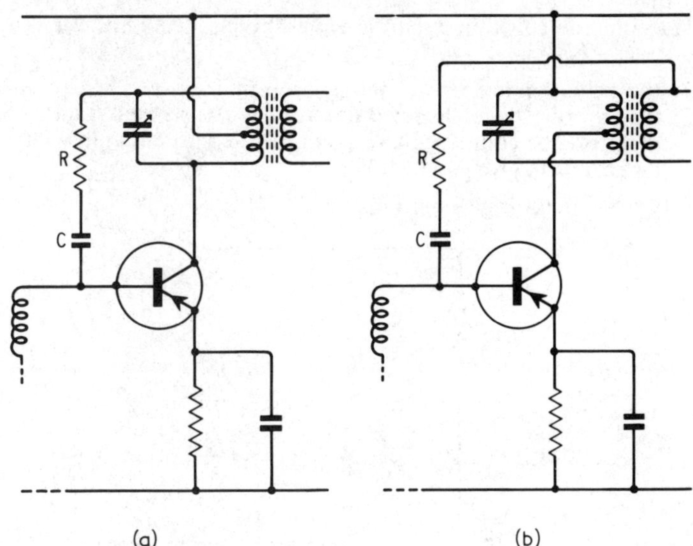

Fig. 3.3 Unilateralisation

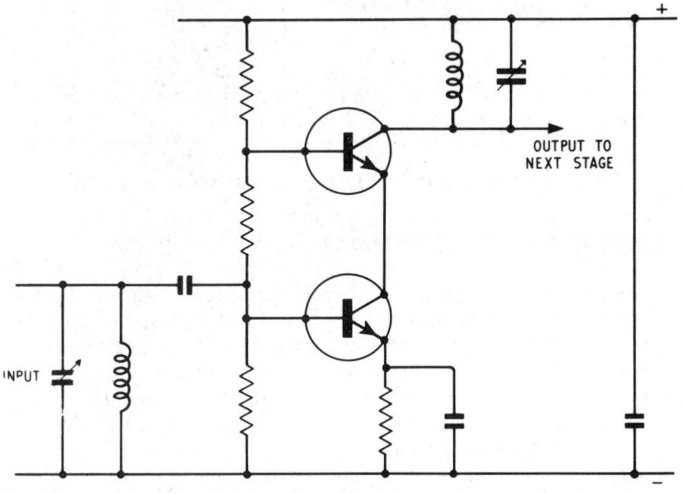

Fig. 3.4 Transistor wide-band v.h.f. cascode circuit

circuit. In a cascode circuit the transistors are connected in series as shown in Fig. 3.4 for two npn transistors. The input is to the lower transistor and the output is drawn from the collector of the upper one. The cascode circuit gives good isolation between input and output, and also provides a convenient point for the application of automatic gain control.

The common collector circuit is sometimes used at v.h.f. Two advantages are:

1. the reduction in stray capacitance, and
2. the fact that the collector is not at r.f. potential and it may, therefore, be connected to a point of fixed potential in such a manner as to give good cooling.

A typical circuit is given in Fig. 3.5.

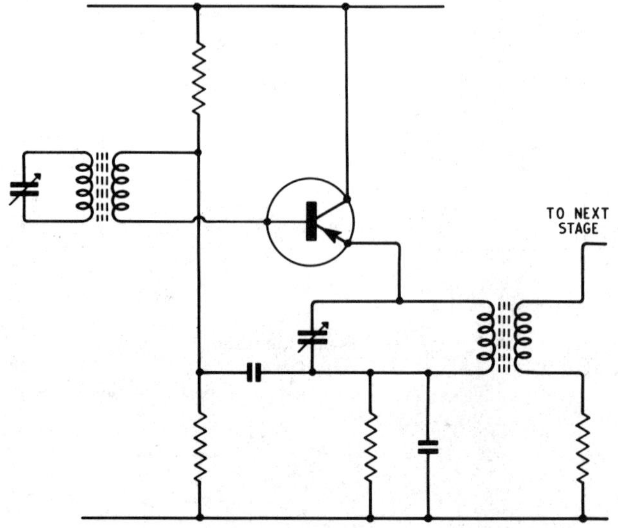

Fig. 3.5 Transistor v.h.f. amplifier stage (common collector configuration)

3.1.2. Signal frequency amplification using valves

Normally-connected triodes are useless as amplifiers above a few tens of kilohertz because of the coupling between input and output circuits through anode-grid capacitance C_{ag}. This is of a few picofarads only but, because of the Miller effect, it becomes effectively $(A + 1)\ C_{ag}$, where A is the gain. Depending on the nature of the load, this results in either instability or degeneration. For tuned high-frequency amplifiers, however, there is always instability.

Pentodes, with an anode–grid capacitance value of only about one-thousandth of that for triodes, may be and are used satisfactorily, with the normal common cathode connection, at frequencies up to some 150 MHz.

A typical basic circuit is shown in Fig. 3.6. The signal from the aerial, or previous signal frequency amplifier, is applied between grid and cathode; grid bias is provided as usual by the *CR* combination in the cathode lead. The screen grid supply may be obtained as shown, with the capacitance between screen grid and earth in order to earth the screen at r.f. A potential divider may alternatively be used to fix the screen d.c. potential (the capacitor being retained to fix the r.f. potential of the screen at zero).

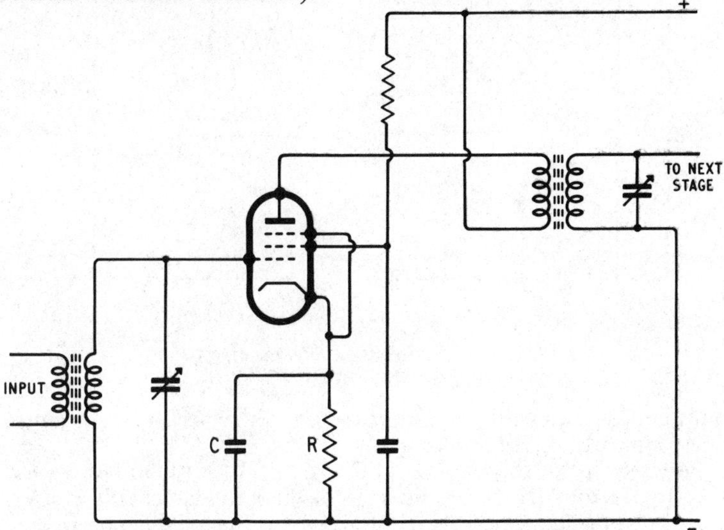

Fig. 3.6 Pentode signal-frequency amplifier stage

The anode load may be a coupled circuit with tuned secondary. For valve circuits this is usually a better arrangement than that employing a tapped parallel *LC* circuit because coupling is easier to adjust than a tap, and the relatively high anode supply voltage is isolated from the tuned circuit. There is likely to be an effective step down of about 3 to 1 between the primary and secondary windings but the overall voltage gain may nevertheless be about 100. Decoupling (see Chapter 4) is not shown in the circuit diagram.

With increase of frequency the gain falls (and the relative noise level rises) because of the increase:

1. in transit time of the electrons moving across the valve,
2. in the effect of the valve input capacitance, and
3. in the inductance of the cathode lead (with a reduction, sometimes a considerable reduction, in the input impedance).

The increasing effect of mutual inductance coupling between leads and of capacitance coupling may cause a reduction in gain or may lead

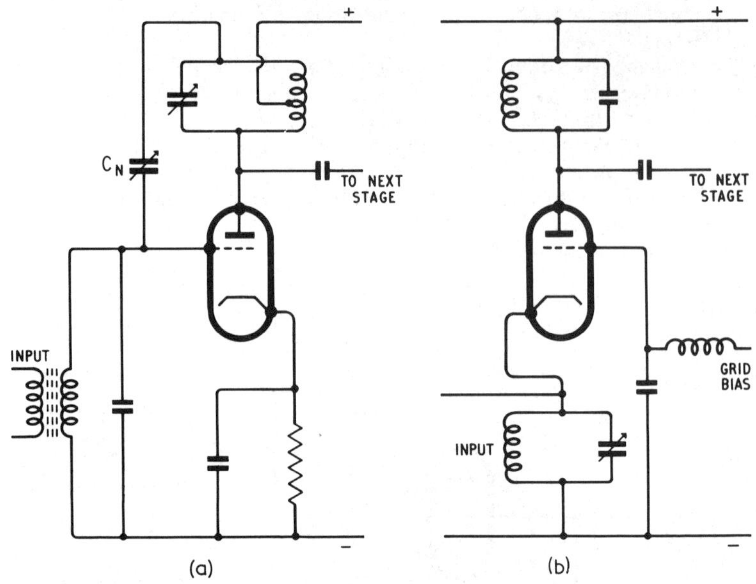

Fig. 3.7 (a) Neutralisation triode amplifier;
(b) common grid amplifier

to instability, depending on circumstances. Other factors, mentioned in Sections 3.1 and 3.1.1, also apply.

For frequencies above about 150 MHz, pentode r.f. amplifiers are often replaced by triode amplifiers (which are fundamentally less noisy). The feedback must be neutralised, or, the coupling between input and output circuits may be considerably reduced. When this is done, suitable triodes can be used as very successful amplifiers at very high frequencies. To assist in the performance, valves of very small physical size may be used—there are some types which have several internal cathode leads, in parallel, in order to reduce materially the lead inductance. External to the valve, the negative feedback effect due to the cathode lead inductance may be neutralised by the inclusion of an inductance between the cathode pin and the common negative H.T. line, and a small capacitance between the grid and the common negative line. (This creates a parallel resonance circuit of high impedance.)

Using a triode in the common-cathode configuration (Fig. 3.7(a)) it is necessary to cancel the effect of the anode–grid capacitance coupling by the inclusion of a neutralising circuit, as explained in Section 5.7. Alternatively the earthed-grid arrangement described in Section 5.6 may be used, which removes the need to neutralise. The basic circuit is repeated for convenience in Fig. 3.7(b).

Because it has fewer electrodes, the triode is much less noisy than the pentode. This is a major advantage which at very high frequencies, outweighs the fact that the input impedance of the circuit is low and so not only causes a reduction in gain but also renders the cascading of two stages not very fruitful. (But it is common to find an earthed-grid stage followed by a neutralised triode stage which has a much higher input impedance).

The cascode circuit, which we have already met in Section 3.1.1 using two transistors in a wideband v.h.f. amplifying circuit, is also very useful with triode valves, and gives useful amplification with low noise level at frequencies higher than those at which the pentode gain/noise level becomes unduly small. The circuit is shown in Fig. 3.8. The input is to *V1* which is in series with a similar valve *V2*. The grid of *V2* is earthed to the radio-frequency signals by capacitor *C* and so acts as a screen between input and output. The output is taken from the anode of *V2*.

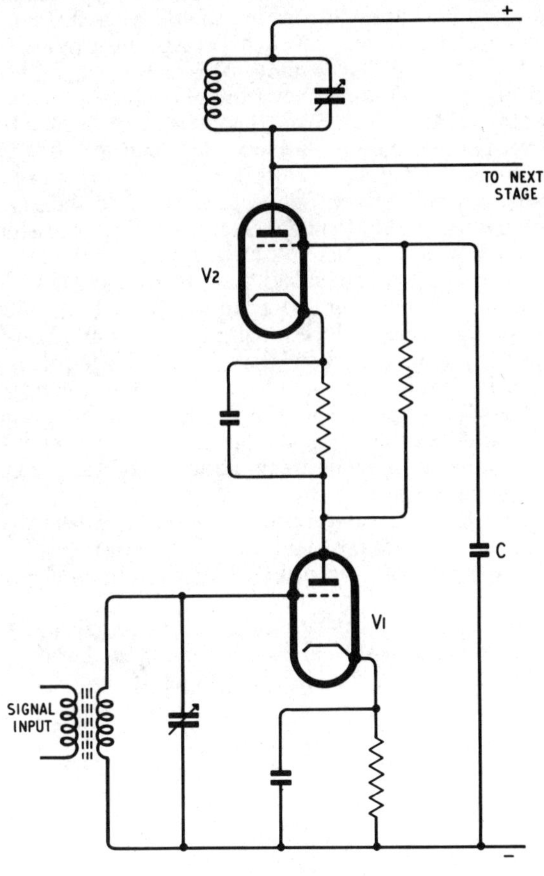

Fig. 3.8 Triode cascode circuit

3.2. FREQUENCY CHANGERS

The function of the frequency changer in the superheterodyne receiver is to mix the signal and a locally generated oscillation so as to yield the intermediate-frequency signal.

Generally the oscillator frequency is higher than the signal frequency ($f_0 - f_s = f_i$). When the signal frequency is fairly high, the oscillator frequency may be below that of the signal ($f_s - f_0 = f_i$) so that a stable oscillator is more easily obtained. The side frequencies associated with the signal carrier also beat with the oscillator signal, with the result that the sidebands are also translated to the i.f. range.

The frequency changer should provide the maximum signal-to-noise ratio output for a given input signal. This requires that the conversion power gain (or the conversion conductance for a valve) should be large and that the amount of noise introduced should be as small as possible.

The output from the frequency changer stage must not vary in frequency (i.e. $f_0 - f_s$ must be constant). This calls for a stable oscillator frequency. A highly stable frequency is more readily obtained when using a separate oscillator (Fig. 3.10) than when a self-oscillating mixer is used (Fig. 3.9). There should be the very minimum coupling between the oscillator and signal-frequency circuits. In particular, the oscillator frequency must not pull in (a self-descriptive term) to the signal frequency: this is more likely to happen at the higher frequencies, and with a relatively low intermediate frequency, because the ratio of oscillator/signal frequency is here less than the lower signal frequencies.

For transistor circuits the oscillator voltage needs to be about 0·1 V, for valve circuits the value must be much higher–round about 10 V for the triode-hexode circuit described later. Unless a harmonic is to be used, the output should contain minimum harmonic content so as to reduce the interference whistles produced. For this the coupling in the oscillator circuit itself should be as small as possible, and the loading of the oscillator, that is the coupling between it and the frequency changer, should be a minimum.

The amplitude of the oscillator output should be substantially constant over the tuning range. The inclusion of a resistor of a few hundred ohms in series with the grid lead in valve circuits helps, as does auto bias.

For frequency changing to take place, the two signals must be suitably mixed. They may be applied in series to a non-linear circuit, or in such a manner as to act multiplicatively in a linear circuit.

Example 3.1

If the signal voltage is $V_c \sin \omega t$, and the oscillator voltage is $V_0 \sin \omega_0 t$ the varying component of resulting anode current is proportional to $V_c \sin \omega t \times V_0 \sin \omega_0 t$. Thus:

$$i_a = KV_cV_0 \sin \omega t \sin \omega_0 t$$
$$= \frac{KV_cV_0}{2} \left[\cos (\omega_0 \sim \omega) t - \cos (\omega_0 + \omega) t \right]$$

and so has a component at the difference frequency, i.e. the intermediate frequency $(\omega_0 \sim \omega)/2\pi$.

At v.h.f. it is particularly important that the oscillator frequency is stable, because the i.f. amplifier bandwidth must allow not only the acceptance of the sidebands of the wanted signal but also for any anticipated drift in the transmitter frequency and in the receiver oscillator frequency.

If the i.f. bandwidth requires to be appreciably greater than that required by the signal, the result is a lower signal-to-noise ratio and greater likelihood of interference. One factor which contributes to variation in oscillator frequency is the level of the signal input to the frequency changer. For good stability this should vary very little. The application of a.g.c. to the s.f. stages is helpful. Where possible crystal oscillators are used.

Above about 150 or 200 MHz it is necessary, or at least desirable, to include a buffer amplifier between the oscillator and the mixer. The importance of screening increases with frequency. At v.h.f. the tendency of the oscillator frequency to drift increases, so do the difficulties arising from the drift. For this reason, as well as others noted earlier, there are advantages, when the frequency is very high, in running the oscillator at a lower frequency than the desired one and using a harmonic to mix with the signal frequency. In some double superheterodyne receivers, a single oscillator is used for both the mixing stages, a relatively high harmonic being used for the first mixer, and the fundamental for the second.

Example 3.2

A 14-MHz crystal oscillator is used in a double superheterodyne receiver which is receiving a signal on 86 MHz. The fifth harmonic of the oscillator is used as the heterodyne voltage in the first mixer—giving a first i.f. of (86 – 70) MHz = 16 MHz. The fundamental of the oscillator is used in the second mixer, giving a (second) i.f. of (16 – 14) MHz = 2 MHz

Particularly when an oscillator harmonic is being used to develop the i.f., the tuned circuit must be of high Q value so that the correct harmonic may be selected and the others fully rejected, otherwise the unwanted harmonics may beat with other incoming transmissions and cause interfering signals at the intermediate frequency.

Sometimes in double superheterodyne receivers a single crystal oscillator is used, at its fundamental harmonic, to provide the injection frequency for both the first and second mixer.

Example 3.3

If the incoming signal is centred on a carrier of 92 MHz and the first i.f. amplifier stages are tuned to 47 MHz, the local oscillator may generate a frequency of 45 MHz. This provides the first i.f., (92 – 45) MHz = 47 MHz, and, used again to heterodyne the output of the first i.f. amplifier, gives (47 – 45) MHz = 2 MHz, which is a convenient value for the second i.f.

For broadcast v.h.f. receivers, and for some others, the oscillator frequency is often chosen to be equal to the signal frequency minus the intermediate frequency (instead of equal to their sum) so that the second harmonic of the oscillator will not, if radiated, interfere with Band III television reception.

Transistors, pentodes, or triodes are usually used as frequency changers for v.h.f. circuits.

3.2.1. Frequency changers using transistors

In the simpler receivers, a single transistor is used both for the generation of oscillations and for mixing. A typical basic circuit is shown in Fig.3.9.

R_1, R_2 and R_3 function as usual to fix the base and emitter potentials. C_p is the padder and is included so that the capacitor C, which is probably identical with the other tuning capacitors in the receiver (Chapter 4), is able to tune L over the required oscillator frequency range while the signal frequency capacitors are tuning over the signal frequency range. The requirement is, of course, that the oscillator frequency shall always equal $f_s + f_i$. C_T is a pre-set trimmer (Section 3.1.1) to allow for variations in stray capacitance between circuits, and slight differences in the minimum capacitance of the different tuning capacitors in the receiver.

Inductors L_1, L_2 and L are coupled. The feedback enables oscillation to be set up and maintained at a frequency f_o which is determined mainly by the circuit L, C, C_p, C_T. In series with L_2 in the base-emitter circuit is L_3 carrying the signal frequency f_s. Provided that the transistor is operating over a non-linear characteristic the collector current contains components of signal varying not only at f_o and f_s but also at $f_o - f_s$ $(= f_i)$ and $f_o + f_s$. The tuned transformer (or maybe a simple tuned LC circuit), resonant at frequency f_i, constitutes the collector load so that the f_i component of collector current and the associated side frequencies set up across the transformer secondary an output which is taken to the first i.f. stage for amplification. The other components of collector current (at f_o, f_s and $f_o + f_s$) are too remote from f_i to create any response.

As noted in the previous section, there is always a danger that coupling between oscillator and signal circuits may result in degradation of the quality of the generated oscillation or even in its frequency being

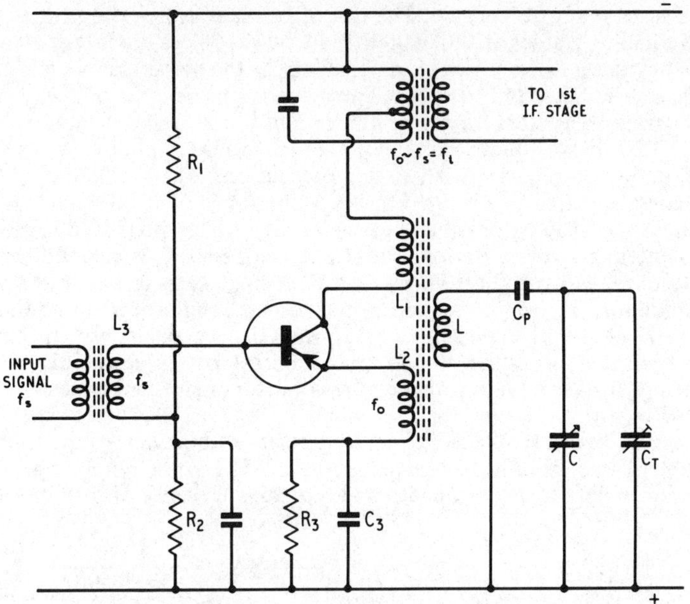

Fig. 3.9 Oscillator-mixer circuit

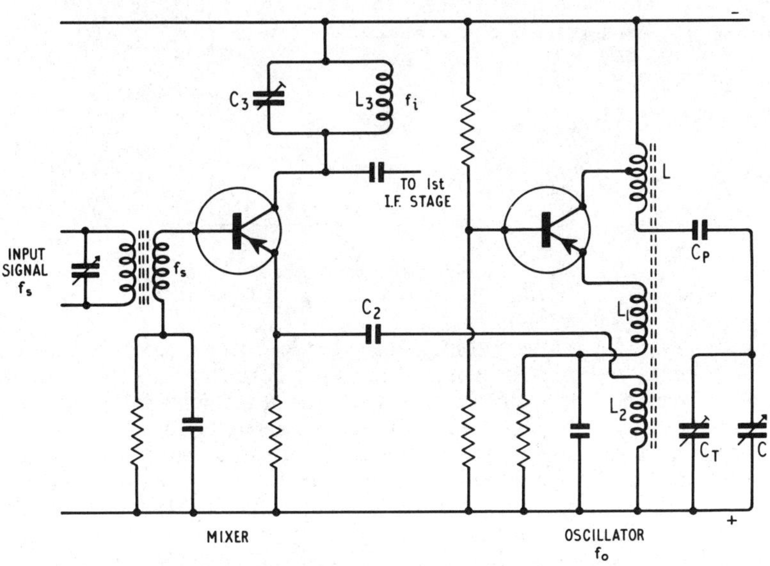

Fig. 3.10 Separate oscillator and mixer circuits

pulled in to that of the signal. The risk of this happening, other things being equal, is greater at the higher frequencies (unless the intermediate frequency is very high). Hence, particularly at the higher frequencies, care has to be exercised to reduce coupling to a minimum–one way is to use separate transistors for the oscillator and for the mixer functions.

Fig. 3.10 shows a basic circuit suitable for use up to about one or two hundred megahertz–with all due precautions. The oscillator frequency is mainly determined by the values of L, C, C_p and C_T, while coupling, the smallest possible, between L and L_1 provides feedback for the maintenance of oscillation. Voltage at frequency f_0 is also induced in L_2 and taken from there via C_2–which is small so as to put the minimum loading on the oscillator–to the base-emitter circuit of the mixer. In series with the oscillator output is the signal. Hence, subject to the necessary non-linearity of the mixer characteristic, intermediate frequency signals are developed in the collector circuit and appear across L_3C_3.

For v.h.f. work (and at u.h.f. and microwave frequencies), the tunnel diode performs usefully as a mixer-oscillator. The tunnel diode is a pn junction with heavily doped p and n regions. Because of this

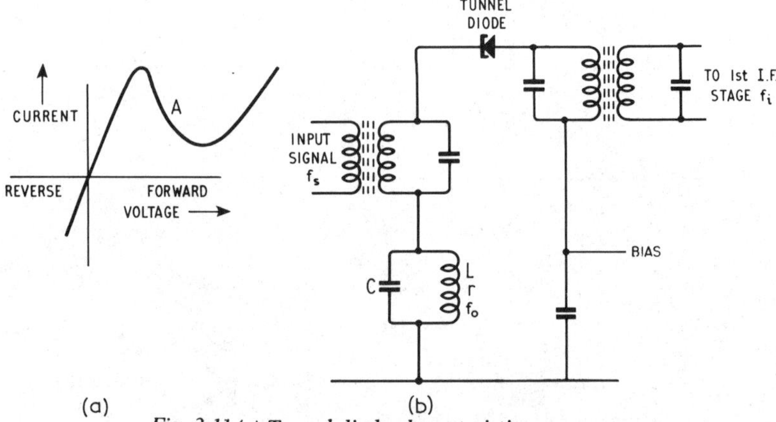

Fig. 3.11 (a) Tunnel diode characteristic;
(b) self-oscillating mixer using tunnel diode

the depletion layer is very thin, and the breakdown voltage, which is normally in the reverse current region of the characteristic, is moved into the forward region. The characteristic is thus of the general form shown in Fig. 3.11(a) with a negative resistance characteristic for a range of forward voltage, usually extending over about half a volt, in the region A. When the diode is operating in this region the charges move across the junction at very high speed (not far short of that of light) and the diode is thus able to operate at very high frequency.

A tunnel diode in series with an LC circuit (Fig. 3.11(b)) is able, provided that the negative resistance of the diode is numerically less than the positive resistance of the LC circuit (L/Cr), to nullify the LC circuit resistance and so enable oscillations to be maintained. Having a non-linear characteristic, it is also able to function as a mixer and so to act as a self-oscillating mixer.

3.2.2. Frequency changers using valves

There are many types of valved frequency changers. They employ diodes, triodes, pentodes, octodes and various combinations of these. One that has been in successful and extensive use in different types of receivers for a long time is the triode-hexode frequency changer. The arrangement provides for mixing in the multiplicative manner, the signal and the oscillator voltages being applied to two different grids in the hexode. In addition to carrying out the mixing process, the hexode is able to function as an amplifier and thus provide considerable gain.

The basic circuit is of the form shown in Fig. 3.12. The signal (f_s) is applied between grid and cathode and so directly influences the anode

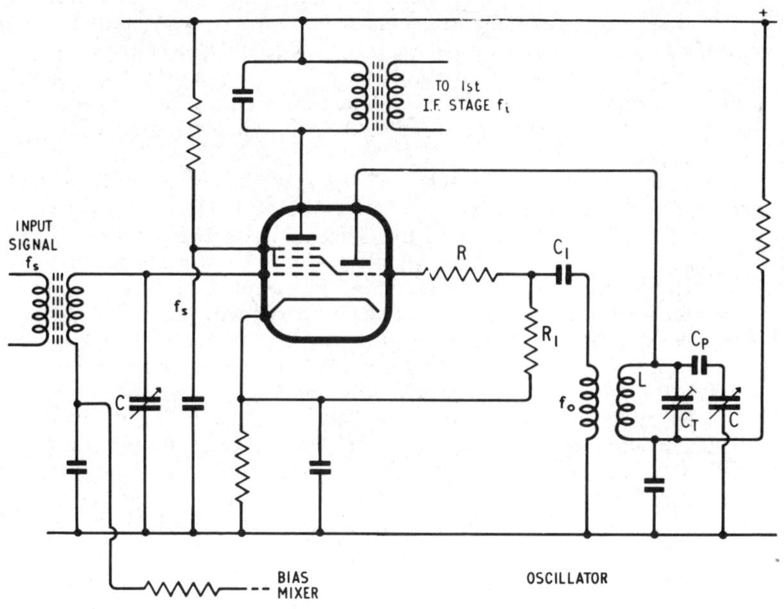

Fig. 3.12 Triode-hexode frequency changer

current. The triode is used in a tuned anode reaction type oscillator circuit, the oscillatory frequency of which is largely determined by L, C, C_p and C_T. The oscillator output, f_0, is taken (internally) from the triode to grid 3 of the hexode and so also directly influences the anode current. Multiplicative mixing, therefore, results and the anode current contains components of frequency at f_0, f_s and $f_0 \pm f_s$. The difference frequency ($f_0 - f_s = f_i$) is selected in the tuned circuits and passed to the i.f. amplifier.

C_1 R_1 provides grid bias and automatic limitation of the amplitude of the oscillation. Resistor R (of the order of 1 kΩ) helps to maintain the oscillation amplitude substantially independent of frequency.

The presence of grid 4 provides pentode characteristics with high anode impedance and good conversion conductance; it also enables grid 3 to have a fairly close mesh without unduly reducing the anode current, so enabling the frequency changer to work well with fairly low oscillator voltage.

Grid 3 is screened from the input by grid 2 thus keeping to a minimum the coupling between signal and oscillator circuits, and aiding oscillator stability.

The higher the frequency the more intense the frequency changer problems, but valve mixers may be used with due care at frequencies up to about 300 MHz. The precautions necessary include those already outlined. The oscillator and mixer valves should be separated and screened, and the smallest possible degree of coupling employed. For the oscillator, a circuit which oscillates readily, such as a Hartley or a Colpitts, should be chosen. Careful design is needed to prevent the occurrence of oscillations *(parasitic oscillations)* at unwanted frequencies.

Multi-electrode frequency changers prove unduly noisy in the v.h.f. range of frequencies so that triodes are often used. The separately generated oscillator voltage, and the signal, are injected in series in the grid-cathode circuit. The valve, therefore, needs to work over a non-linear part of its characteristic and this fact, coupled with the much lower output impedance of the triode compared with valves having pentode-like characteristics, means that the gain is less than with a pentode or similar type of valve. However, the signal-to-noise ratio, which is the important consideration, is greater. Neutralisation may not be necessary because of the different frequencies at the input and the output, but may prove to be essential at the intermediate frequency.

Diodes are sometimes used for frequencies higher than those at which transit time effects render triodes unsuitable (from perhaps 100 MHz upwards in some instances). The smaller the diode, the smaller the capacitance and the smaller the transit time, the higher the frequency up to which it may be used. Point contact diodes, with minute capacitance and transit time, are used even in the microwave region. The oscillator and signal voltages are applied in series with the diode and the i.f. tuned circuit. The discontinuity in the diode characteristic enables mixing to be effected.

3.3. INTERMEDIATE–FREQUENCY AMPLIFICATION

Intermediate-frequency amplifiers operate over a fixed band of frequencies which clearly simplifies the design and manufacture of the tuned circuits. The amplifier must possess ample gain so that the voltage at the output of the frequency changer may be raised to the level needed for the satisfactory operation of the detector (several volts). The tuned circuits must have adequate selectivity so that unwanted adjacent channel transmissions still remaining at the frequency changer output are eliminated, while the frequency components of the wanted signal are preserved. The tuned circuits must also provide the impedance needed to enable the desired degree of amplification to be obtained.

In some receivers provision is made for the bandwidth to be variable–either in two or three steps or continuously. This matter is further referred to in Section 4.1.5.

The tuning of the i.f. stages is pre-set. It may be by variable capacitor, or by variable inductance (using magnetic cores, or non-magnetic metallic cores which may be screwed axially into the coils– the inductance increases with penetration for magnetic cores, decreases

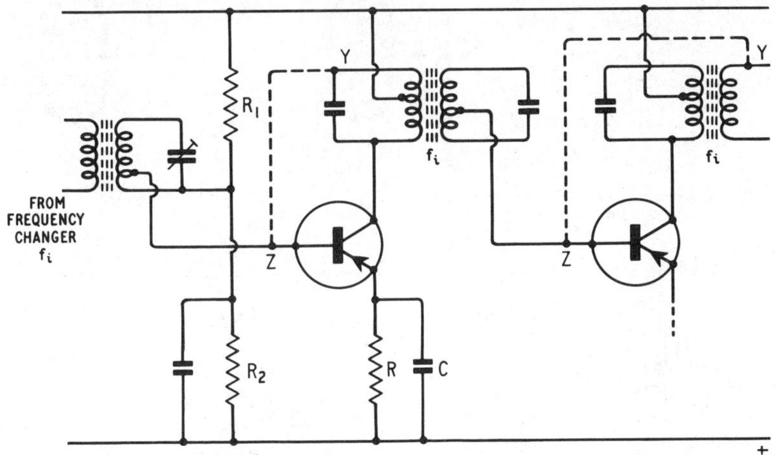

Fig. 3.13 I.F. stage using transistor

for metallic non-magnetic ones). The gain is greater with large inductance and small capacitance. However, stray capacitances are not negligible and apart from varying somewhat between one receiver and another of nominally the same type, they vary with time and also depend on the ambient conditions, particularly temperature, in any given receiver. It is necessary, therefore, to include fixed capacitance, of known and substantially invariable value, which is sufficient to constitute a fairly high proportion of the total stray capacitance at any particular part of the circuit. To nullify the effect of variations in capacitance with temperature, it is helpful to include some capacitance which has the opposite sign of temperature coefficient so that the total capacitance varies hardly at all with temperature change.

The individual i.f. stages are carefully screened to prevent feedback, and the transformers are contained in screening cans. For transistor circuits, alloy diffused transistors with a screen between input and output are often used. The internal feedback of such transistors is very small.

At v.h.f. the usual considerations apply concerning choice of intermediate frequency. Frequency stability is particularly important. For this, and for good gain and selectivity, a low intermediate frequency is an

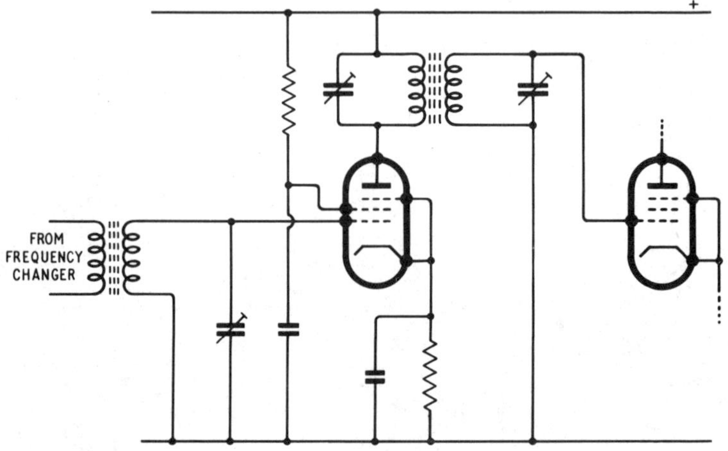

Fig. 3.14 I.F. stage using pentode

advantage, but image frequency rejection may be difficult, as may the obtaining of a wide bandwidth, if required. For multichannel work on v.h.f., bandwidths up to 2 MHz are used. Sometimes simple resistive or inductive loads are used for the i.f. amplifiers—selectivity being obtained in a single multisection bandpass filter.

3.3.1. Intermediate-frequency amplification using transistors

A typical basic transistor circuit stage for i.f. amplification is shown in Fig. 3.13. The input transformer, which is the output transformer of the frequency changer, selects the narrow band of frequencies centred on the intermediate frequency and rejects the other frequencies in the frequency changer output. The wanted band is applied between base and emitter of the first i.f. transistor. The transformer provides matching between the high impedance presented to its primary winding and the low input impedance of the first i.f. transistor. There are usually two or three stages of i.f. amplification depending on the type of receiver. The gain may be about 30 dB per stage.

R_1, R_2, R and C provide bias and stabilisation as usual. Unilateralisation may be applied by including series or parallel C and R between Y and Z in the figure in the position occupied by the dotted lines (see Section 3.1.1 and Fig. 3.3). The series arrangement is nearly always used because the capacitor constitutes a simple means of keeping the collector supply voltage from the transistor base.

The output transformer is tapped in both primary and secondary windings to provide proper impedance values for the output of the first transistor and the input of the second respectively.

3.3.2. Intermediate-frequency amplification using valves

A typical basic amplifying stage using a pentode is given in Fig. 3.14. The transformers are often tuned in both primary and secondary, and are identical. The action is similar, apart from the fixed tuning, to that of the pentode s.f. amplifier described in Section 3.1.2.

3.4. OTHER SUPERHETERODYNE RECEIVER CIRCUITS

Certain circuits, like those for automatic gain control and for heterodyning a c.w. signal, are described in the next chapter. Basic circuits for detection and low-frequency amplification are described in the two earlier volumes in this series (*Radio and Line Transmission 1 and 2*). Detection and noise limiting are referred to in the next section.

3.4.1. Detector and noise limiter

For v.h.f. reception, diode detection is nearly always used because of its simplicity. A noise limiter, also a feature of many communication receivers, usually follows. The limiter is designed to eliminate pulses of short duration, such as those from motor car and similar ignition systems, and of amplitude greater than some pre-determined level, often corresponding to about 50-85% modulation.

The operation is usually by the action of the noise pulse either cutting off a normally conducting diode which is in series with the signal circuit, or bringing into a state of conduction a diode (which is normally cut off) connected in shunt with the signal circuit. Thus the noise pulse is caused either to break the circuit to the a.f. amplifiers, or to short circuit the input to these amplifiers.

The noise limiter sometimes includes both series and shunt connected diodes.

The circuit of Fig. 3.15 shows the use of a series diode. In normal operation the voltage across the capacitor *C* follows the mean detector output voltage (the time constant is about one-fortieth of a second). If the mean output voltage is *V*, the voltage with respect to earth at each of the points *A, F* and *E* is $-V$. The voltage at *D* is $-0{\cdot}5$ V. The noise diode, therefore, conducts and the signal is taken through the low-resistance diode to the first a.f. amplifier in the usual way.

A spike of noise voltage of magnitude *V*, or more, volts immediately depresses the voltage at *A* by *V*, or more, volts and that at *D* by 0·5 V, or more, volts. The voltage at *E* cannot instantaneously change because the capacitor *C* needs a little time to charge and discharge, so the diode cuts off. No signal is now transferred to the a.f. amplifier.

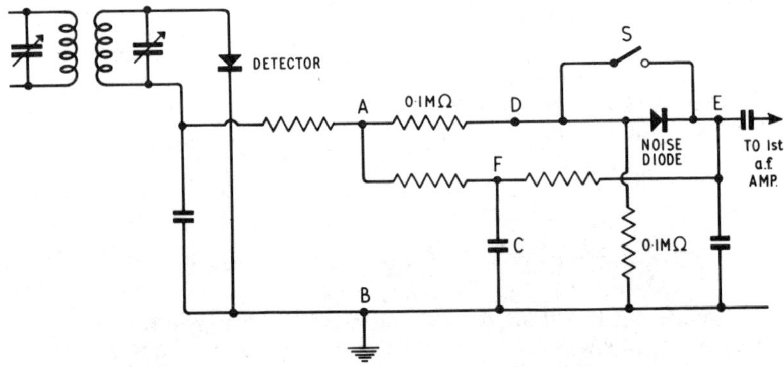

Fig. 3.15 Noise limiter

Normal operation is resumed when the noise voltage terminates—or, if the duration is unduly prolonged, when the voltage across *C* increases sufficiently to render the noise diode conducting.

The noise limiter is cut out of circuit by simply short circuiting the noise diode; the switch *S* is included for this purpose.

QUESTIONS

1. Draw the circuit diagram of one stage of a transistorised intermediate frequency amplifier normally used for a superheterodyne broadcast radio receiver. Neutralising and stabilising circuits should be shown. Describe the function of each of the components in the design.

   Why is this particular transistor circuit configuration normally chosen for this application? *(C & G)*

2. Discuss the advantages and disadvantages which result from the provision of considerable amplification at the signal frequency in front of the first frequency changer in a superheterodyne receiver.

3. State concisely the precautions necessary in the design and construction of amplifiers for use at v.h.f.

4. Explain why, for a given transistor, there is a frequency above which it is better to employ the common-base instead of the common-emitter configuration.

5. Why is neutralisation often necessary in transistor amplifier circuits? Sketch a circuit and explain how neutralisation may be applied for an i.f. transistor amplifier.

6. Sketch a triode valve circuit suitable for amplification at v.h.f. Why is a 'normally connected' triode unsuitable? Are pentodes suitable for use as v.h.f. amplifiers? Do they need neutralising?

7. Sketch cascode circuits using (a) transistors, (b) thermionic valves. What advantages are associated with the use of cascode circuits?

8. Sketch a circuit and explain carefully the action of a noise limiter suitable for use in a communication-type radio receiver.

9. Sketch a circuit and explain the action of a frequency changer and oscillator for a transistorised superheterodyne receiver.

10. Discuss the problems associated with oscillator circuits for superheterodyne receivers for use at v.h.f.

4

Coupled Circuits
Ganging and Tracking
Automatic Gain Control

4.1. COUPLED CIRCUITS

A requirement for a radio-frequency amplifier is that there should be a constant level of amplification over a given range of frequencies—sufficient to accept the carrier and side frequencies necessary for the proper conveyance of the intelligence contained in the signal. Outside the range of this passband the gain should fall as steeply as possible to zero. A simple tuned *LC* circuit does not achieve this, because over all it is not selective enough—the sides of the response curve fall too gradually, and it does not usually provide a sufficiently flat response close to the resonance frequency.

For circuits with variable tuning (and constant Q), the absolute selectivity decreases as the resonance frequency increases. With capacitor tuning the gain increases with frequency—permeability tuning is better in this respect (i.e. the gain changes less with change of frequency). There is a tendency to instability, or to distortion, when the frequency is below resonance, that is for all lower sideband frequencies. This may, however, be reduced by tapping down the inductor of the *LC* circuit. A transformer is easier to adjust than a tap, its use also has the advantage of removing the d.c. supply voltage from the tuned circuit. As we shall now see there are other weighty advantages to be derived from the use of coupled circuits (i.e. transformers).

When using inductively coupled circuits, it may be necessary to tune only one of the windings in order to reduce the number of variable capacitors (Fig. 4.1 (a)). Best results, however, are generally obtained when both primary and secondary coils are tuned (Fig. 4.1 (b)). The diagrams show both the input and the output connected across the whole of the appropriate winding—for proper matching it is often necessary to tap down the winding (see for example Fig. 3.13).

With the single tuned circuit, best results are obtained with L_1 as small as possible and the coupling as tight as possible. A large value of L_1 shifts the secondary response curve and gives a maximum response at a frequency above the natural frequency of the secondary tuned circuit. Since stray coupling modifies the selectivity characteristics, the primary winding should be at the earthy end of the secondary. With

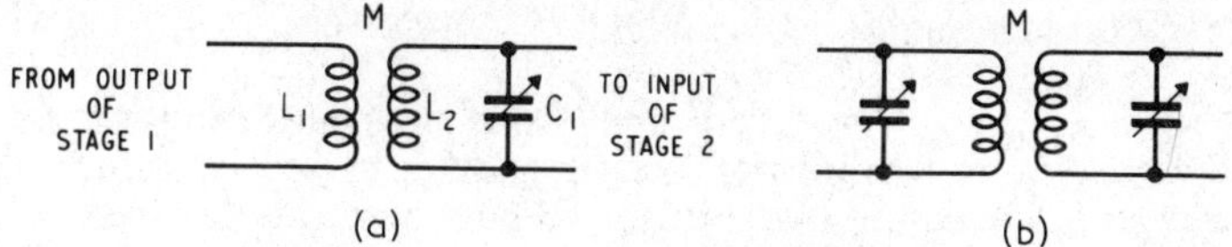

Fig. 4.1 Inductively coupled circuits

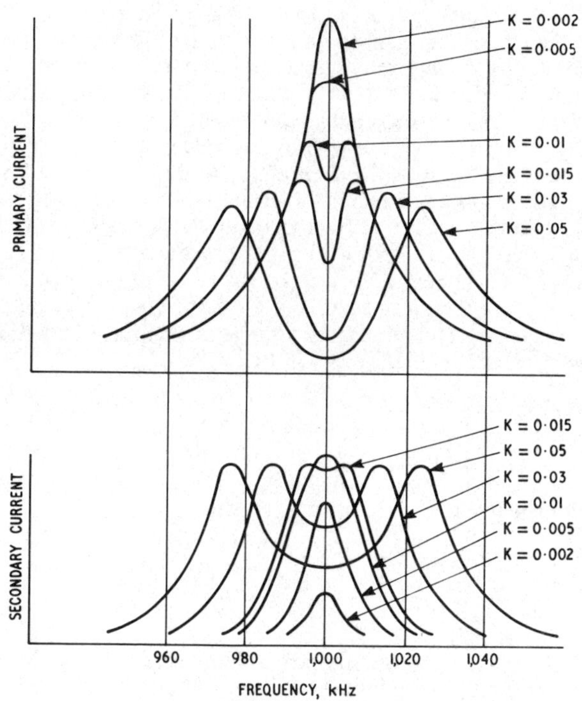

Fig. 4.2 Effect of value of coefficient of coupling

this arrangement the overall effective Q is rather less than that of the secondary alone.

With primary and secondary windings both tuned, as for i.f. transformers, it is usual with narrow band receivers, such as broadcast receivers, to tune both windings to the same frequency. (To obtain wide bandwidth with fairly steep cut-off to either side the different windings are tuned to different frequencies a few kilohertz apart.)

When the coupling is very loose the output from the secondary is very small but the selectivity is great (Fig. 4.2). The overall response curve is given by the product of the individual ones. As the coupling is increased the response increases and the selectivity decreases as described in the following sections.

4.1.1. Critical coupling–overcoupling

Increase of coupling from a very small value gives more output and less selectivity. At a certain value of coupling, called *critical coupling,* the output is a maximum (and this maximum output is obtained at the resonant frequency). For greater values of coupling the single hump of the resonance curve breaks into two to form double peaks which move further and further apart as the coupling is increased. In this condition the circuits are said to be *overcoupled.* The output voltage at the peaks remains substantially constant for quite a wide range of overcoupling values. What happens in fact is that the peak value of current in the primary winding decreases steadily as the coupling is increased from zero. At very low coupling values very little of the primary magnetic flux cuts the secondary winding and the secondary output voltage is small but increases with coupling. For quite a wide range of coupling, from the critical value upwards, the increase in the proportion of primary flux which cuts the secondary just about offsets the fall in primary current and flux so that the amount of flux which cuts the secondary remains practically constant at the frequencies of maximum response. The secondary voltage at the peaks likewise remains practically constant. This is shown in Fig. 4.2.

4.1.2. Hump spacing

The critical coupling value is

$$k_c = 1/Q$$

Hence, if the Q values for the primary and secondary circuits (assumed identical) are each 100, critical coupling is $1/Q = 0{\cdot}01$.

When the primary and secondary circuits are both tuned to the same frequency but the Q values are different (Q_1 and Q_2) the critical coupling is

$$k_c = 1/\sqrt{(Q_1 Q_2)}$$

For coupling k several times greater than critical, the spacing of the peaks of the double hump may be determined from the relationship

$$f_p \triangleq \frac{f_0}{\sqrt{(1 \pm k)}}$$

where the two values obtained for f_p are the approximate frequencies at which the peaks occur.

Example 4.1

Two circuits are resonant at 1 MHz; they are coupled with a coupling factor of 0·03. If critical coupling for the pair of circuits is 0·01 the the peaks of the humps occur at frequencies of about

$$\frac{10^6}{\sqrt{(1+0{\cdot}03)}} = 985 \text{ kHz}$$

and
$$\frac{10^6}{\sqrt{(1-0{\cdot}03)}} = 1\,015 \text{ kHz}$$

When k is much less than unity, as in these calculations, application of the binomial theorem shows that the hump peaks are at about

$$f_p = f_0\,(1 \pm \frac{k}{2})$$

The spacing of the humps, Fig. 4.3, is

$$2(f_p \sim f_0) = kf_0$$

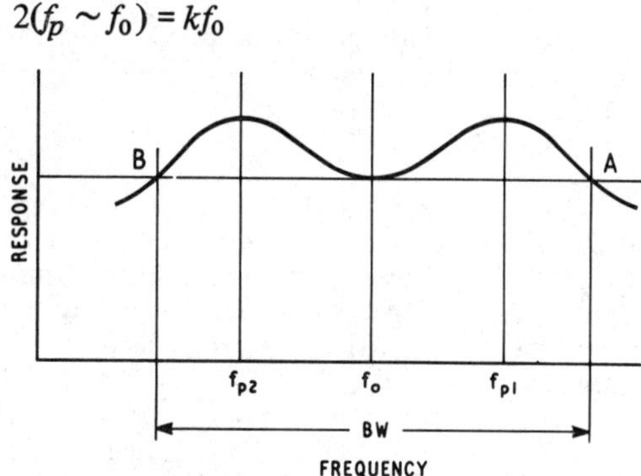

Fig. 4.3 Overcoupled circuit–response

If the bandwidth is regarded as that between frequencies for which the gain is never less than it is at the trough of the curve, that is between A and B in Fig. 4.3, it can be shown that this is

$$BW = \sqrt{2}\,kf_0$$

The serious drop in amplification which exists in the vicinity of the middle frequency for overcoupled circuits can be overcome to a large extent by introducing into the amplification chain a single LC circuit having a Q value about half that of either the primary or secondary of the tuned transformer, and tuned to be resonant at the middle frequency of the overcoupled circuit. The response due to the peak of the single circuit just about makes up for the trough in the response of the overcoupled circuit.

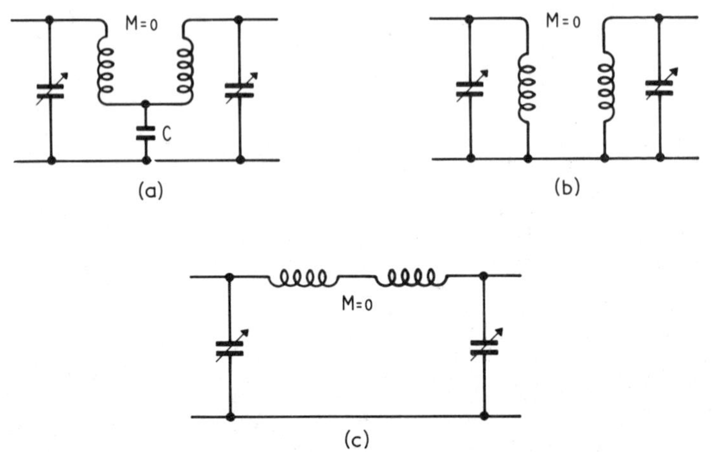

Fig. 4.4 Shunt capacitance coupling

4.1.3. Methods of coupling

We have discussed circuits with mutual inductance coupling. Other methods of coupling giving similar response curves are possible and are sometimes used. Fig. 4.4 (a) shows *shunt capacitance coupling.* There is no mutual inductance between the coils. The degree of coupling depends on the value of C, the smaller the value the greater the coupling. The equivalent circuit for $C = \infty$ is given in Fig. 4.4 (b) (C is virtually a short circuit). This shows that in this conditions the circuits are divorced from one another—apart from the single conductor at the base. The coupling is zero. Fig. 4.4 (c) gives the equivalent circuit for $C = 0$, i.e. an open circuit, and the coupling is manifestly maximum. Values of C intermediate between these two extremes give corresponding intermediate values of coupling. For a frequency of about 1 MHz, a typical value for C is 0·02 μF.

Fig. 4.4 (a) shows *series capacitance coupling*—or *top capacitance coupling.* Again there is no mutual inductance. C is the coupling capacitance and the coupling increases from zero to the maximum value as C is increased from zero to infinity. Fig. 4.5 (b) shows the equivalent circuit for $C = 0$ (i.e. an open circuit) which leaves the circuits entirely uncoupled. When C is infinity (i.e. effectively a short

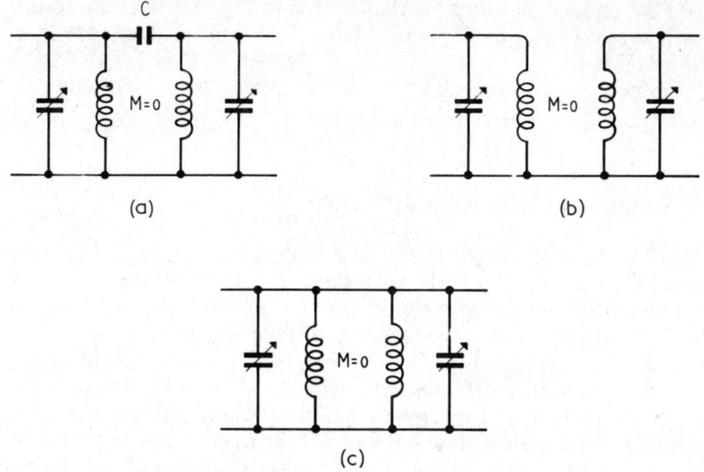

Fig. 4.5 Series (or top) capacitance coupling

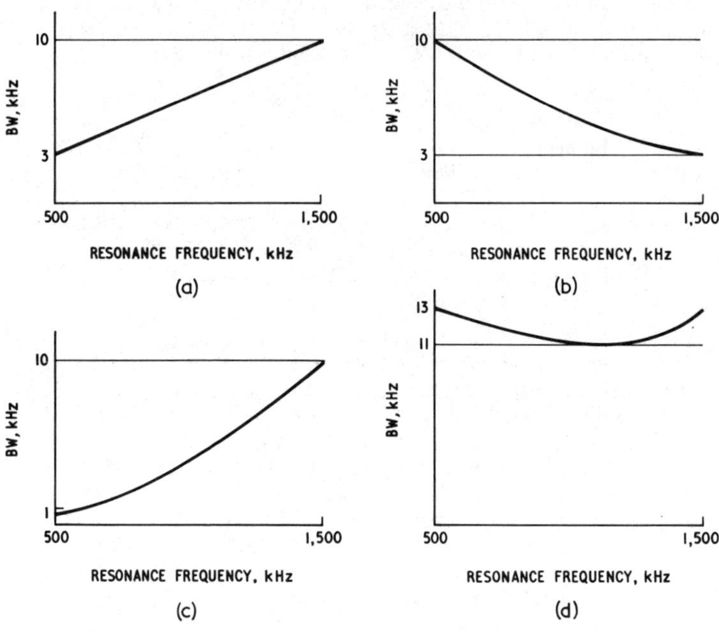

Fig. 4.6 Variation of bandwidth with resonance frequency

circuit) the equivalent circuit is that shown in Fig. 4.5 (c). The coupling is clearly a maximum. C is typically 1 μF for resonance frequencies of about 1 MHz.

Combinations of types of coupling are referred to in the next section.

4.1.4. Variation of bandwidth with tuning

For fixed tuned circuits the bandwidth for a given circuit configuration is fixed. When the tuning is variable, the bandwidth also varies.

For the mutually coupled circuit of fixed M the bandwidth increases linearly with rise in resonance frequency. This is shown in Fig. 4.6 (a).

For the shunt capacitance circuit, with fixed value of coupling capacitor, bandwidth decreases with increase of resonance frequency (Fig. 4.6. (b)) while for top capacitance coupling the bandwidth increases with resonance frequency (Fig. 4.6 (c)).

The shunt capacitance curve changes in the opposite direction to the other two. Judicious combination of (b) with either (a) or (c) leads to a much smaller variation of bandwidth with resonance frequency than results with the use of only one type of coupling.

The usual combination is of mutual inductance with shunt capacitance. Typically this gives a fairly small bandwidth variation as shown in Fig. 4.6 (d) for the 500–1 500 kHz frequency range.

4.1.5. Provision of variable bandwidth

For fixed tuned i.f. circuits it is possible to make provision for the manual variation of bandwidth in order to cater for the different operating conditions encountered at different times. This provision can be made in several ways, including:

1. Mechanically varying the mutual inductance coupling.
2. Variation of capacitance coupling (but this simultaneously changes the middle frequency of the response curve).
3. By coupling a third coil, shunted by a variable resistor, to the primary and secondary windings. When the resistance is set to a high value the third coil has little effect and the coupling between primary and secondary gives maximum selectivity, i.e. minimum bandwidth. With reduced resistance the third coil absorbs power thus loading the tuned circuits and widening the bandwidth. This automatically reduces the gain and sometimes other circuits are included to make good this deficiency.

It can also be arranged that the selectivity varies automatically depending on the strength of the received signal, or on the amount of interference, or both.

4.2. CRYSTAL FILTERS

To improve selectivity the slopes of the skirts of the response curves must be steepened, ideally until the response curve is a rectangle (centred on the carrier for double sideband reception, or with the carrier at one edge if a single sideband only is to be accepted), and the acceptance band narrowed. For telephony the bandwidth must not be reduced too much or first quality and then intelligibility will be lost. For c.w., for example morse code transmissions, a quite narrow bandwidth, perhaps only 1 kHz or even 100 Hz, may be resorted to in cases of extreme interference.

4.2.1. Narrow band reception

For extreme selectivity the Q of an ordinary LC circuit is too low. A fairly simple way of obtaining a very narrow passband is to use a quartz crystal filter in conjunction with the usual i.f. circuit tuning arrangements. The use of a crystal to couple the output of one amplifier to the input of the next (Fig. 4.7) causes a virtual open circuit to exist between the two circuits (that is between A and B) except at and very

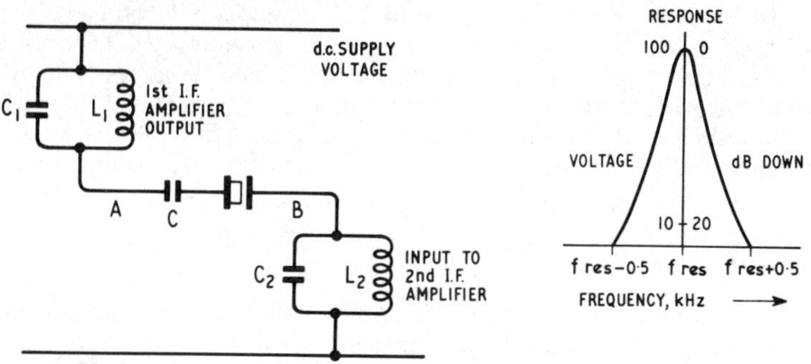

Fig. 4.7 Crystal filter–basic principle

close to the series resonance frequency of the crystal. Over this very narrow frequency band there is effectively a short circuit between A and B and maximum coupling between the first amplifier and the second.

The narrow peaked type of response which is ideally obtained is shown in the figure. Selectivity at and near resonance can further be sharpened, at the expense of a slight widening of the response curve at its skirts, by mistuning the two LC circuits in opposite directions.

The ideal performance of the circuit of Fig. 4.7 is upset by the capacitance across the crystal, and for the circuit to function as described it is necessary to cancel the effect of the capacitance. This can be done by coupling from B into L_2 a voltage which just equals and opposes the effect of the capacitance.

More usually the crystal capacitance is balanced by using two crystals, or a crystal and a capacitor, one at each end of a centre-tapped i.f. transformer winding. Fig. 4.8 shows a suitable arrangement using two crystals. The series resonance frequencies of the two crystals are arranged to be different. If the frequency spacing is two or three

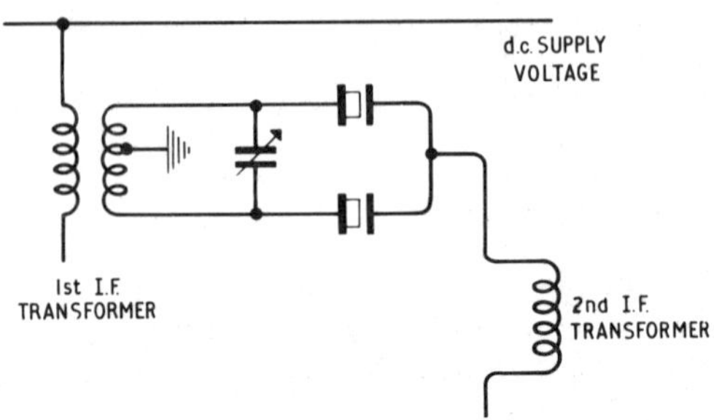

Fig. 4.8 Crystal filter circuit

hundred hertz, a good narrow bandpass characteristic is obtained, suitable for code reception. With a wider spacing–a few kilohertz–a bandpass characteristic may be obtained which is suitable for telephony reception. With the two crystal arrangement there is greater attenuation outside the passband than with the single crystal unit–also the slope of the sides of the response curves is steeper.

4.2.2. Transfilters

There are ceramic filters, called *transfilters,* which are sometimes used in addition to, or instead of, i.f. transformers. The filter may be in the form of a disc of piezo-electric ceramic material with an electrode on either side, or, as shown in Fig. 4.9 (a), with an electrode A covering one face and two electrodes, an outer ring B and a central disc C, on the other face. The disc is cut so as to resonate at the desired intermediate frequency or at a sub-harmonic. In the form shown in Fig. 4.9 (a) the impedance between C and A is several times greater than that between B and A. Hence, if the output from an i.f. amplifying transistor is applied between C and A, and if B and A are used as the input source to a further amplifying transistor, both operating in the usual common-emitter configuration, an approximate impedance match is obtained.

Fig. 4.9 (b) shows a basic circuit employing transfilters. The inclusion of the single crystal filter shown in the cathode lead may be necessary to preserve the stability of the circuit.

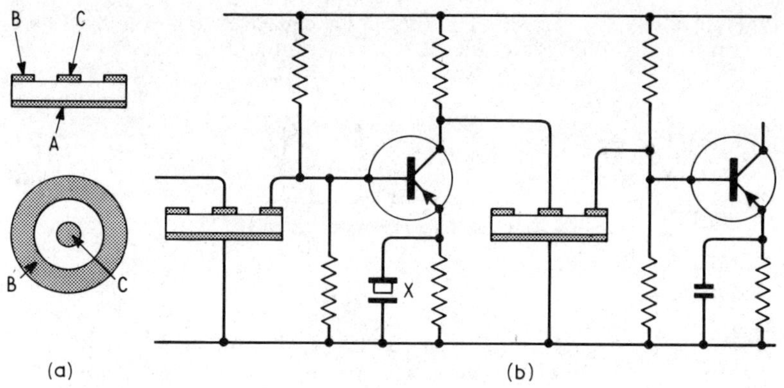

Fig. 4.9 Transfilter

4.3. GANGING AND TRACKING

Signal-frequency circuits normally need to cover a range of frequencies. The oscillator circuit needs to be tuned through an equal frequency difference.

When the oscillator operates above the signal frequency, the ratio maximum/minimum frequency is less for the oscillator than it is for the signal frequency circuits. If both circuits are to be tuned by capacitance and if the capacitors are to be ganged, that is driven on the same shaft by the same control, it is necessary either that the oscillator capacitor plates are specially shaped, or that some other suitable arrangements are adopted to ensure that the frequency difference between oscillator and signal remains substantially the same throughout the tuning range.

For a receiver which covers a number of wavebands, the oscillator tuning capacitor plates can be specially shaped to suit only one of the bands–for the other bands other arrangements would need to be made. In practice the oscillator tuning capacitor is usually of the same type and characteristics as the signal frequency capacitors.

When the oscillator operates above the signal frequency, the oscillator tuning inductance is lower in value than those of the signal-frequency circuits. To tune over the smaller oscillator frequency range, the oscillator capacitance variation needs to be less than that of the signal-frequency circuit capacitance. This is usually arranged by putting a fixed capacitor C_P (the *padder*) in series with the tuning capacitor (Fig. 4.10). The capacitance of the padder is fairly large, comparable with the maximum value of the tuning capacitance: its presence therefore has little effect when the tuning capacitance is at a minimum but it has a marked effect on the total maximum capacitance.

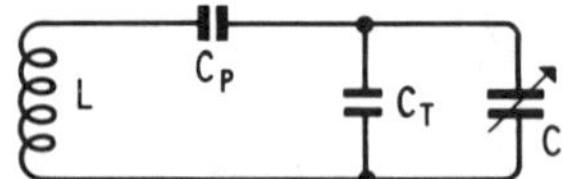

Fig. 4.10 Padder and trimmer capacitors

Example 4.2

If the tuning capacitance (Fig. 4.10) has maximum and minimum values respectively of 460 pF and 15 pF, and if the stray capacitances and the trimmer C_T total 40 pF, the overall capacitance range (with no padder) is 500 pF to 55 pF. If the required oscillator frequency range is, say 2:1 the necessary capacitance range is 4:1. Putting a padder C_P = 290 pF in series with C gives a maximum capacitance of

$$\frac{500 \times 290}{790} = 184 \text{ pF}$$

and a minimum capacitance of

$$\frac{55 \times 290}{345} = 46 \text{ pF}$$

giving the required maximum/minimum ratio of 4:1.

The mere choice of a value for C_P which gives the correct maximum/minimum capacity ratio does not ensure that the oscillator frequency is always the correct amount above the signal frequency,

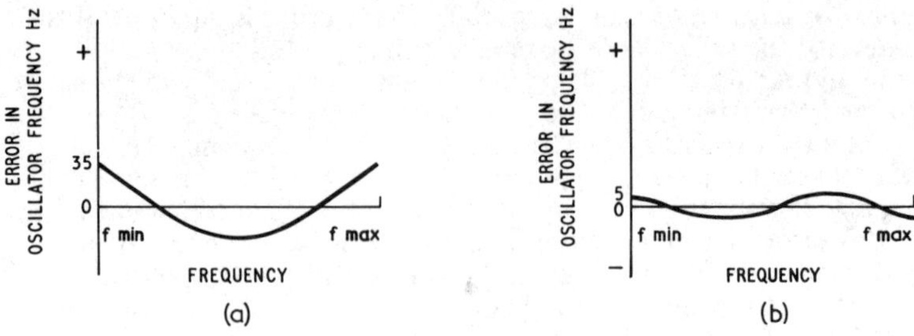

Fig. 4.11 Tracking error

i.e. it does not ensure that the *tracking* is perfect. In fact, tracking can not be perfect, but it can be made exact at as many frequencies as there are controllable fixed components.

Without C_T there are only two fixed reactances at our disposal (L and C_P) and, therefore, only two frequencies at which the tracking is exact. The error in the oscillator frequency is likely to be quite large in places (Fig. 4.11 (a)).

Introducing the third control C_T results in there being three points at which the tracking error can be made to be zero (Fig. 4.11 (b)). This reduces the maximum error considerably—in the ratio of possibly five or six.

The bulk of the selectivity is in the i.f. circuits. In using the receiver, therefore, the operator, who is seeking the strongest and least distorted output, automatically tunes so that the signal entering the i.f. stages is at the correct (i.e. intermediate) frequency. With tracking error present this means that the signal-frequency stages are mistuned to the extent of the tracking error. If the signal-frequency range is high, the s.f. circuits are not likely to be very selective and the effect of slight mistune is probably of little consequence. At lower frequency ranges, however, the s.f. tuning may be more selective and the effect of mistune more serious. The points of zero tracking error can be placed at will by correct choice of C, C_P and C_T. If the s.f. circuits are not very selective, the points of zero error may be chosen to be equally distributed along the frequency scale so that the four points of maximum error are all equal and as small as possible. If the effects of error may be important it can be arranged that the minimum error occurs where the signal-frequency selectivity is greatest—often at the low-frequency end of the band (if Q is approximately constant, the bandwidth is least at the low frequencies).

Having chosen the frequencies of zero error, and after ensuring that the i.f. circuits are correctly aligned, the receiver is set up by:

1. applying a signal at the highest frequency of zero error, and adjusting C_T for maximum output, and
2. applying a signal at the lowest frequency of zero error and adjusting C_P for maximum output.

This is because C_P has little effect at the high frequency and C_T has little effect at the low frequency. Both procedures are repeated once or twice until the best results are obtained. There is no point in adjusting for the middle frequency.

If L is decreased (and C_P and C_T, therefore, increased), the error curve shape is altered and gives less error at the lower frequency end of the band and more error at the higher frequency end. The middle zero error frequency (if the other two are not altered) is shifted towards the low-frequency end of the band.

If one part of the band is used much more than, or to the exclusion of, the remainder, tracking adjustments are made to give good tracking in the desired sector even if this results in bad tracking elsewhere.

Example 4.3

A receiver having an intermediate frequency of 465 kHz is required to tune over a range of 600 kHz to 1 800 kHz with a ganged variable capacitor having a range of 320 pF per section. Calculate the values of:

1. the minimum capacitance needed in the r.f. circuit,
2. the inductance required in the r.f. circuit,
3. the padding capacitance required in the local oscillator circuit assuming that the minimum value of the variable capacitance is the same as that found in (1) above,
4. the inductance required to tune the local oscillator.

We may proceed thus:

The frequency range for the signal is 1 800 : 600 = 3 : 1, and the corresponding capacitance range is thus 9 : 1.

If the minimum capacitance is C_m, and assuming that the maximum capacitance is $320 + C_m$, we have

1. $$\frac{320 + C_m}{C_m} = 9$$

therefore $C_m = 40$ pF

2. From $f = 1/[2\pi\sqrt{(LC)}]$ and substituting $f = 600$ kHz

and $C = 360$ pF (or $f = 1\,800$ kHz and $C = 40$ pF)

$L = 193\ \mu$H

3. The local oscillator frequency needs to be 2 265 kHz when the signal-frequency is 1 800 kHz. The local oscillator frequency is 1 065 kHz when the signal-frequency is 600 kHz. The ratio is 2 265/1 065 = 2·13 : 1. The capacitance ratio thus needs to be $2{\cdot}13^2 : 1 = 4{\cdot}54 : 1$.

The maximum capacitance is

$$\frac{360\ C_p}{360 + C_p}$$

where C_p is the padder capacitance in picofarads.

The minimum capacitance is

$$\frac{40\,C_p}{40 + C_p}$$

Hence

$$\frac{9\,(40 + C_p)}{360 + C_p} = 4{\cdot}54$$

and

$$C_p = 285 \text{ pF}$$

4. Substitute $f = 2\,265$ kHz and $C = 40 \times 285/(40 + 285)$ pF in $f = 1/[2\pi\sqrt{(LC)}]$ to obtain the value of L_{osc} for the oscillator circuit.

$$L_{osc} = 140\ \mu\text{H}$$

4.4. AUTOMATIC GAIN CONTROL

Automatic gain control (a.g.c.) is designed to give the condition that, for a given degree of modulation, any variation in carrier amplitude shall produce but little change in the amplitude of the receiver output. The output level may of course be adjusted to any desired level using manual gain control.

4.4.1. Advantages of using a.g.c.

The use of a.g.c. minimises the effects of fading, prevents the output level from changing violently when tuning from weak to strong transmissions, and enables the detector and following stages to be designed on the basis of a fairly constant and known input.

4.4.2. Requirements

The a.g.c. must normally be controlled by the amplitude of the radio-frequency carrier and be independent of the amplitude of the signal at the output. If it is not so independent, with a sound receiver, for instance, the action is to even out the difference between loud and soft passages.

Automatic gain control operates by reducing the signal level. It should, therefore, not come into operation below the minimum level of signal strength deemed adequate–i.e. its action should be delayed.

The application of a.g.c. should not introduce distortion, nor should it cause the frequency of the frequency changer oscillator to shift. It should act reasonably rapidly so that normal fading may be counteracted.

4.4.3. Method of application

The control may be effected by applying a voltage to some of the amplifying devices–if these are valves the voltage is applied in the grid–cathode circuit and its polarity must be such as to make the grid more negative to reduce the signal level. For npn transistors a negative voltage to the base is needed. For pnp transistors the opposite polarity is required. In transistor circuits additional control is often applied as will be described later.

The control voltage is derived, by rectification, from the r.f. output; it may be taken from the detector or a separate rectifier may be used. The voltage is carefully filtered to ensure that only pure d.c. is fed back to the earlier stages of the receiver, otherwise instability results.

In valve receivers a.g.c. may be applied to any or all of the valves handling r.f. signals–i.e. to the s.f. amplifier, frequency changer(s) and the i.f. stages (Fig. 4.12). In transistor receivers, with no circuit element

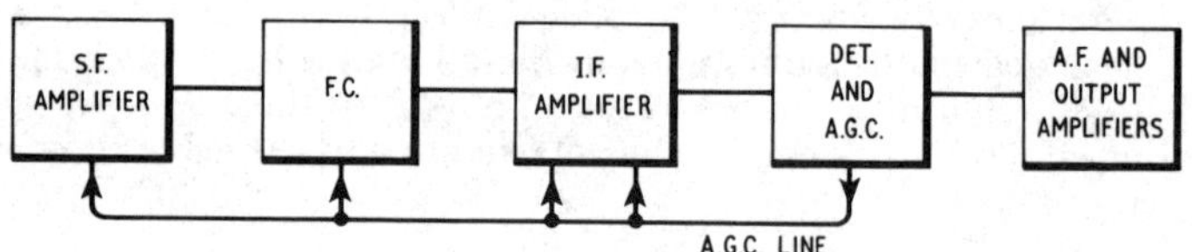

Fig. 4.12 Application of a.g.c.–valve receiver

quite comparable with the variable mu valve, the a.g.c. is often applied only at the first i.f. stage–but there are many exceptions to this as will be seen in Chapter 10.

4.4.4. Types of a.g.c.

There are two general types of a.g.c.–*non-amplified* and *amplified.* The controlling voltage in the former cannot be greater than the rectified output from the last i.f. amplifier, for the latter it can be.

In addition, the control may be *simple,* i.e. come into operation at all signal levels, or it may be *delayed* so that it does not come into effect until the signal reaches some minimum level. Conditions and needs vary with the type of receiver, but in general it can be said that it is preferable if the a.g.c. does not come into operation when the input signal is less than 1 mV.

The curves of Fig. 4.13 give an idea of the output level which results from input voltages between a few microvolts and one volt with different types of a.g.c. for a given receiver. It is clear that simple a.g.c.

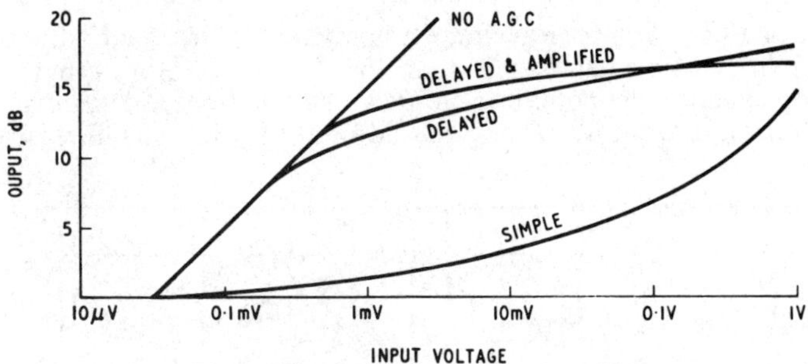

Fig. 4.13 Performance of different types of a.g.c.

unduly reduces the output for small and medium level inputs.

The control, of course, can be adjusted so that the output scale is with reference to any desired level.

4.4.5. Transistor receivers

The application of a.g.c. in transistor receivers is not so straightforward as it is for valve receivers. There are two general methods of direct control. One is to use the a.g.c. voltage to control the gain of some of the transistors, the other is to cause the signal to introduce an amount of damping of the first i.f. transformer. Less directly, the a.g.c. voltage may be used to control the magnitude of the voltage output of the local oscillator(s) so that the latter, and consequently the mixer output in relation to the signal input, falls with increasing signal.

4.4.6. Gain control by variation of transistor gain

There are two ways in which the variation in the carrier level can be caused to change the gain of one or more of the transistors:

1. The increase in a.g.c. voltage brought about by an increase in signal is used to alter the transistor base voltage in such a direction as to reduce the emitter current and hence the gain. That is, to shift the voltage in a positive direction for a pnp transistor and in a negative direction for an npn transistor.

2. The increase in a.g.c. voltage brought about by an increase in signal is used to alter the base voltage in such a direction as to increase the emitter current (and hence the collector current) and so, with the aid of a resistor in series with the load in the collector circuit (i.e. in series with the transformer primary in an i.f. circuit), to reduce the collector voltage, and hence the gain. This method is in less common use than the first.

Fig. 4.14 shows a typical circuit employing the first method. Two i.f. amplifiers and a detector D are included. The detector output is developed across the potentiometer P which functions as a manual gain control in front of the next (audio frequency in a sound receiver)

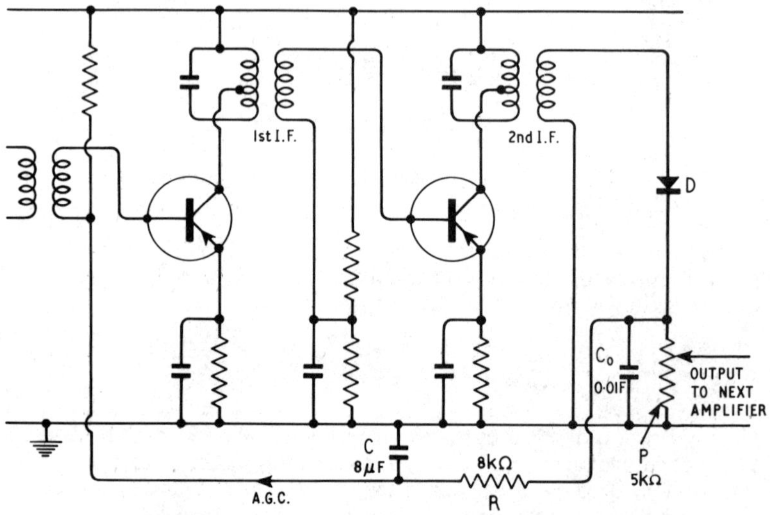

Fig. 4.14 A.G.C. by control of emitter current

amplifying stage. The voltage across P is also taken, through R, to the base of the first transistor. The greater the rectified output from D (and this depends on the carrier level and not on the setting of P), the greater is the voltage applied to the transistor base. The voltage polarity must be such as to reduce the gain.

Typical measurements using this simple system show an input signal change of 30 dB resulting in an output change of 6 dB.

The use of this method gives maximum emitter current when the signal is weakest. The greater the emitter current the greater the damping of the preceding transformer and the greater the bandwidth. On the other hand, a strong signal leads to minimum emitter current, minimum damping and the narrowest bandwidth. This is the opposite of what is usually desired, but the effect is seldom serious. The operation of the a.g.c. leads to a small amount of mistuning but with good design this effect also is not serious.

Fig. 4.15 shows a typical basic circuit for the type of control in which increasing signal strength results in decreasing voltage at the collector of the controlled transistor. The polarity of the a.g.c. voltage is the opposite of that needed for the previous method (i.e. it is negative for pnp, positive for npn). If a common emitter configuration is used for the detector, the a.g.c. voltage must be taken from the emitter; if a diode is used, it must be connected in the reverse polarity to that of Fig. 4.12. Increase of signal strength results in the base voltage of the first transistor moving negatively, and this increases the emitter current and the collector current. The voltage drop across R_g (probably about 1 kΩ) increases and the collector voltage falls, as does the gain. C_d must be included as its function is to maintain the the bottom of R_g at earth potential to r.f.

The increased emitter current which results from a strong signal gives greater damping of the i.f. transformer and a wider bandwidth— and vice versa. This is usually an advantage. The action of the a.g.c. does, however, introduce some mistuning.

Fig. 4.16 shows the waveforms for both methods. At (a) is shown the output from the last i.f. stage for each of three signal levels during

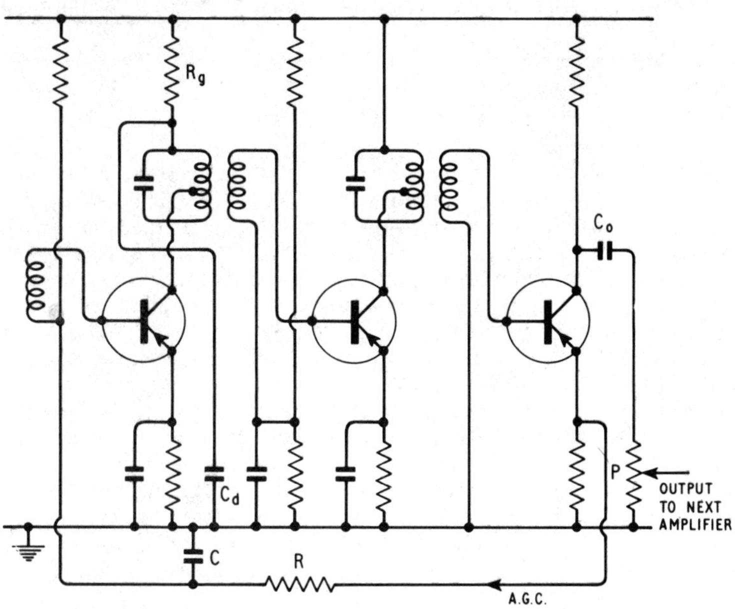

Fig. 4.15 A.G.C. by control of collector voltage

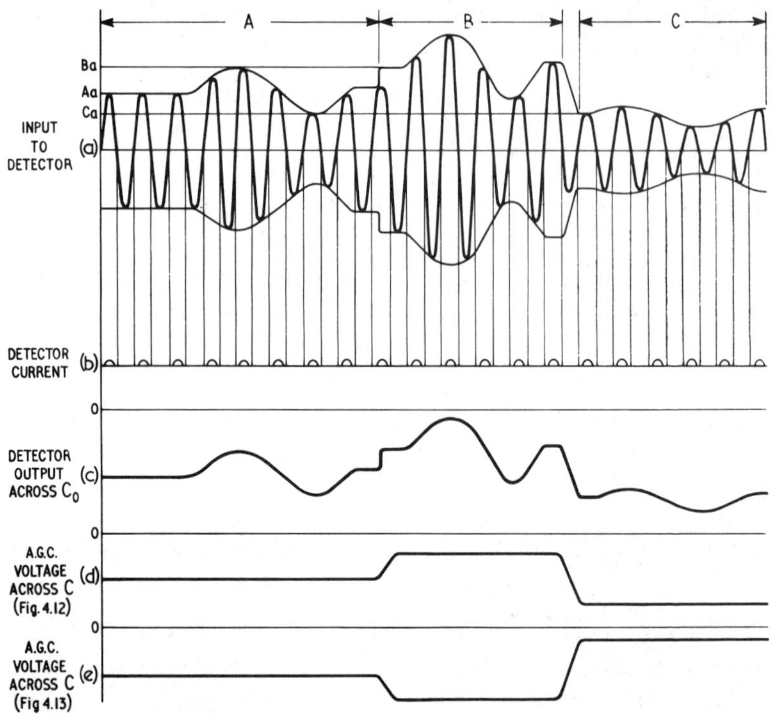

Fig. 4.16 A.G.C. waveforms

periods *A*, *B* and *C*. Referring also to Fig. 4.14 it can be seen that the waveform of Fig. 4.16 (a) is applied to the diode *D* which passes only positive pulses (Fig. 4.16 (b)) sufficient to keep C_0 charged. The time constant of PC_0 is about 50 μs (*P* is about 5 kΩ, C_0 is about 0·01 μF). This is long compared with the periodic time of the i.f. variations, and the voltage across C_0 is, therefore, not able to follow the i.f. changes; instead it traces the peak value (Fig. 4.16 (c)). There is very little current drain; if there were the voltage across C_0 would be less than the peak value of the i.f. waveshape.

Subject only to any necessary further amplification, the voltage tapped off at the slider of *P* is the output voltage. The a.g.c. voltage (the whole voltage across *P*) is applied across *RC* and thence to the base of the first i.f. transistor. For a sound receiver the time constant of *RC* is about 0·1 *s* ($R \triangleq 8$ kΩ, $C \triangleq 12$ μF), which is long compared with the a.f. variation of a.g.c. voltage, so that the voltage across *C* is a smoothed version of the a.f. voltage (Fig. 4.16 (d)) and its value approximates to the peak carrier voltage of Fig. 4.16 (a).

The a.g.c. voltage is at its most negative during period *C*, thus giving maximum gain when the signal is at its weakest. It is less negative during period A when the signal strength is greater, and the gain is thus reduced. During period B–maximum signal strength–the a.g.c. voltage is at its least negative value, and this gives the least gain.

Note that the manual gain control P may be used to vary (ultimately) the strength of the output signal, but that it does not affect the a.g.c. circuit. Thus the manual gain control can be set to give the desired output level and all transmissions, subject to the limitations of the a.g.c. system, will be received at this general level provided that the transmission has the necessary minimum amplitude at the input.

For the a.g.c. system of Fig. 4.15, the a.g.c. voltage changes need to be reversed (Fig. 4.16 (e)).

Because of the element of distortion and of restriction of available range of characteristic introduced by these forms of a.g.c., their use is generally restricted to the early stages of a receiver where the signal level is low.

Both these forms of a.g.c. operate at all signal levels however low. The type now to be described, does not come into operation until the carrier reaches a certain, pre-determined level.

4.4.7. Gain control by diode damping

For this method of gain control the a.g.c. voltage, with the aid of an additional diode, is used to damp the first i.f. transformer thus reducing its output. This method is often used in addition to that employing control of the transistor gain in accordance with the strength of the received carrier. When this is done a closer, and often wider, control of output is obtained. Measurements previously quoted for a.g.c. without damping diode showed a 6 dB change in output resulting from a 30 dB change in signal input. With the damping diode added to the same circuit, an input change of 50 dB resulted in an output change of only 4 dB.

The diode is shown at A in Fig. 4.17. Its left-hand end is taken through L_1 to a point B of fixed potential (say −2·5 V) in the emitter circuit of the second i.f. transistor. The right-hand end of the diode goes to the emitter of the first i.f. transistor. With a weak signal, the emitter of the first i.f. transistor is at a voltage of, say, −0·60 V and there is no current in A. As the signal increases, the a.g.c. line brings an increase in the base voltage of the first i.f. transistor, i.e. it becomes less negative. The emitter voltage must follow, and it also becomes less negative. At some signal level the emitter voltage becomes −0·25 V and the diode A begins to conduct thus damping L_1C_1 somewhat. Since L_1 is coupled to the other transformer windings the effect is to reduce the transformer output. The stronger the input signal, the greater is the reduction in the negative voltage at the emitter of the first i.f. transistor, the greater is the forward voltage across the diode A, and the greater is the damping and reduction in transformer output.

The signal level at which the diode damping becomes operative depends on the voltage selected for point B. Thus it is possible to

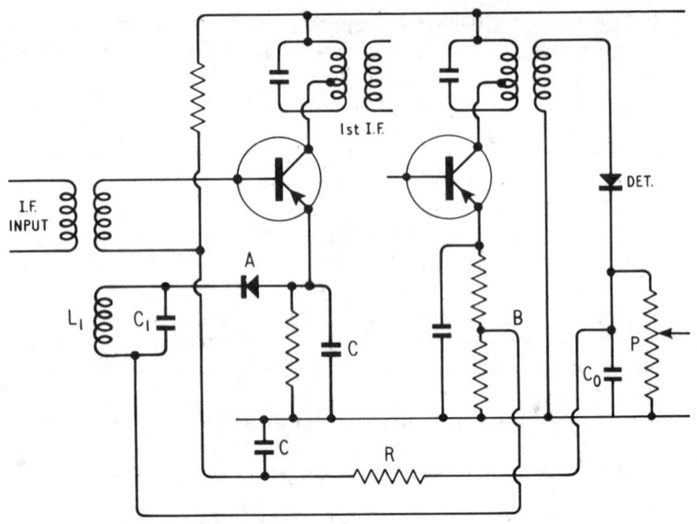

Fig. 4.17 A.G.C. by diode damping

select the input level at which the a.g.c. comes into operation.

When the diode damping control is in operation, the transformer bandwidth is increased, as has been mentioned. This is usually desirable. The considerable damping of a powerful local signal is also very advantageous as it helps to avoid overloading of later stages.

4.4.8. Valve receivers

For valve receivers, valves are available in which the g_m value varies smoothly over a very wide range of grid bias values. This makes the provision of a.g.c. relatively simple and trouble free. The control may be applied to all the valves between the aerial and the detector (Fig. 4.18 (a)).

In its simplest form (compare Fig. 4.14) the output across the detector *CR* circuit is used as the a.g.c. voltage and taken to the grids of the earlier stages. The greater the detector output, the greater is the output across PC_0 (negative at the top with respect to the bottom), the greater is the bias applied to the grids of the variable–mu valves, and the greater is the reduction in gain.

A separate diode may be used for a.g.c. and it may derive its input from its own i.f. stage. This i.f. stage need not be subject to a.g.c. action and it is thus able always to supply maximum input to the a.g.c. system. The a.g.c. diode may with advantage be fed from the primary of the transformer (Fig. 4.18 (b)) because here the tuning is

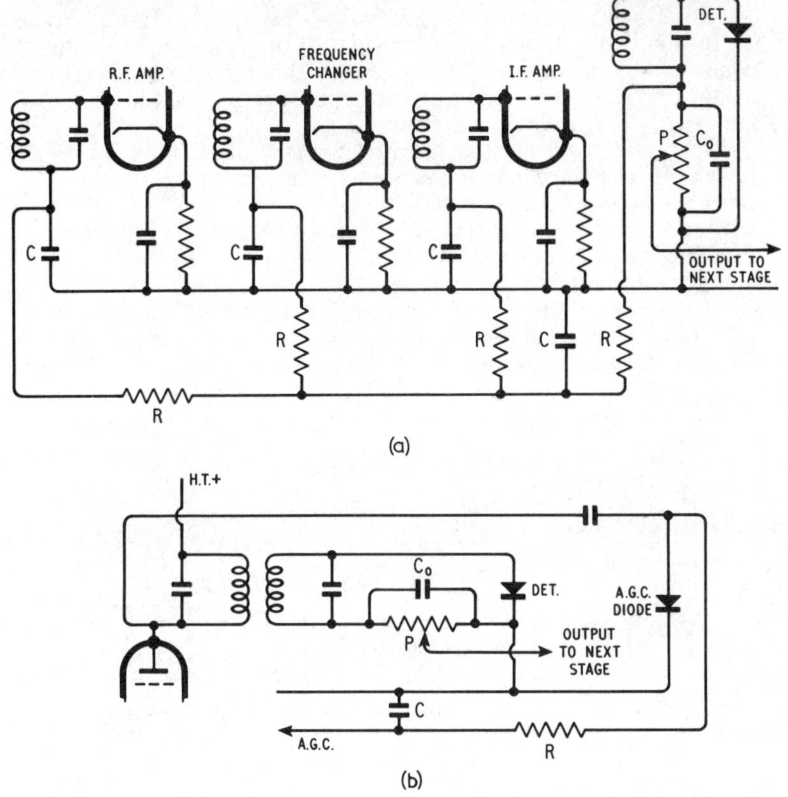

Fig. 4.18 A.G.C. for valve receivers

less selective than at the secondary and the change in output is less sudden when tuning to, or through, a station. The action of the a.g.c. is, therefore, less noisy.

The various time constants are the same as for transistor circuits, but because of the much higher impedance in valve circuits the reactance values can be greater and the capacitances correspondingly less.

The damping and hence loss of amplification introduced by the diode increases as the resistance of P is reduced. For sound receivers P is made about 0·5 MΩ, with C_0 about 100 pF. The time constant is about one-twentieth of a millisecond.

The RC filters in the a.g.c. leads need to have time constants of about one-tenth of a second and, whereas in a transistor circuit this is obtained by R = 8 kΩ and C = 12 μF for the valve circuit, we can have much greater R (1 MΩ) and smaller C (0·1 μF). These filters remove any remaining r.f. and a.f.

The r.f. must be removed to prevent the possibility of instability or the introduction of interference whistles. If the a.f. voltages are not removed they provide their own a.g.c. and thus reduce the effective modulation ratio. This reduces the amplitude at the peaks of modulation and (relatively) increases it at the troughs.

Separate C and R must be provided for each stage supplied with a.g.c. otherwise the common impedance to two or more circuits increases the possibility of interference whistles.

In order that the a.g.c. shall not come into action until the signal amplitude has a certain minimum value, the a.g.c. diode may be biased to prevent conduction until the signal is of the required strength. Fig. 4.19 shows a way of doing this. D_2 is the detector the rectified output of which, from the manual volume control P, is taken to the first post detector amplifier. The cathode resistor of this amplifier is in

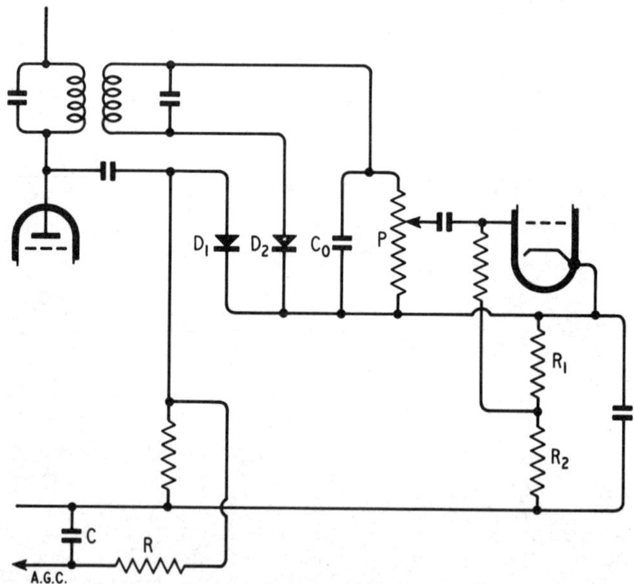

Fig. 4.19 Delayed a.g.c.

two parts of which R_1 provides the bias for the amplifier. The voltage across R_1 and R_2 together acts as a bias in series with the a.g.c. diode (positive at the top of R_1), so that the diode cannot conduct until the signal peak is equal to this bias. There is no a.g.c. signal developed, therefore, below this level. Some distortion can result from this type of a.g.c. because, if the positive envelope of the signal is such as to sometimes exceed and sometimes be less than the delay voltage across R_1 and R_2, the a.g.c. diode switches on and off. Since it causes damping when it is on it distorts the crests of the modulation envelope and not the troughs. This results in the generation of a.f. harmonics.

For reception in the higher frequency ranges the a.g.c. is often not applied to the frequency changer because changes in current here cause changes in valve capacitance and may give rise to frequency drift of the oscillator. This effect is not serious at the lower radio frequencies.

For reception in the higher frequency ranges it is also often important to obtain a good signal-to-noise ratio at the front of the receiver. The a.g.c. is often, therefore, not applied to the first s.f. amplifier so that maximum gain may be obtained (and, therefore, the best signal-to-noise ratio). On the other hand, excessive signal leads to overloading and cross-modulation and to avoid this the provision of manual and/or automatic gain control early in the receiver is desirable.

Two a.g.c. troubles are:

1. The tendency of the last i.f. amplifier to overload despite the action of the a.g.c.
2. The already mentioned tendency for delayed a.g.c. to introduce distortion for signals between certain levels.

These difficulties are removed or considerably reduced by the use of *amplified a.g.c.* The general principle of operation is that the level of the a.g.c. voltage is raised above that of the detector output. This is accomplished in one of two ways:

1. The a.g.c. diode may be fed from an additional stage of amplification to which a.g.c. is not applied, or
2. d.c. amplification may be applied between the a.g.c. voltage source (i.e. the detector or the a.g.c. diode output) and the controlled valves.

Example 4.4

A receiver has two i.f. amplifier valves and one r.f. amplifier, each having an a.g.c. characteristic of 3·5 dB/V, and a frequency changer valve with a control of 2·5 dB/V. A change from 10 μV to 100 mV in input produces a 6 dB change in output; find the corresponding change in a.g.c. voltage.

The r.f. and the two i.f. amplifiers provide a control of 3·5 dB/V–this is a total of 10·5 dB/V.
The frequency changer provides 2·5 dB/V.
The total control is 13 dB/V.
The input level ratio is from 0·1 V to 10^{-5} V, i.e. 10^4.
This is 20 log 10^4 dB = 80 dB.
The output change is 6 dB.
The action of the a.g.c. system eliminates 74 dB.
The change in bias voltage is 74/13 = 5·7 V.

Example 4.5

The output of a receiver is required to vary by not more than 6 dB for a 60 dB change in input. Three amplifying stages are controlled in each of which the gain varies by 3 dB/V of grid bias. Calculate the delay voltage, assuming the a.g.c. voltage is directly proportional to the amplitude of the carrier.

The 60 dB signal change referred to is presumably to be considered as taking place above the level at which the a.g.c. comes into operation.

The control exercised is (60 – 6) dB = 54 dB.

The voltage change is, therefore, 54/(3 × 3) = 6 V.

The change is one of 6 dB, i.e. a 2:1 ratio.

The initial voltage, therefore, must be at least 6 V. This is the required delay voltage.

4.5. MANUAL GAIN CONTROL

In simple receivers manual gain control is effected by use of a potentiometer in the output of the detector stage–by the potentiometer *P*, for instance, in Figs. 4.18 and 4.19. In a sound receiver this is the a.f. gain control.

More sophisticated receivers may have two, or even three, manual gain controls–the a.f. control being supplemented by others in the s.f. and possibly i.f. circuits. This enables adjustment to be made for best signal-to-noise ratio, minimum distortion, absence of overloading etc. for all conditions of input signal.

4.6. DECOUPLING

Power supplies for transistor and valve stages have a finite impedance. Unless precautions are taken the collector, anode and screen currents, varying in accordance with the signal, pass through the power supply and set up voltage drops across the supply impedance. Because the impedance is common to all stages the voltage drop variations are transferred to all the amplifiers except the input to the first. Feedback of this sort can cause all sorts of difficulties from instability to degeneration of the signal.

It is not a practical possibility to reduce the supply impedance to so low a value as to remove the possibility of troublesome feedback. What can be done easily and cheaply is to ensure that current variations are bypassed so that they do not pass through the power supply. For this, filter circuits are introduced into the supply leads. Instead of connecting the collector or anode supply direct to the load R_L (Fig. 4.20 (a) and (b), a filter circuit *RC* is interposed as shown in Fig. 4.20 (c) and (d). For valves with screen grids the screen must also be supplied through a filter circuit. With correct choice of *R* and *C* a large percentage

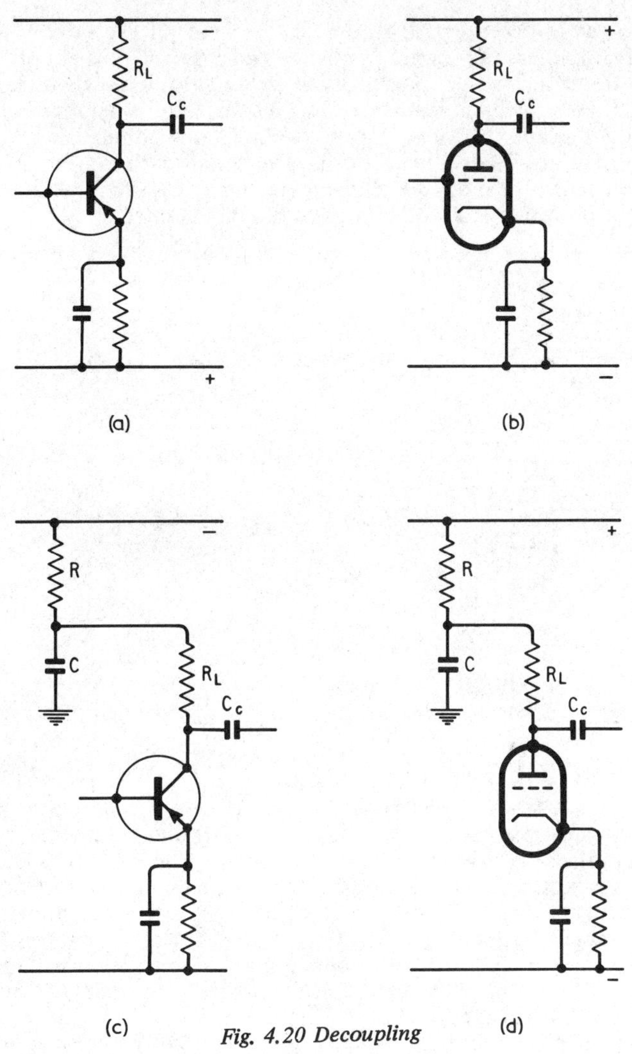

Fig. 4.20 Decoupling

of the current variations in the valve or transistor output circuit is bypassed through *C*. The fraction of the current which passes through *R*, and so into the power supply, is only (approximately) X_c/R– assuming that X_c is small compared with *R*. This ratio can be made as small as is necessary to prevent trouble due to feedback: the process is called *decoupling*.

The d.c. supply voltage to the active device is reduced by the amount of the voltage drop in the decoupling resistor *R*. Some compromise is necessary, therefore, in the choice of the valve of *R*. For a.f. amplifiers the lowest frequency to be amplified is about 50 Hz. At this frequency a 4-μF capacitor has a reactance of a little under 1 000 Ω. To reduce the feedback to a few per cent of its original value a decoupling capacitor of 4 μF needs to be used in conjunction with a resistor of about 20 or 30 kΩ. For a valve amplifier running at an anode current of one or two milliamperes such a resistance value gives a voltage drop which is about as large as can be tolerated.

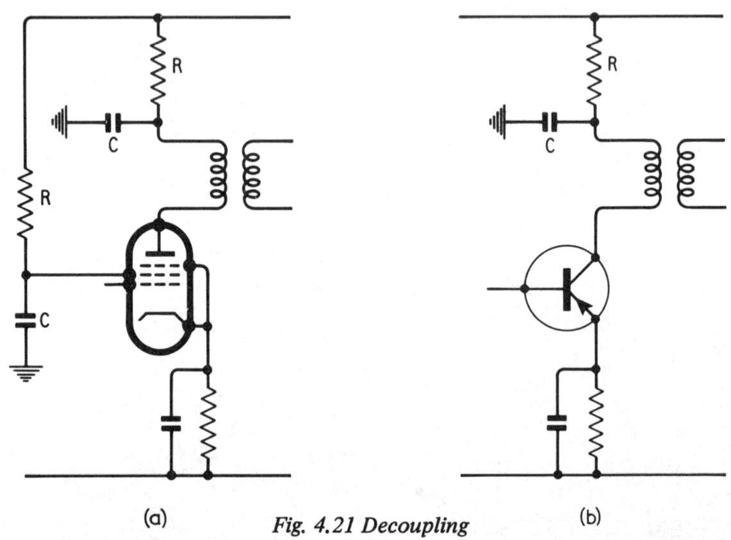

Fig. 4.21 Decoupling

For an s.f. or i.f. amplifier the lowest frequency to be amplified is much higher so that *C* and *R* may be proportionately smaller. Typical values for valve circuits are $C = 0{\cdot}1$ μF and $R = 1\,000$ Ω.

For transistor amplifiers the impedances are much lower. A given *X*/*R* ratio, therefore, must be obtained by using a much larger value of *C*. The values of *R* are correspondingly reduced. For a.f. circuits *C* is typically 100 μF and *R* a few hundred ohms. For i.f. circuits *C* may be about 0·1 μF and *R* about 100 Ω for higher frequencies *C* may be reduced to 0·01 μF or lower.

Fig. 4.21 shows decoupling *RC* applied to anode and screen of a pentode, and to the collector supply of a transistor.

4.7. BEAT FREQUENCY OSCILLATOR

For the reception of continuous-wave signals an additional unit must be added to the circuits we have so far discussed. This is because a continuous-wave transmission consists of an on-off sequence of waves at the signal frequency—and this frequency is too high to be audible.

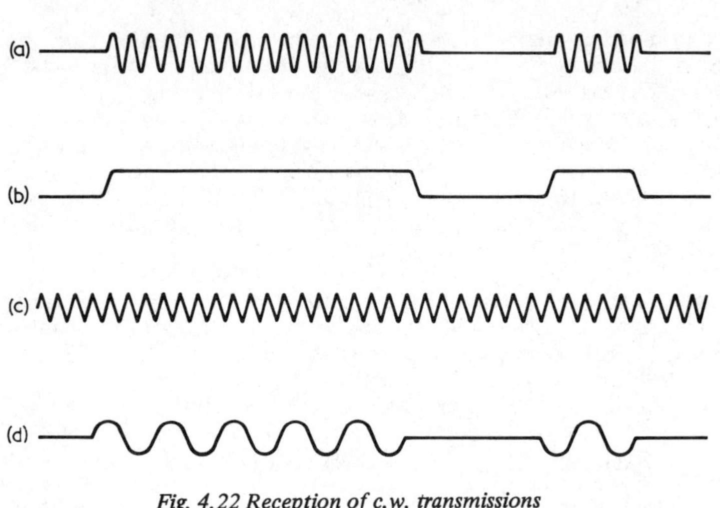

Fig. 4.22 Reception of c.w. transmissions

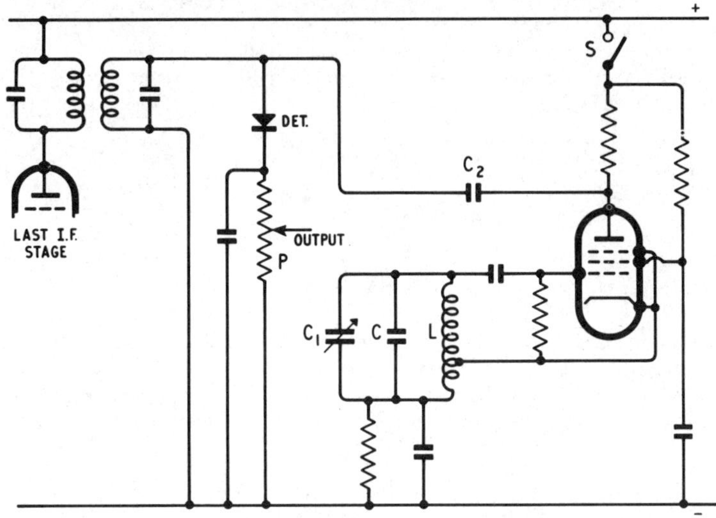

Fig. 4.23 Beat-frequency oscillator

Fig. 4.22(a) shows the letter N in morse code as it would exist at the output of an i.f. amplifier stage. (This takes about half a second to transmit manually.) After detection the waveform is as shown in Fig. 4.22 (b). At the best this is audible only as a series of clicks as the wave starts and stops. To provide proper audibility the wave of Fig. 4.22 (a) is heterodyned by an oscillation the frequency of which is about 1 kHz removed from the intermediate frequency. Thus, if the intermediate frequency is 1 600 kHz, the heterodyning oscillator (the *beat frequency oscillator–b.f.o.*) has a frequency of about the same value and is variable through 2 or 3 kHz on to either side.

It is not possible to draw these frequencies to scale within the compass of a small page, but Fig. 4.22 (a) is representative of the intermediate frequency, say 1 600 kHz, Fig. 4.22 (c) shows the frequency of the beat-frequency oscillator, say 1 601 kHz, and Fig. 4.22 (d) is to be understood to represent the resulting beat frequency (1 kHz) which is audible through the loudspeaker.

Fig. 4.23 shows a typical beat-frequency oscillator and its coupling to the signal circuits. It is a Hartley oscillator and the capacitor C_1, in shunt with the main frequency determining circuit LC, enables the oscillator frequency to be set to values within the range of 2 or 3 kHz to either side of the intermediate frequency.

The b.f.o. output is taken via C_2 to join the output from the last i.f. stage in the detector. It follows that the detector output contains components at the two input frequencies (b.f.o. and intermediate) and at the sum and difference frequencies. All these frequencies except the difference frequency are relatively high and are removed in the detector filter circuits. The difference-frequency component is developed across P and passed on to the next stage for amplification.

Opening the switch S removes the H.T. supply from the b.f.o. and so stops it functioning. It is necessary to do this when the receiver is to be used for telephony.

QUESTIONS

1. How may the bandwidth of a single stage, or of a number of stages, be widened? What is the effect on the gain?

   Explain how coupled and single LC circuits may be arranged to give a substantially flat response curve over a range of frequencies (which is small compared with the middle frequency).

2. Explain how a type of coupled circuit, with variable tuning, may be developed to give a bandwidth which is fairly constant irrespective of the tuning.

3. Why is the provision of variable bandwidth often desirable in a receiver? Indicate in outline two ways in which variable bandwidth may be provided in an i.f. amplifier.

4. Explain meaning of the terms *ganging* and *tracking* as applied to superheterodyne receivers.

   What is *tracking error?* How may it be reduced? Can it be eliminated completely?

5. Why is automatic gain control normally provided in radio receivers? What is the function of a.g.c.? Quote some typical performance figures.

Distinguish between simple and delayed a.g.c. What is the advantage of using delayed a.g.c.?

6. Are the general methods of obtaining a.g.c. the same for transistor and valve receivers? If not, give the main differences.

Sketch a basic circuit and briefly explain the action of a delayed a.g.c. system for a valve receiver.

7. At what stage(s) and why is manual gain control provided in:

(a) simple broadcast receivers,
(b) sophisticated communication receivers?

8. Why is decoupling necessary in radio receivers? Explain why the respective component values are different as between a.f. and r.f. stages and as between transistor and valve receivers.

9. Explain why a beat-frequency oscillator must sometimes be provided in a radio receiver. Sketch a typical basic circuit and show and explain the relationship of this stage of other adjacent stages in the receiver.

10. Why must the beat-frequency oscillator be switched out of circuit for the reception of certain types of signals?

Explain in detail how you could use a variable beat-frequency oscillator, in conjunction with the receiver tuning controls, to separate a wanted weak c.w. signal from an unwanted strong c.w. signal on a closely adjacent radio frequency.

5

R.F. Amplification

5.1 INTRODUCTION

The amplifying valve or transistor is a converter of d.c. power to a.c. power. An h.t. supply or battery is the source of d.c. power, while the variation of anode or collector current at the drive frequency produces an alternating component of voltage across a load impedance and thus provides output power. The ratio of a.c. power produced to the d.c. power supplied is the conversion efficiency. It is usually desirable to operate with a maximum possible efficiency having regard to the amount of tolerable distortion of the waveform of the voltage being amplified. This is especially true for large transmitting valves where power consumption is high and electricity costs are important.

5.2. CLASS A

5.2.1. Amplifier Efficiency

Fig. 5.1 is a simplified circuit of an amplifier biased as shown in Fig. 5.2 into linear class-A operation by a negative bias on the grid. The valve is assumed to have an anode rating of the order of 100 W.

If the drive voltage V_g is switched off, the anode current has a steady value of 100 mA. If the resistance of the transformer primary for d.c. is negligible, the anode potential is the h.t. voltage of 1 kV. Under these conditions the h.t. supplies a power of $1\,000 \times 100 \times 10^{-3}$ W or 100 W. This power is dissipated as heat at the anode of the valve. The production of heat at the anode is the chief reason why the conversion efficiency is inevitably less than 100%.

If a grid drive voltage of 36 V peak value is applied as shown in Fig. 5.3 (c), the grid potential varies between the limits of −14 V and −86 V. The dynamic mutual conductance of the valve is assumed to be 2·5 mA/V so that the grid voltage swings the anode current in accordance with the drive voltage waveform between a minimum of 10 mA and a maximum value of 190 mA. The matching transformer T (Fig. 5.1) has a primary input impedance of 10 kΩ at the drive frequency. Z_L represents the input impedance of a transmission line of characteristic impedance 600 Ω. T in this instance has a step down ratio of $\sqrt{(10^4/600)}$.

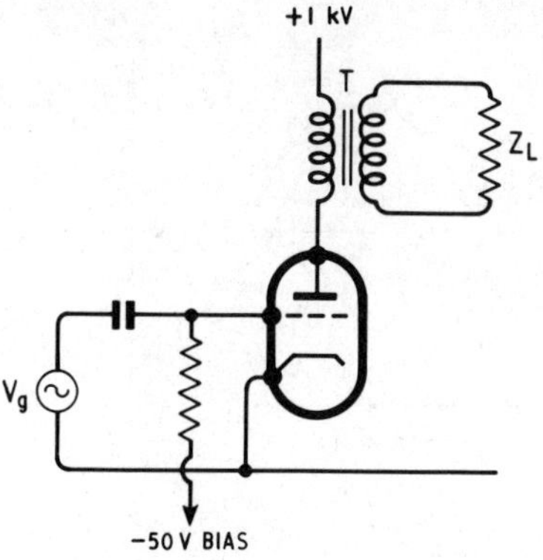

Fig. 5.1 A transformer-coupled output valve

The changes of valve current in the transformer primary cause changes of anode voltage of the value shown in Fig. 5.3 (a). The peak change in anode voltage is given by: $90 \times 10^{-3} \times 10^4$ V or 900 V. The

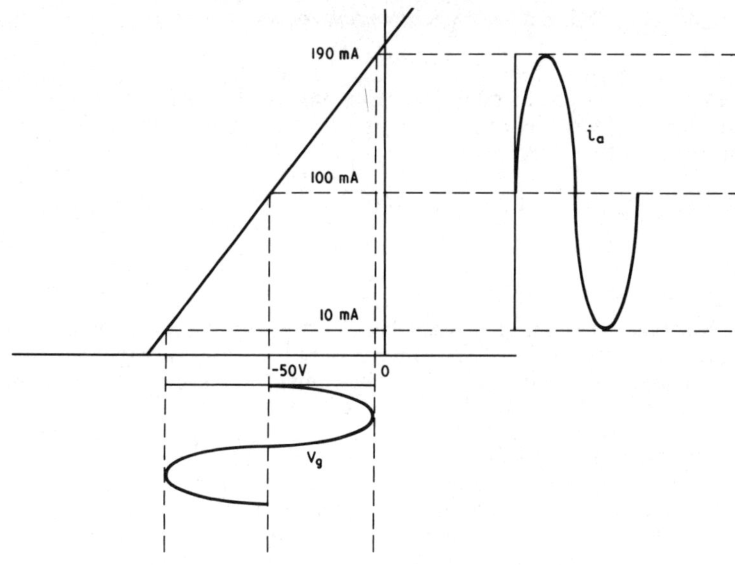

Fig. 5.2 Class-A operation

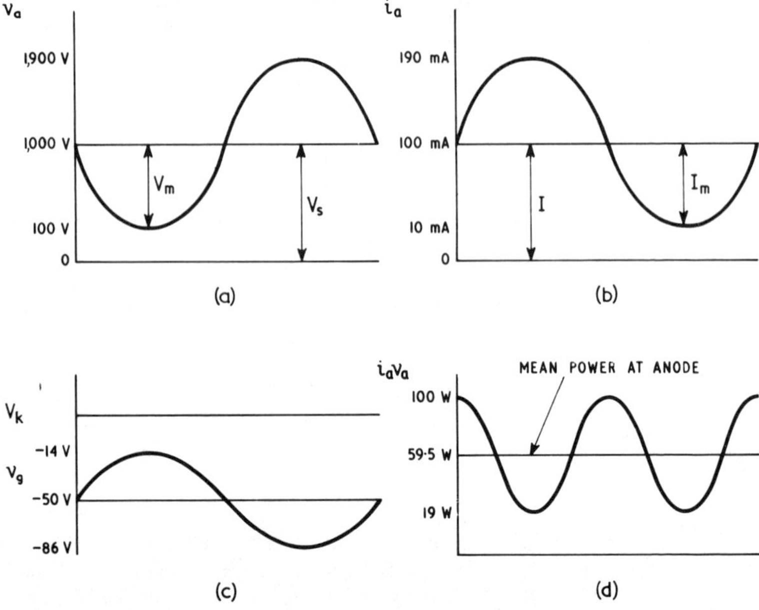

Fig. 5.3 Phase relationships for a resistive load

peak a.c. power produced is, therefore, $900 \times 90 \times 10^{-3}$ W or 81 W. The mean power is half this value or 40·5 W. In this example the efficiency is therefore:

$$\frac{\text{A.C. POWER PRODUCED}}{\text{D.C. POWER SUPPLIED}} = \frac{40{\cdot}5}{100} = 0{\cdot}405$$

or 40·5%.

It should be noted that with class-A operation the mean anode current is unchanged by the application of the drive so that the d.c. power supplied by the h.t. source also remains the same. It follows that as more a.c. power is developed, less power is dissipated at the anode of the valve. The instantaneous power at the anode is given by the product of the instantaneous values of anode current and voltage. Fig. 5.3 (d) shows how the power dissipation at the anode varies throughout a cycle between a maximum value of 100 W and a minimum value of 18 W. The mean power level is thus 59·5 W. The power dissipation falls to a minimum when either the anode voltage or the anode current is at a minimum. It will be noted that v_a and i_a vary in opposite phase and this accounts for the lowered value of power loss at the anode.

The efficiency is improved if the swing of anode current and voltage can be increased or if the d.c. power supplied can be reduced for the same level of a.c. output power. If the waveform of anode current and voltage are to be preserved, the maximum swing of anode voltage and current in the example above are 1 000 V and 100 mA respectively. The peak a.c. power is then $1\,000 \times 100 \times 10^{-3}$ W or 100 W and the mean a.c. power is 50 W. As the d.c. power remains unchanged at 100 W, the maximum theoretical efficiency is 50/100 or 50%.

5.2.2. Class A–relationships

The following symbols with the meanings stated below are used in the text:

V_m = maximum swing of anode voltage.
V_s = H.T. supply voltage.
I = d.c. or mean anode current (the supply current).
I_m = maximum swing of anode current (peak a.c.).
v_a = instantaneous anode voltages.
i_a = instantaneous anode current.
Z_L = effective load impedance.
ω = $2\pi \times$ the drive frequency.
t = time in seconds from the beginning of cycle.

The following relationships apply:

Power taken from the h.t. supply = $V_s I$ W
Peak alternating power produced = $V_m I_m$ W
Mean alternating power produced = $V_m I_m/2$ W

$$\text{Efficiency} = V_m I_m/2 \times 1/V_s I \times 100\%$$

$$Z_L = V_m/I_m \ \Omega$$

$$\text{and } I_m = V_m/Z_L \text{ A.}$$

The instantaneous power dissipation at the anode is

$$i_a v_a = (I + I_m \sin \omega t)(V_s - V_m \sin \omega t)$$

$$= IV_s + V_s I_m \sin \omega t - IV_m \sin \omega t - I_m V_m \sin^2 \omega t$$

But $\sin^2 \omega t = \frac{1}{2}(1 - \cos 2\,\omega t)$

$$\therefore i_a v_a = IV_s + V_s I_m \sin \omega t - IV_m \sin \omega t - I_m V_m/2 + V_m I_m/2 \cos 2\,\omega t$$

The second, third and fifth of these terms have an average value of zero over any whole number of cycles–as is the case with any sine or cosine alternating quantity—so that the average power dissipated at the anode is:

$$IV_s - I_m V_m/2$$

This is the d.c. power supplied minus the a.c. power.

5.3. CLASS C

5.3.1. Improved Efficiency by Increased Bias

If the drive voltage for a power amplifier is of constant amplitude, a large bias voltage may be applied to the grid as shown in Fig. 5.4. (a). A bias such as this, which is greater than the cut-off voltage of the valve, is termed a class-C bias. Without the drive voltage no anode current flows. The presence of the drive voltage raises the grid potential for a portion of the positive half cycles to a level which permits a pulse of anode current to flow. The duration of anode current flow is measured in electrical degrees and is called the *angle of anode current flow.*

Usually the drive voltage swings the grid positive for a small fraction of the input cycle and during this time a grid current will flow in addition to the anode current. The duration of grid current flow expressed in electrical degrees is the *angle of flow of grid current.* The average grid current multiplied by the value of the grid resistor indicates the bias secured by the grid current biasing. A typical angle of anode current flow is 120° but the angle of grid current flow may be approximately 100°.

The use of a large grid bias which suppresses the anode current for a large fraction of the input cycle reduces the anode current and thereby the heat loss at the anode while still permitting large changes of anode current and voltage. In this way efficiencies of 75% or so may be obtained. Because the waveform of the anode current has been severely distorted, harmonics of the input frequency are present in the anode circuit. This difficulty can be overcome by using a tuned anode load which selectively amplifies only the fundamental frequency and offers negligible impedance to the unwanted harmonics. Alternatively, the anode load may be tuned to one of the harmonics if the valve is to function as a harmonic generator. See Section 5.6.

Before the efficiency of a class-C amplifier can be calculated it is necessary to know the following values:

1. The d.c. h.t. supply voltage.
2. The mean anode current, which is the supply current.
3. The peak value of the anode current component at the drive frequency and the anode voltage swing at the drive frequency; *or* the peak anode current component at the drive frequency and the anode load impedance at the drive frequency; *or* the anode voltage swing and the anode load impedance at the frequency to be amplified.

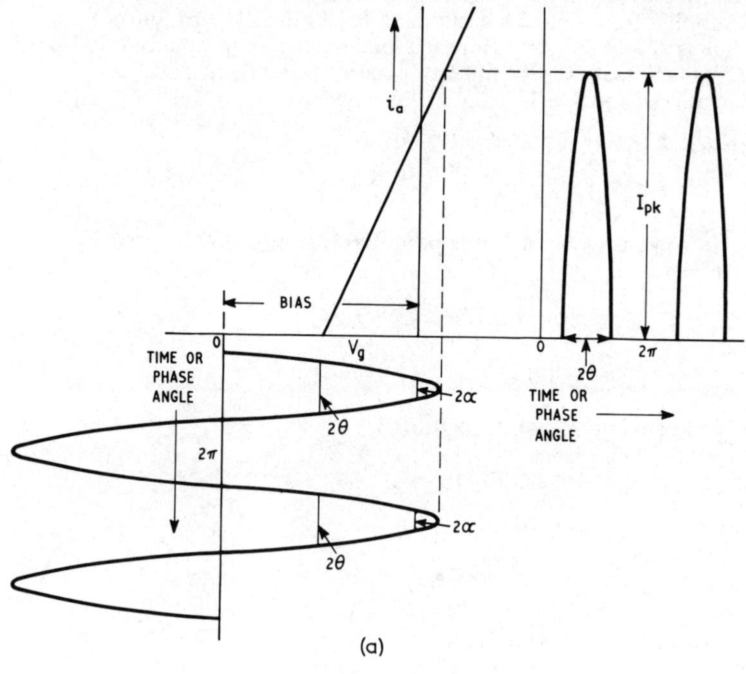

(a)

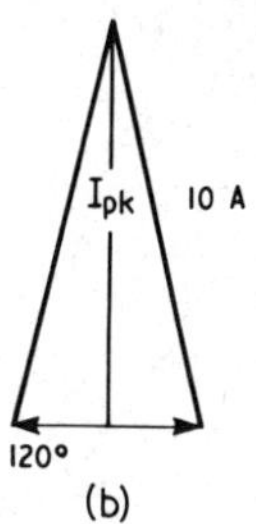

(b)

Fig. 5.4(a) Class-C operation;

(b) approximate current pulse shape

Example 5.1

A class-C power amplifier operates at a frequency of 5 MHz using an h.t. supply of 2 kV. The d.c. supplied is 120 mA. The r.f. output voltage is developed across a tuned load having a capacitance of 31·8 pF and a Q factor of 10. The minimum anode potential is 100 V. Calculate the anode efficiency of the valve.

The d.c. power supplied is IV_s watts or

$$2\,000 \times 120 \times 10^{-3}\text{ W} = 240\text{ W}$$

The downward swing of anode potential is from 2 000 V to 100 V. Therefore

$$V_m = (2\,000 - 100)$$

$$= 1\,900\text{ V}$$

The anode load impedance is $Q/\omega C$.

$$\therefore Z_L = 10/2\,\pi\,5 \times 10^6\;.\;31{\cdot}8 \times 10^{-12}$$

$$= 10^4\ \Omega$$

The peak output power is $V_m^2/10^4$

$$= (1\,900)^2/10^4$$

$$= 361\text{ W}$$

The mean power is ½ × 361 W, or 180·5 W.

The efficiency $=$ r.f. power/d.c. power

$$= 180{\cdot}5/240$$

$$= 0{\cdot}75 \text{ or } 75\%$$

Example 5.2

The anode current flow of a class-C power amplifier is described by the equation:

$$i_a = 0{\cdot}25 + 0{\cdot}4\cos\omega t + 0{\cdot}28\cos 2\,\omega t + \text{——etc.}$$

(higher terms being neglected).
This current flows through a tuned circuit which at the drive frequency has a resistive impedance of 5 kΩ. The h.t. voltage is 2 kV. Explain the

significance of each term in the equation and calculate the efficiency of the amplifier.

The first term indicates a component of zero frequency which is the d.c. drawn from the h.t. source. The second term states that a current of 0·4 A flows at the fundamental frequency of $\omega/2\pi$ Hz.

The third term relates to a second harmonic current the peak value of which is 0·28 A.

The power drawn from the h.t. source is IV_s.

$$IV_s = 0{\cdot}25 \times 2\,000$$

$$= 500 \text{ W}$$

The peak output power at the fundamental frequency is equal to $I_m^2 Z_L$, assuming a power factor of unity for the load. Thus output power is $(0{\cdot}4)^2 \times 5\,000$ W peak = 800 W and the mean power is 400 W.

The amplifier efficiency = r.f. power/d.c. power

$$= 400/500 \times 100\%$$

$$= 80\%$$

Example 5.3

A class-C power amplifier has a total filament emission current of 10 A and uses an h.t. supply of 10 kV. The amplitude of the anode current at the fundamental frequency is 3·0 A. Find the r.f. power output and the efficiency if the angle of anode current flow is 120° and the anode current rises to saturation value at its peak when v_a is 1 kV.

The peak swing of anode voltage is from 10 kV down to 1 kV.

$$\text{Thus } V_m = (10 - 1) = 9 \text{ kV.}$$

The peak r.f. power $= I_m V_m$

$$= 3 \times 9 \text{ kW}$$

$$= 27 \text{ kW}$$

The mean r.f. power = 13·5 kW

An approximate value for the mean anode current may be calculated by assuming that the graph of anode current plotted against the angle of flow is triangular in shape as shown in Fig. 5.4.(b).

The base of the triangle represents a time equal to 120/360 parts of the period of the drive frequency. Thus the current flows for $120/360 \times 1/f$ seconds where f is the drive frequency in hertz. The area of the triangle represents the charge conveyed across the valve in this time. From the geometry of the figure, the charge carried by one pulse of current is $1/2 \times 1/3f \times 10$ coulombs. The average flow of current during a complete period is this quantity of charge divided by the time taken by a complete cycle.

Hence

$$I = \frac{1/2 \times 1/3f \times 10 \text{ C/s}}{1/f}$$

$$= 1{\cdot}67 \text{ A}$$

The d.c. power supplied $= IV_s$

$$= 1{\cdot}67 \times 10 \text{ kW}$$

$$= 16{\cdot}7 \text{ kW}$$

The efficiency of the valve = r.f. power/d.c. power

$$= 13{\cdot}5/16{\cdot}7$$

$$= 81\%$$

In all these examples it will be noted that the efficiencies are much greater than those possible with class-A operation. The increase in efficiency is due to the reduction in mean anode current and to the fact that the pulses of anode current flow at times when the anode potential is at or near to its minimum value.

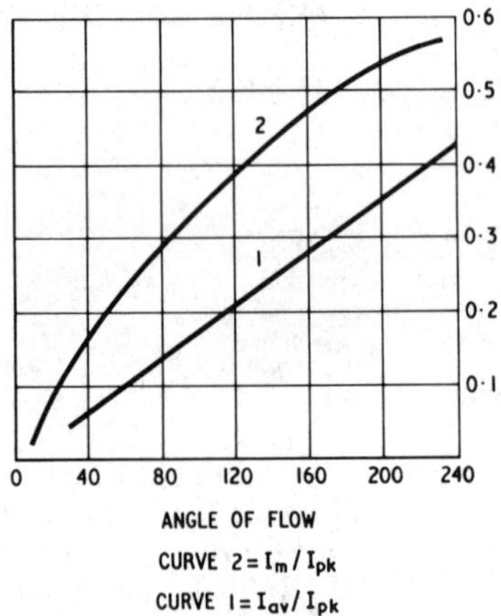

Fig. 5.5 Graphs relating I_m *and* I_{av} *to* I_{pk}

5.3.2. The Effect of a change in Bias

If the bias is altered so that the angle of flow of anode current is changed this affects:

1. the ratio of average anode current to the peak pulse value and,
2. the ratio of the amplitude of anode current at the fundamental frequency to the peak pulse value.

Fig. 5.5 shows graphically how these two ratios are related to the angle of anode current flow. The graphs are not precise for all valves, as the ratios depend to a small extent upon the valve characteristics. The derivation of formulae from which the graphs are drawn is outside the scope of this volume but appropriate formulae are stated below for reference:

$$I_m = \frac{I_{pk}}{\pi(1 - \cos\theta)} \cdot (\theta - \sin 2\theta/2)\ \text{A}$$

and

$$I_{av} = \frac{I_{pk}}{\pi(1 - \cos\theta)} \cdot (\sin\theta - \theta\cos\theta)\ \text{A}$$

The angle θ is half the angle of anode current flow. Where θ appears in the above formulae its value must be inserted in radians as shown in Example 5.4.

Example 5.4

A class-C power amplifier has a peak current pulse of 2·0 A and the angle of anode current flow is 110°. The h.t. supply is 5 kV and the anode load has a resistive impedance of 5·7 kΩ at the frequency of the drive voltage. Find the power taken from the supply, the r.f. power produced, and the anode efficiency of the amplifier.

It is first necessary to find the amplitude of the fundamental component of the anode current at the drive frequency and also the mean anode current. The equations given above are used:

$$\theta = \pi \times 55/180 \text{ rad}$$

$$= 0{\cdot}96 \text{ rad}$$

$$I_m = 2{\cdot}0\,(0{\cdot}96 - \frac{\sin 1{\cdot}92}{2})/\pi(1 - \cos 0{\cdot}96)$$

$$= 2{\cdot}0\,(0{\cdot}96 - 0{\cdot}4698)/\pi(1 - 0{\cdot}5736)$$

$$= 2{\cdot}0 \times 0{\cdot}4902/\pi \times 0{\cdot}4264$$

$$= 0{\cdot}7 \text{ A}$$

Also,

$$I_{av} = 2{\cdot}0\,(\sin 0{\cdot}96 - 0{\cdot}96 \cos 0{\cdot}96)/(1 - \cos 0{\cdot}96)\,\pi$$

$$= 2{\cdot}0\,(0{\cdot}8192 - 0{\cdot}96 \times 0{\cdot}5736)/(1 - 0{\cdot}5736)\,\pi$$

$$= 2{\cdot}0\,(0{\cdot}2692)/\pi(0{\cdot}4264)$$

$$= 0{\cdot}5384/1{\cdot}34$$

$$= 0{\cdot}4 \text{ A}$$

The power drawn from the h.t. supply is:

$$0{\cdot}4 \times 5 \text{ kW}$$

$$= 2{\cdot}0 \text{ kW}$$

The peak r.f. power at the drive frequency is:

$$(0{\cdot}7)^2 \times 5{\cdot}7 \text{ kW}$$

$$= 2{\cdot}8 \text{ kW}$$

The mean value of this power is 1·4 kW. The valve efficiency is 1·4/2·0 = 0·7 or 70%.

In Fig. 5.5, the graph relating the amplitude of the fundamental component of anode current, I_m, to the peak value of the anode current pulse, I_{pk}, decreases noticeably in gradient beyond the co-ordinate for an angle of flow of 120°. This is approximately the angle of flow which provides the best operating conditions for class-C operation. Greater angles of flow involve too great a value of d.c. while smaller angles of flow include too small a component of the fundamental frequency in the anode current.

5.3.3. The Grid Circuit

Bias for a class-C amplifier is provided by the rectification of the drive voltage in the grid circuit of the valve. The previous stage must provide the power which is dissipated as heat in the grid resistor and also the power which is dissipated at the grid itself.

If the peak value of grid current and its angle of flow are known, then the component of grid current at the drive frequency can be found by reference to graphs similar to those drawn in Fig. 5.5. The mean value of grid current can similarly be found. Knowing the bias value required for the valve and the expected d.c. grid current, the grid resistor value may be calculated:

$$R_g = \frac{\text{BIAS VOLTAGE}}{\text{D.C. GRID CURRENT}}$$

The grid resistor often has too small a value for it to be connected directly in shunt with the output impedance of the previous stage. If necessary, therefore, an r.f. choke is connected in series with the grid resistor as shown in Fig. 5.6. C_2 decouples the r.f. component of current from the grid resistor so that r.f. power loss does not occur in the resistor.

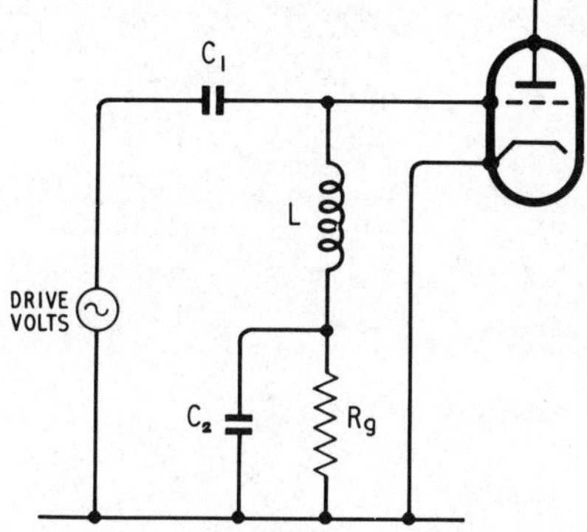

Fig. 5.6 Grid circuit bias arrangements

The peak r.f. drive power is equal to the product of r.f. drive voltage and the amplitude of the fundamental component of grid current at the drive frequency. The average drive power is half of this. The power dissipated in the grid resistor is V_b^2/R_g watts, where V_b is the bias voltage and R_g the bias resistor value. The difference between the drive power supplied and the power dissipated by the grid resistor is the power dissipated by the grid electrode. The drive must not be excessive or the grid dissipation is too big and the grid mesh may melt.

Example 5.5

A grid drive voltage for a class-C amplifier has a peak value of 90 V. The grid bias due to grid current flow is –60 V and the peak grid current is 50 mA with an angle of flow of 100°. Find the following:

1. the maximum and minimum values of grid potential
2. the value of grid resistor required
3. the power dissipated by the grid resistor
4. the drive power
5. the grid dissipation.

1. The maximum grid potential is –60 + 90 = 30 V
   The minimum grid potential is –60 – 90 = –150 V

2. From graph (1) of Fig. 5.5, for an angle of flow of 100° of a sinusoid, the ratio $I_{av}/I_{pk} = 0{\cdot}17$

$$I_{av} = 0{\cdot}17 \times 50 \text{ mA} = 8{\cdot}5 \text{ mA}$$

$$R_g = V_b/I_{av} = 60/8{\cdot}5 \text{ k}\Omega$$

$$= 7{\cdot}05 \text{ k}\Omega$$

3. The power dissipated in the grid resistor is,

$$V_b\, I_{av} = 60 \times 8{\cdot}5 \text{ mW}$$

$$= 510 \text{ mW}$$

4. From graph 2 of Fig. 5.5,

$$I_m = 0{\cdot}35\, I_{pk}$$

$$= 0{\cdot}35 \times 50 \text{ mA}$$

$$= 17{\cdot}5 \text{ mA}$$

The peak drive power thus equals $90 \times 17{\cdot}5$ mW and the average drive power = $\frac{1}{2} \times 90 \times 17{\cdot}5$ mW

$$= 785 \text{ mW}$$

5. The grid dissipation = drive power – loss in R_g

$$= (785 - 510) = 275 \text{ mW},$$

which is the loss in the grid mesh.

5.3.4. Design Features

The following are relevant points in the design of class-C amplifier stages

1. A valve must be chosen of sufficient anode rating to dissipate the power not converted to r.f. output power. The anode dissipation may be between 30% and 40% of the required output power—e.g. for 100-W output the anode dissipation may be 30 W which is equivalent to an efficiency of 77%.

2. The peak anode current is limited by the emission current from the filament or cathode of the valve.
3. The valve design sets an upper limit to the maximum anode–cathode potential difference which can occur without damage to the valve.
4. The peak grid potential is limited by the maximum safe grid dissipation.
5. The minimum anode potential must not be less than the maximum grid potential or excessive grid current will flow. Optimum conditions sometimes make the maximum grid potential equal to the minimum anode potential.
6. The value of H.T. voltage lies midway between the maximum and minimum anode potentials.
7. A grid bias is selected which provides for an angle of flow of approximately 120°.
8. The amplitude of grid drive voltage must be equal to the sum of the bias and the positive excursion of grid potential.
9. A tuned load is provided with a variable coupling to the valve anode circuit. As the coupling to the load is made tighter, the Q of the tuned anode circuit falls and the feed current of the valve increases. The coupling must not be increased beyond the value at which the maximum safe feed current is flowing to the anode. The dynamic impedance of the load is $Q/\omega C$ where C is the capacitance across the coil and ω is the angular frequency in rad/s.

5.4. LINEAR CLASS-B AMPLIFIERS

If the grid drive voltage to an r.f. amplifier is amplitude modulated, then the high efficiency of class-C operation is no longer practicable because with a modulated drive the amplitude during the troughs of modulation may be insufficient to drive the valve into conduction. The output would thus cease except during part of the positive half cycles of modulation. A valve which is to amplify an amplitude modulated waveform uses a projected cut-off bias as indicated in Fig. 5.7. The valve current flows in a series of half sinusoids and the amplitude of each one is proportional to the amplitude of the drive voltage. The angle of anode current flow is approximately 180°. As with a class-C bias, this class-B bias gives rise to harmonics of the drive frequency in the anode current waveform. A tuned load is again necessary to attenuate the harmonics relative to the amplitude of the fundamental frequency.

Reference to the graphs of Fig. 5.5 or a calculation using the appropriate equation, shows that a valve with an angle of flow of 180° has an average anode current which is 0·318 or $1/\pi$ of the peak current pulse, and that the amplitude of the fundamental component of anode current is half the peak current pulse. If it is assumed that the anode voltage is allowed to swing to zero, and thus vary by a peak swing equal to the h.t. voltage, the efficiency is $\pi/4$ or 78·5%.

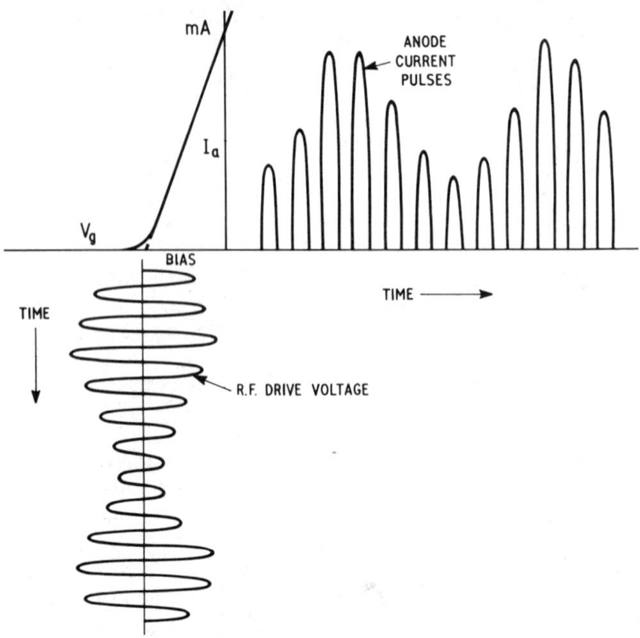

Fig. 5.7 Class-B operation with modulated drive

Example 5.6

A linear class-B tuned r.f. power amplifier has an H.T. supply of 800 V and a peak anode current of 100 mA at the instant when the anode potential is zero. Calculate the r.f. power produced, the d.c. power supplied by the H.T. source and the efficiency.

The d.c. power supplied $= V_s I$

$= 800 \times 100 \times 0{\cdot}318$ mW

$= 25{\cdot}4$ W

The r.f. power produced $= \frac{1}{2} \times I_m V_m$

$= \frac{1}{2} \times 50 \times 800$ mW

$= 20$ W

The efficiency $= 20/25{\cdot}4$

$= 0{\cdot}785$ or $78{\cdot}5\%$

In the above example ideal and impracticable conditions are assumed which lead to an efficiency rather bigger than that attainable in practice. The necessity of keeping the minimum anode potential higher than the maximum grid potential and also of using a bias a little less than the cut-off bias results in efficiency values in practice of the order of 60–65%.

As the efficiency of a class-B amplifier depends on the amplitude of the grid drive voltage, the efficiency is less if the modulation is removed from the drive voltage. The anode current pulses with only the carrier present are half the amplitude they reach at the positive peak of modulation when the depth of modulation is 100%. The efficiency with carrier only applied at the input is approximately 30%. With modulation applied to the input, the anode current pulses increase in amplitude on the positive half cycle of modulation and decrease to an equal extent on the negative half cycle of modulation. The mean anode current remains almost unchanged, but the output contains side frequencies. The anode dissipation is decreased by the power output at the side frequencies. Thus efficiency is least with the drive unmodulated and increases as the depth of modulation of the drive voltage is increased.

Example 5.7

A class-B linear amplifier has an H.T. of 1 kV and a tuned anode load of dynamic impedance 16 kΩ. With only the carrier voltage applied, the anode current pulses have a peak value of 50 mA. Find the efficiency of the valve, (a) with the drive unmodulated and (b) with a sinusoidal modulation to a depth of 80%.

The mean anode current is $I = 50 \times 0{\cdot}318$ mA

$$= 15{\cdot}9 \text{ mA}$$

The d.c. input power is $V_S I = 1\,000 \times 15{\cdot}9 \times 10^{-3}$ W

$$= 15{\cdot}9 \text{ W}$$

The mean carrier power is $\frac{1}{2} I_m^2 Z_L$ where $I_m = 50 \times 0{\cdot}5$ mA

Thus carrier power $= \frac{1}{2} \times (25 \times 10^{-3})^2 \times 16 \times 10^3$

$$= 5 \text{ W}$$

The efficiency $= 5/15{\cdot}9 = 0{\cdot}314$ or $31{\cdot}4\%$.

For a modulation factor m, the power in the side frequencies is $m^2/2 \times$ the carrier power. If $m = 0{\cdot}8$ and the carrier power is 5 W, the side frequency power is, $0{\cdot}64/2 \times 5 = 1{\cdot}6$ W. The total power with modulation $= 5 + 1{\cdot}6$

$$= 6{\cdot}6 \text{ W}$$

The efficiency $= 6{\cdot}6/15{\cdot}9$

$$= 0{\cdot}415 \text{ or } 41{\cdot}5\%$$

The bias voltage for a linear class-B amplifier must be held constant and independent of the drive amplitude. This indicates the need for a separate fixed bias source. A possible arrangement is to apply a positive cathode bias obtained from a potential divider across the H.T. source. This would apply only to small power circuits. For valves with directly heated cathodes, a separate negative bias supply for the grid is preferable.

Although an output valve having a modulated drive voltage must use class-B bias, an output valve having modulation applied in its anode circuit employs class-C bias. The anode modulated class-C amplifier is dealt with in Chapter 6.

5.5. Harmonic Generators

The distortion of the anode current waveform in class-B and class-C amplifiers involves the production of components of anode current which are harmonics of the drive frequency. It is the usual function of the anode load to select the fundamental frequency (the first harmonic) and to develop a large output voltage at this frequency, to the exclusion of the other harmonics to which the load is not resonant. If the valve is to operate as a harmonic generator, the load is made resonant to the particular harmonic which is to be selected and amplified. With the load tuned to the second harmonic, the stage is a frequency doubler. Tuning the load to the third harmonic makes the stage into a frequency tripler. Higher harmonics could be tuned and amplified but the amplitude becomes smaller as the order of the harmonic becomes higher. *Doublers* and *triplers* are, therefore, the most common types of harmonic generators.

The angle of flow of the anode current must be adjusted to suit the order of harmonic required as an output. In the interests of efficiency the anode current should flow only while the anode voltage of the valve is swinging below its mean value. In the frequency doubler this will be for half a cycle of the output frequency but only for a quarter of a cycle of the input frequency. Hence the angle of flow of anode current for a frequency doubler should not be more than 90° for best efficiency. It may be increased some 50% beyond this to give a larger output power but only with a reduced efficiency. In a frequency tripler the anode voltage falls below its mean value for half the output cycle but for one-sixth of the drive period. Hence 60° is the angle of anode current flow for best efficiency, but 90° of anode current flow provides increased output power at the expense of a reduction in efficiency.

If frequency multiplication by a large factor is necessary, then several stages of frequency multiplication are arranged in cascade. For example, a phase modulator in a small f.m. transmitter may be followed by frequency multiplication by 24 in order to raise the frequency to the v.h.f. band and to multiply the frequency swing by a similar factor. This multiplication is in four stages including one tripler and three doublers.

5.6. Grounded-Grid Amplifiers

Power amplifiers in large transmitters may be pentodes, beam tetrodes or triodes. As there are some difficulties in providing adequate cooling for the screen grids of pentodes and tetrodes, triode output valves are often favoured despite their larger value of anode to control grid capacitance. One way of reducing the unwanted coupling between the anode and grid circuits of a triode valve is to use an earthed or grounded-grid connection. In such a circuit the grid is connected

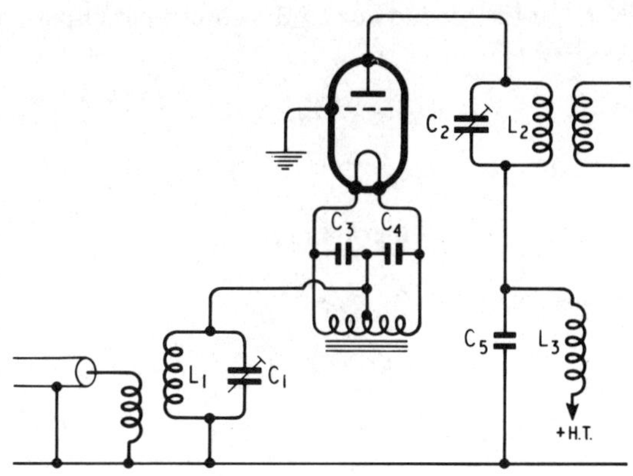

Fig. 5.8 An earthed-grid amplifier

directly to earth and the input signal is applied between earth and the cathode of the valve. The cathode potential varies with respect to earth and the grid at the signal frequency. The control grid acts as an electrostatic screen between the input and output circuits. Fig. 5.8 is a possible circuit diagram and Fig. 5.9. is an equivalent circuit for the sinusoidal components of anode current and voltage. In Fig. 5.8, L_1 and C_1 are the tuned input circuit. The input impedance of the circuit

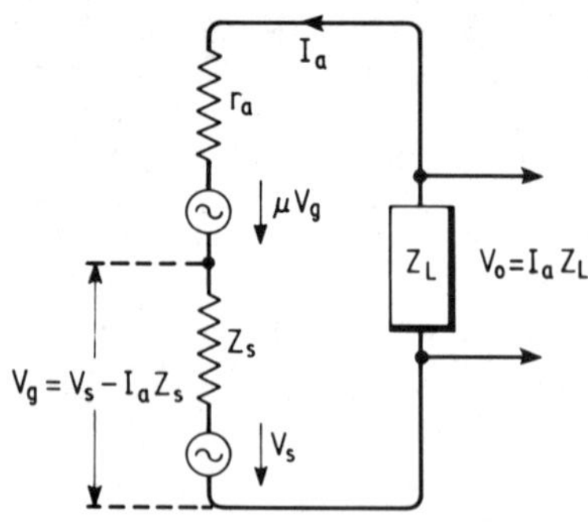

Fig. 5.9 Equivalent circuit of earthed-grid amplifier

is matched to the characteristic impedance of the input feeder by a suitable matching transformer. C_3 and C_4 are the by-pass capacitors across the secondary of the filament supply transformer. C_2 and L_2 form the tuned output circuit with transformer coupling to a following stage. L_3 and C_5 decouple the signal frequency components of current from the H.T. source.

Applying Kirchhoff's second law to the equivalent circuit, gives the following equations:

$$\mu V_g + V_s = I_a r_a + I_a Z_s + I_a Z_L \tag{5.1}$$

and
$$V_g = V_s - I_a Z_s \tag{5.2}$$

Substituting equation (2) in equation (1)

$$V_s + \mu V_s - \mu I_s Z_s = I_a (r_a + Z_s + Z_L)$$

whence
$$I_a = V_s(1 + \mu)/r_a + Z_L + Z_s (1 + \mu) \tag{5.3}$$

$$V_0 = I_a Z_L \tag{5.4}$$

From equations (5.4) and (5.3)

$$V_0 = V_s(1 + \mu)/r_a + Z_L + Z_s (1 + \mu)$$

The gain is the ratio of V_0/V_s and thus the gain

$$= (1 + \mu)Z_L/r_a + Z_L + Z_s (1 + \mu)$$

The appearance of the symbol for the source impedance in the denominator of the expression for the gain shows the desirability of feeding the earthed-grid amplifier from a low impedance source.

An expression for the input impedance is one which relates the anode current to the voltage applied between grid and cathode since this is the voltage at the input terminals of the amplifier. This expression is not, therefore, affected by the source impedance. From Fig. 5.9 Kirchhoff's Law gives:

$$\mu V_g + V_g = I_a (r_a + Z_L)$$

Thus $$V_g (\mu + 1) = I_a (r_a + Z_L)$$

and $$V_g/I_a = (r_a + Z_L)/(\mu + 1)$$

This is a low value of input impedance when compared with that of a grounded-cathode stage. The physical reason for this is that the anode current of the valve flows through the input circuit of the grounded-grid stage, whereas the input current of the grounded-cathode stage is merely that accepted by the input capacitance of the valve. It follows that the signal source of a grounded-grid amplifier must supply a considerable amount of drive power. Using the notation of Fig. 5.9 this is $V_s I_a$ volt-amps. Some of this power is available, however, in the anode load and there is a transfer of drive power to the load circuit. The essential features of a grounded-grid amplifier may be summarised thus:

1. There is no Miller Effect and hence the input capacitance is low.
2. The input impedance is low.
3. Power is transferred from the drive source to the load.
4. With resistive source and load impedance the input and output voltages are in phase.

Example 5.8

A grounded-grid amplifier has a tuned load of Q factor 40 and a shunt capacitance of 25 pF. The valve has an amplification factor of 24 and anode slope resistance of 5 kΩ. Assuming that the sources impedance matches the input impedance, find the voltage gain at the tuned frequency of 45 MHz.

The load impedance $Z_L = Q/\omega C$

$$= 40/2\pi \times 45 \times 10^6 \times 25 \times 10^{-12}$$

$$= 5{\cdot}65 \times 10^3$$

$$= 5{\cdot}65 \text{ k}\Omega$$

The input impedance $= (r_a + Z_L)/(\mu + 1)$

$$= (5 + 5{\cdot}65)/(24 + 1)$$

$$= 0{\cdot}426 \text{ k}\Omega$$

As the source impedance is matched to the input impedance,

$$Z_s = 0{\cdot}426 \text{ k}\Omega$$

$$\text{The gain} = (1 + \mu)Z_L / [r_a + Z_L + Z_s (1 + \mu)]$$

$$= 25 \times 5{\cdot}65/[5 + 5{\cdot}65 + 0{\cdot}426 (1 + 24)]$$

$$= 141/21{\cdot}3$$

$$= 6{\cdot}6$$

5.7. THE NEUTRALISED TRIODE

Tuned r.f. amplifiers in transmitters which are not operated as grounded-grid valves usually require a neutralising circuit for the unwanted feedback which takes place through the anode-grid capacitance. A circuit is provided external to the valve through which current may flow from anode to grid circuits in equal amplitude but opposite in phase to that which is admitted by the anode-grid capacitance.

A number of circuits are available, one of which is the tapped output circuit of Fig. 5.10. The neutralising capacitor C_n is adjusted so that the current passing to the grid from point *n* of the tuned anode load is equal in amplitude to that passing via the anode-grid capacitance from end *a* of the tuned load. *k* is the r.f. earth point of the tuned circuit so that the points *a* and *n*, being on opposite sides of the earth point, have r.f. potentials which are opposite in phase. Fig. 5.11 is the equivalent circuit for the circuit of Fig. 5.10. In the equivalent circuit a bridge arrangement is shown. The circulating current in the tuned circuit, consisting of L_{ak}, L_{kn} and C in series, is Q times the r.f. component of the anode current of the valve. If the Q factor of the circuit is high, the r.f. valve current may be neglected compared with the circulating current. With this approximation, balance is achieved if:

$$\frac{1/\omega C_{ag}}{1/\omega C_n} = \omega L_{ak}/\omega L_{kn}$$

$$\text{or } C_n/C_{ag} = \omega L_{ak}/L_{kn}$$

With the bridge balanced, the r.f. voltage across the tuned circuit $a - n$ produces no p.d. between the grid connection *g* and the cathode point *k*. Hence the p.d. across the tuned circuit does not contribute to the input.

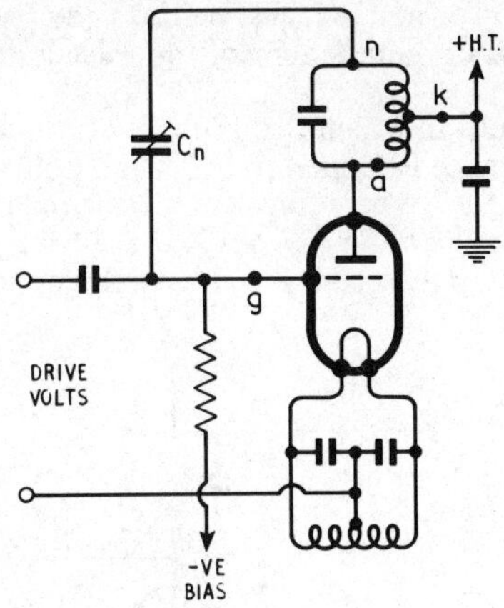

Fig. 5.10 Tapped-output neutralised circuit

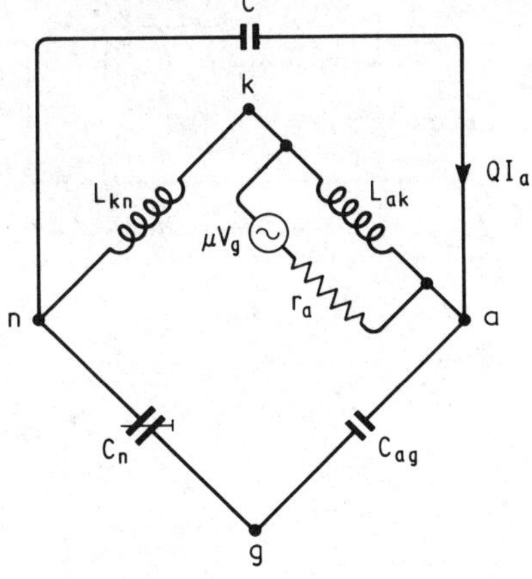

Fig. 5.11 Equivalent circuit of tapped-output circuit

The practical adjustment of C_n can be made as follows: The H.T. voltage to the valve is switched off but the filament supplies are run at normal voltage and the drive voltage is applied to the grid from the previous stage. C_n is then adjusted until a sensitive r.f. detector loosely coupled to the anode load shows that no r.f. voltage exists across the load. This indicates that there is no resultant feed forward of power via C_n and C_{ag}. Conditions of no feed forward via anode–grid capacitance are also those required for no feedback via the same path when the H.T. is restored and the a.c. component of the valve current is providing output power.

An alternative circuit, in which the grid circuit inductance rather than the anode inductance is tapped to earth at a point between the ends, is shown in Fig. 5.12. The equivalent circuit for the tapped input neutralisation circuit is shown in Fig. 5.13. If the bridge is balanced, the valve equivalent alternator connected between points *a* and *k*

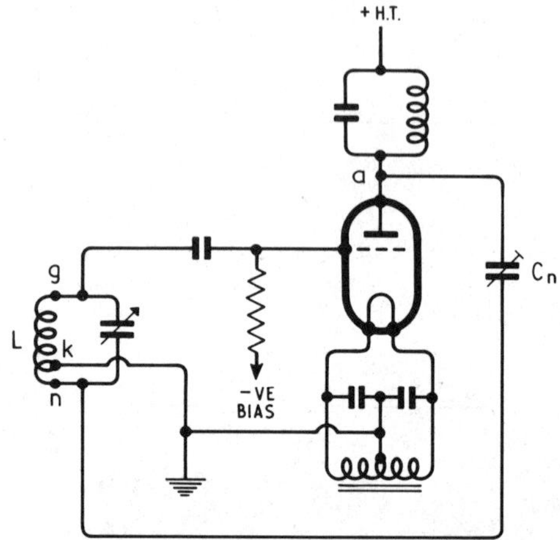

Fig. 5.12 Tapped input neutralisation

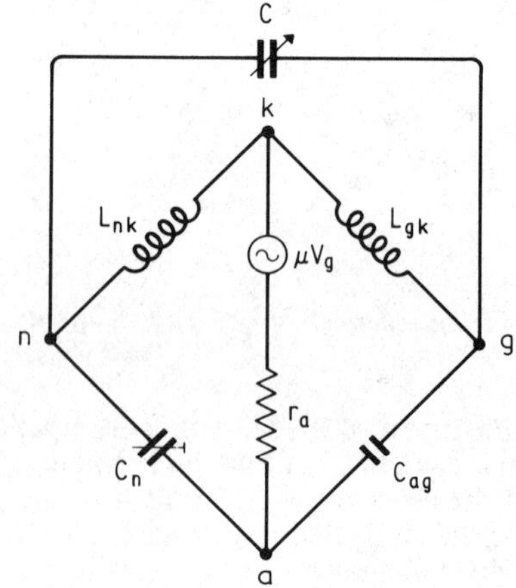

Fig. 5.13 Equivalent circuit for tapped-input circuit

produces no voltage across the input circuit. There is no p.d. between points n and g except that due to the drive voltage coupled from the previous stage. For balance, the following conditions apply:

$$\frac{1/\omega C_{ag}}{1/\omega C_n} = \frac{\omega L_{gk}}{\omega L_{nk}}$$

$$\text{and } C_n/C_{ag} = L_{gk}/L_{nk}$$

The circuit which lends itself most easily to neutralisation is the push-pull amplifier. In this circuit the opposite ends of the output

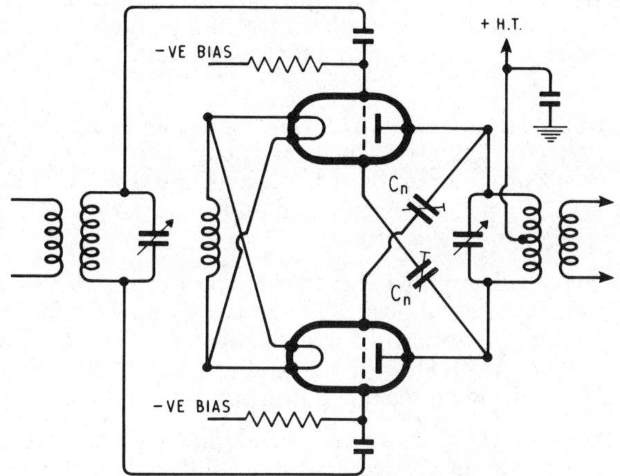

Fig. 5.14 A neutralised push-pull circuit

circuit have antiphase voltage variations which can be coupled back to the opposite valve grids as shown in Fig. 5.14. Feedback through each anode–grid capacitance is thus balanced by opposite feedback from the opposite end of the output circuit.

QUESTIONS

1. State briefly the meaning of the term amplifier efficiency. A water-cooled triode has an anode rating of 20 kW and can provide a maximum output power of 70 kW. State the maximum efficiency of which it is capable.

2. The tank circuit of a class-C r.f. triode amplifier has an effective dynamic resistance of 5 000 Ω when coupled to the output load. The amplification factor of the valve is 35, the H.T. supply is 2 000 V and the minimum voltage at the anode is 200 V. The mean anode current is

250 mA, the angle of flow is 120° and the anode current pulses may be assumed to be of triangular waveform. Calculate (a) the peak anode current, (b) the anode and grid voltages at which the anode current starts to flow, (c) the power delivered to the tank circuit and (d) the anode efficiency.

3. Explain why a class-C biased valve is capable of a higher efficiency than a class-A biased amplifying valve and state the limitations on the use of class-C bias.

4. The grid drive voltage for a class-C amplifier has a peak value of 60 V and causes a peak grid current pulse to flow of 30 mA. The grid bias is –40 V and the angle of flow of grid current is 100°. Find the grid resistor value required, the power dissipated in the grid resistor, the drive power and the power dissipated by the grid (see Fig. 5.5 for appropriate graphs).

5. A silicon transistor operating in class-A conditions as a power output device has a collector potential which varies between 10 V and 50 V for a variation of collector current between 1·25 A and 0·25 A. The collector supply is at 30 V. Find the dynamic load impedance of the amplifier, the a.c. load power and the efficiency.

6. A class-B push-pull amplifier uses two similar triodes which are coupled by a transformer of negligible loss to a load of 5·0 Ω. The ratio of the total primary turns to the secondary turns is 40. The H.T. supply is 600 V and the minimum anode potential is 100 V. Find the power coupled to the load, the d.c. power supplied and the efficiency.

7. Explain the principle of operation of a harmonic generator and state why the angle of flow should decrease when the order of harmonic required as an output is increased. A frequency tripler has an input frequency of 2 MHz and the anode load has an inductance of 10 μH. What value of capacitance is required for the anode load?

8. State the advantages to be found in an earthed grid connection circuit for a v.h.f. triode amplifier. An earthed grid amplifier has a tuned parallel load with a Q value of 25 and with a 25 pF capacitor which resonates with the inductance at 50 MHz. The valve has a μ of 25 and an r_a of 5 kΩ. Calculate the gain of the amplifier and its input impedance. Assume the source impedance to be negligible.

9. Explain briefly the purpose of neutralisation in r.f. power amplifiers and draw a circuit diagram to illustrate the principle. Say how the adjustment of the neutralisation circuit can be carried out.

10. Referring to Fig. 5.12, assumed tha L_{gk} = 20 μH, L_{nk} = 5 μH, and that C_n is set to 8 pF for correct neutralisation. What is the anode-grid capacitance of the valve?

6

Amplitude Modulation

6.1. FUNDAMENTALS OF AMPLITUDE MODULATION

Modulation of a signal of constant amplitude V_c and frequency f (angular velocity $\omega = 2\pi f$), Fig. 6.1(a), by a signal of constant amplitude V_a and frequency f_a (angular velocity $p = 2\pi f_a$), Fig. 6.1(b), results in a modulated wave

$$v = (V_c + V_a \sin pt) \sin \omega t \text{ and, if } m = V_a/V_c$$

$$= V_c \sin \omega t \,(1 + m \sin pt) \tag{6.1}$$

$$= V_c \sin \omega t + mV_c \sin \omega t \sin pt$$

$$= V_c \sin \omega t + \frac{mV_c}{2} \cos (\omega - p)t - \frac{mV_c}{2} \cos (\omega + pt) \tag{6.2}$$

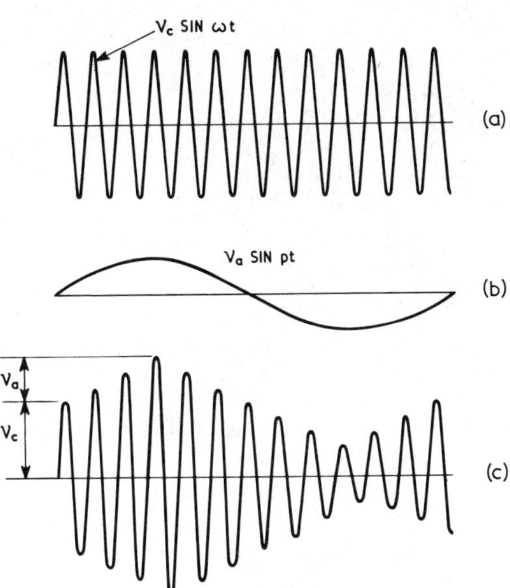

Fig. 6.1 Amplitude modulation

Clearly the amplitude of the modulated wave varies between $V_c + V_a$ and $V_c - V_a$. This can be seen in Fig. 6.1(c).

The *modulation factor* is $m = V_a/V_c$. The percentage modulation is $100m$.

The modulated waveform which results from modulation at a single frequency may be represented by the sum of three constant amplitude, constant frequency, waveforms: the *carrier* (amplitude V_c, frequency f), the *upper side frequency* (amplitude $V_a/2$, frequency $f + f_a$), and the *lower side frequency* (amplitude $V_a/2$, frequency $f - f_a$).

The power in the two side frequencies together, expressed as a fraction of the total power, is $m^2/(2 + m^2)$. It follows that the ratio of total sideband power to carrier power is $m^2/2$.

Example 6.1

The last statement may be readily justified. Let P_s represent sideband power and let P_c represent carrier power.

$$\frac{\text{SIDEBAND POWER}}{\text{TOTAL POWER}} = \frac{m^2}{2+m^2}$$

$$\frac{P_s}{P_s+P_c} = \frac{m^2}{2+m^2}$$

$$P_s = (P_s + P_c)\frac{m^2}{2+m^2}$$

$$P_s\left(1 - \frac{m^2}{2+m^2}\right) = P_c\frac{m^2}{2+m^2}$$

$$\text{and } \frac{P_s}{P_c} = \frac{m^2}{2}$$

Example 6.2

If a transmitter is amplitude modulated 50% (i.e. $m = 0{\cdot}5$) the ratio:

$$\frac{\text{SIDEBAND POWER}}{\text{TOTAL POWER}} = \frac{0{\cdot}25}{2{\cdot}25}$$

$$= 0{\cdot}11 \text{ or } 11\%$$

Thus of the total power developed and radiated only one ninth is in the sidebands, i.e. conveys intelligence.

Similarly, if the percentage modulation is (a) 100 and (b) 25 the respective ratios are 1/3 and 1/33.

When the modulating signal consists not of a single frequency but of a number of frequencies, two *sidebands* are produced consisting, respectively, of the sums of the carrier and modulating frequencies, and the differences between the carrier and modulating frequencies.

A system which results in the transmission of a carrier and an upper and a lower sideband is called a *double sideband* system *(d.s.b.)*. Single sideband *(s.s.b.)* and independent sideband *(i.s.b.)* systems are referred to in Section 6.4.

Example 6.3

A carrier of 1 MHz is modulated by signals at (a) 128 Hz and (b) 256 Hz and harmonics of this latter frequency up to the fifth.

The upper sideband contains the following frequencies (all expressed in kilohertz): 1 000·128, 1 000·256, 1 000·512, 1 000·768, 1 001·024 and 1 001·280.

The lower sideband frequencies, also in kilohertz, are 999·872, 999·744, 999·488, 999·232, 998·976 and 998·720.

The radiated power is greater during modulation than when the carrier alone is being generated. The extra power is delivered by the modulator (and ultimately, of course, drawn from the power supply).

The ratings of the active devices used (valves or transistors) must be adequate for the power levels involved. In particular the devices must individually be able to dissipate the difference between the power supplied to them from the source and the power delivered by them to the adjacent stage. See also Chapter 7.

Example 6.4

A valve modulator stage operates with an anode current of 100 mA from a d.c. supply of 400 V and thus draws 40 W from the supply. If the stage delivers 15 W to the modulated valve, the power difference (40 − 15) W = 25 W must be dissipated at the modulator valve anode.

If for some reason the modulator ceases to deliver power to the modulated valve, but continues to draw 100 mA from the 400-V supply, the whole 40 W is then dissipated at the modulator anode.

6.2. ANODE MODULATION

When a modulating voltage is applied in series with the H.T. supply to the anode of a properly arranged Class-C amplifier, the r.f. output of the

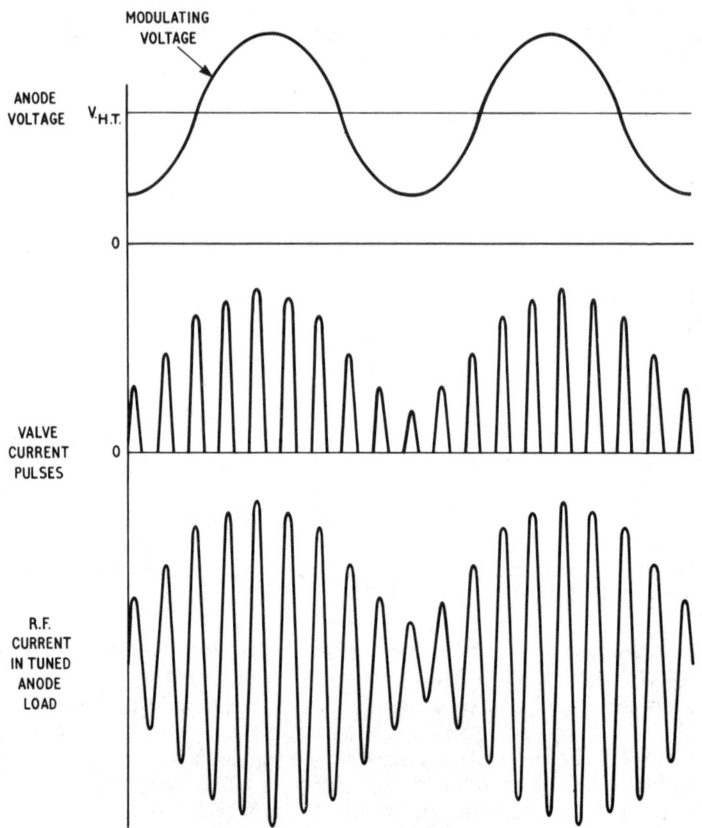

Fig. 6.2 Amplitude modulation: voltage and current waveforms

amplifier varies linearly with the amplitude of the modulating signal (Fig. 6.2). This, basically, is anode modulation.

Fig. 6.3 shows the modulated valve and its anode supply. The latter consists of a steady voltage V_b in series with the modulating voltage $V_a \sin pt$.

The average anode voltage is V_b. The average anode current is proportional to this and is, say, V_b/R where R is the resistance of the modulated valve.

The anode voltage at any instant is $v = V_b + V_a \sin pt$.

Assuming that R is constant, the average power is the average value of v^2/R. This is equal to the average value of $(V_b + V_a \sin pt)^2/R$ which is the average value of

$$\frac{1}{R}(V_b{}^2 + 2V_b V_a \sin pt + V_a{}^2 \sin^2 pt)$$

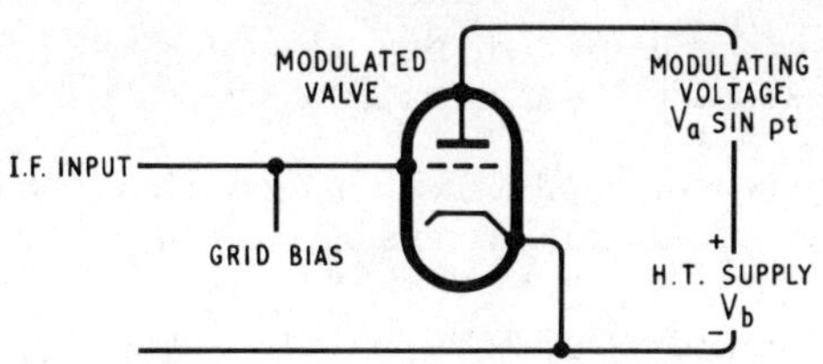

Fig. 6.3 Basic anode-modulation circuit

Remembering that V_b is constant, that the average value of a sine wave is zero, and that that of the square of a sine wave (of peak value unity) is 0·5, we have:

The average value of v^2/R is

$$\frac{V_b^2}{R} + \frac{V_a^2}{2R}$$

The first term (V_b^2/R) is the power supplied by the H.T. supply to the modulated valve–the d.c. power.

The second term $(V_a^2/2R)$ is the power supplied by the source of the modulating voltage, i.e. the modulator circuit–this is the audio power.

Now $V_a/V_b = m$ so that:

$$\frac{\text{AUDIO POWER}}{\text{D.C. POWER}} = \frac{V_a^2}{2V_b^2} = m^2/2$$

and this equals, $\dfrac{\text{SIDEBAND POWER}}{\text{CARRIER POWER}}$

Or, audio power $= \dfrac{m^2}{2} \times$ D.C. POWER TO MODULATED AMPLIFIER.

Example 6.5

The supply to the anodes of the final r.f. modulated amplifier of a transmitter is 1 000 V at 4 A. For eighty per cent modulation, the audio power which the modulator must provide is:

$$\frac{(0{\cdot}8)^2}{2} \times 4\,000 \text{ W} = 1\,280 \text{ W}$$

Example 5.7 may with advantage be referred to again at this stage.

The power requirements may be considerable, efficiency, therefore, is of prime importance. The maximum efficiency attained in practice with class-A working is not more than about 40%, with class-B operation it is about 65% and with class-C operation it is 80% or a little higher.

An ordinary class-A amplifier with inductive load can operate satisfactorily as a modulator but falls far short of the ideal for a number of reasons:

1. The ratio of audio-frequency power output to total power output is small.
2. Unless the power output, and hence the efficiency, is kept to a relatively low level, second and higher order harmonics tend to be generated to an excessive extent.
3. There is the possibility of the introduction of hum.
4. The common supply impedance causes a tendency to instability and the consequent need for precautions, like decoupling, to prevent its occurrence.
5. A very large output inductor is needed if the lower frequencies are to be preserved and if the possibility of saturation of the core due to the steady component of anode current is to be avoided.
6. Little more than 50% modulation may be obtained without serious distortion.

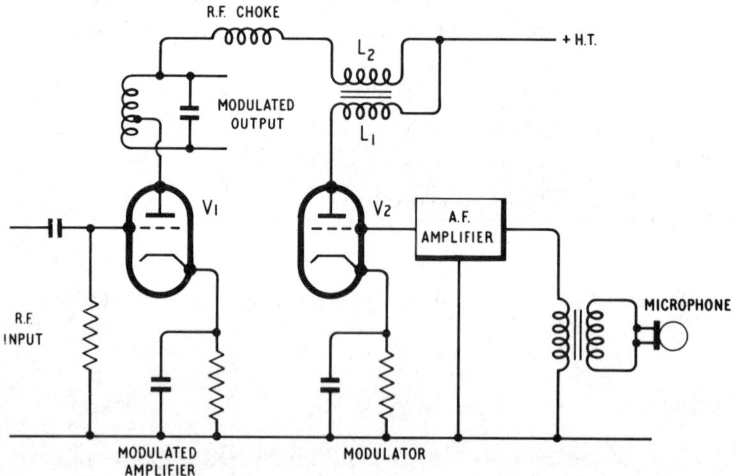

Fig. 6.4 Modulator and modulated stages

A simple basic low-power class-A anode modulator, which suffers from some of the drawbacks just referred to, is shown in Fig. 6.4. (In all but low power equipments such as this one the grid bias voltage is derived from a separate power supply—instead of the cathode resistors shown—in order to prevent the power loss and heat associated with the use of self bias resistors).

A transformer is used for coupling, and such a choice can confer advantages compared with the use of a choke:

1. By using a step-up of voltage between L_1 and L_2 the $V1$ anode voltage swing needed for 100% modulation is easily obtained.
2. Without nullifying advantage (1) the transformer may be used to effect a match between the modulating valve and the modulated valve.
3. By connecting the transformer windings so that the magnetising effects of the anode currents of $V1$ and $V2$ are in opposition the steady magnetisation is much reduced.

Audio input from the microphone (assuming telephony is being employed), amplified as necessary in a.f. amplifiers (not shown), is applied to the modulator grid. The anode current varies similarly, as does the voltage across L_2 which is $V_a \sin pt$. The anode voltage of $V1$ (the modulated amplifier) is thus varied about the H.T. + value by $V_a \sin pt$. If the $V1$ circuit and operating conditions are properly arranged, the amplitude of the r.f. output is proportional to the anode voltage. At 100 per cent modulation V_a is equal to H.T. + and in this condition the anode voltage of $V1$ varies between zero and 2 × H.T. +

For any but low power equipments class-A modulation is too inefficient. Class-C modulation cannot be used because of the distortion which results. Hence push-pull class-B modulators are widely used. (Operation in push-pull is necessary to minimise envelope distortion).

The modulator must provide audio power equal to $m^2/2 \times$ the d.c. power supplied to the anode(s) of the modulated valve(s). For 100% modulation this is one-half of the d.c. power to the anode(s) of the modulated valve(s). Bearing in mind the lower efficiency of the modulating (class-B) circuits compared with that of the modulated (class-C) circuits it is clear that the modulator valves may well need to be capable of dissipating as much power as, or more power than, the modulated valves. Valves (and transistors) must, of course, be able safely to dissipate as heat the total power input from the supply less the power passed on to another stage.

Example 6.6

The final r.f. amplifier of a transmitter develops 4 kW of r.f. power when unmodulated. Its efficiency is 80%. The maximum modulation is to be 95% and it is applied at the final stage anode. A class-B modulator is employed. For purposes of comparison two efficiency values are assumed: (1) 65% and (2) 40%.

The dissipation requirements of the final r.f. and modulator valves can be determined as follows:

At 80% efficiency the final r.f. amplifier needs 4/0·8 kW = 5 kW d.c. input at the anode.

The necessary audio power for 95% modulation is

$$\frac{(0{\cdot}95)^2}{2} \times 5\ \text{kW} = 2{\cdot}26\ \text{kW}$$

1. At 65% efficiency the modulator d.c. input to the anode circuit is 2·26/0·65 kW = 3·48 kW.
2. At 40% efficiency the latter figure becomes 2·26/0·4 kW = 5·65 kW.

The modulator valves must be able to dissipate:

1. (3·48 – 2·26) kW = 1·22 kW
2. (5·65 – 2·26) kW = 3·39 kW

The final r.f. valves must be able to dissipate

$$5 + 2{\cdot}26 - 0{\cdot}8\,(5 + 2{\cdot}26) = 1{\cdot}45\ \text{kW}$$

Example 6.7

An anode modulated class-C amplifier has an anode efficiency of 70% and provides an output power in the absence of modulation of 70 W. Determine the anode dissipation of this valve (a) without modulation applied, and (b) with modulation at 80% and (c) 100%.

If 70 W is 70% of the input power, then the latter must be 100 W, and the anode dissipation without modulation is 30 W.

With 80% modulation the audio power which is required is:

$$\frac{m^2}{2} \times 100 = \frac{0{\cdot}64}{2} \times 100\ \text{W}$$

$$= 32\ \text{W}$$

Of this 32 W, 30% is dissipated at the anode of the power amplifier. Thus the total anode dissipation of the class-C modulated amplifier is:

$$30 + 9{\cdot}6 = 39{\cdot}6\ \text{W}$$

With 100% modulation, the required carrier power is 70 W with a sideband power of 70 × (1/2) = 35 W

The modulator must provide sideband power of

$$\frac{35}{70} \times 100 = 50\ \text{W}$$

Of this, 30 %, i.e. 15 W, is wasted as heat at the output valve anode. The total dissipation at the final anode is thus 30 + 15 = 45 W.

To recapitulate:

The final anode must be capable of dissipating 50% more power than it does with the carrier drive applied without modulation present.

The modulating amplifiers must be chosen so that allowing for only about 60% efficiency they can provide both the sideband power and the fraction of their output power which is wasted at the anode of the the modulated amplifier. In the above example the output power of the modulators is 50 W for 100% modulation. If their efficiency is 60% their anode dissipation must be

$$\frac{50}{60} \times 40 = 33{\cdot}3 \text{ W}$$

This is shared between the modulator valves, probably two, in push-pull.

Finally, if the modulating input is likely to disappear, the modulators must be able to dissipate the whole of the power supplied to them by the H.T. source. However, if push-pull operation is employed the amount of power is relatively small.

6.2.1. High and low level modulation

Modulation may be effected at low power level—prior to the application of most of the r.f. amplification; or, at high level— the final amplifier, or in some instances both the final amplifier and the penultimate amplifier, being modulated. Examples of both types are to be found in Chapter10.

It is not easy to generate economically a.f. power at high level and with little harmonic distortion. Thus an advantage of low level modulation is that it requires little modulation power. Consequently, the modulation circuit is simple and it can be small and compact. On the other hand, with low level modulation, all the amplifiers following the modulated stage must be linear, so that class-C operation, with its high efficiency, cannot be employed. Also, in amplitude-modulation equipments, frequency multiplication may not be effected after modulation.

In general, both high level and low level modulation can give equally good results both in terms of economics and of quality of transmission. The choice is normally determined by the method of frequency control used, the number of frequency multiplication stages, the types of valves available or required to be used, and the traditions of the particular company or designer concerned.

The oscillator is never modulated except in very simple systems and where frequency stability is not important.

6.2.2. Trapezium modulation

In high-frequency broadcasting systems the signal-to-noise ratio at the receiver is often well below the figure regarded as necessary for local broadcast work, and there is often interference. There is clearly no point in transmitting a superbly high quality signal at the expense of power in the sidebands.

The use of trapezium modulation somewhat reduces the quality of the transmission but considerably increases the effective modulation. In most receiver locations this results in a great improvement in the signal-to-noise ratio and this more than compensates for the degradation of the signal quality at the transmitter.

The process consists fundamentally of increasing the amplitude of the higher frequency components of the information signal—with the lower frequencies clipped and relatively reduced.

Transmitters operating with trapezium modulation must be designed to take account of the increased dissipation called for in the final amplifier, modulator and associated equipment, including rectifiers.

A transmitter using trapezium modulation is included in those outlined in Chapter 10.

6.2.3. Push-pull class-B modulator

As already noted, the modulator supplies the sideband power. Particularly in high power, high level modulation equipments, therefore, the efficiency of the modulator is an important economic factor. For this reason push-pull class-B modulators are in wide use as they are capable of developing high power at reasonably high efficiency.

Fig. 6.5 shows the circuit for the modulation of a final r.f. stage by a push-pull class-B modulator. The modulated stage shown in the figure consists of two common cathode neutralised triodes in push-pull. Many other arrangements are common—for example, a single tetrode, or earthed triodes in push-pull.

The r.f. amplifier operates in class-C. The efficiency is about 80%. The valves must thus be able to dissipate one quarter of the power developed. If the bias is removed, the valves are damaged. Hence fixed bias (or at any rate an adequately high proportion of fixed bias) is employed so that in the absence of an input signal the anode dissipation cannot rise to a dangerous level. The series resistors across the filaments are included to enable the effective centres of the filaments to be found for connection to the H.T.—and the earthy part of the grid circuit. If this is not done, the anode current may be modulated at the supply frequency if a.c. is used to heat the filaments.

The grid supply, like that for the anode, is taken to a tap on the appropriate tuning inductor. The tap position is chosen to be as near earth potential as possible—r.f. chokes are included, however, as some r.f. may nevertheless be present. Chokes are preferable to resistors for decoupling because of the heavy currents in both grid and anode circuits.

The modulator, here shown as employing triodes V_1 and V_2, is required to provide the appropriate power. If the modulator circuit is reasonably efficient, the valves normally need to be able to dissipate about the same power as the final r.f. amplifier valve(s) (see Example 6.6). The characteristics of the two push-pull valves must be as near identical as possible. As a means of countering slight differences it may be arranged for the grid bias supplies to be independent (not shown in the figure).

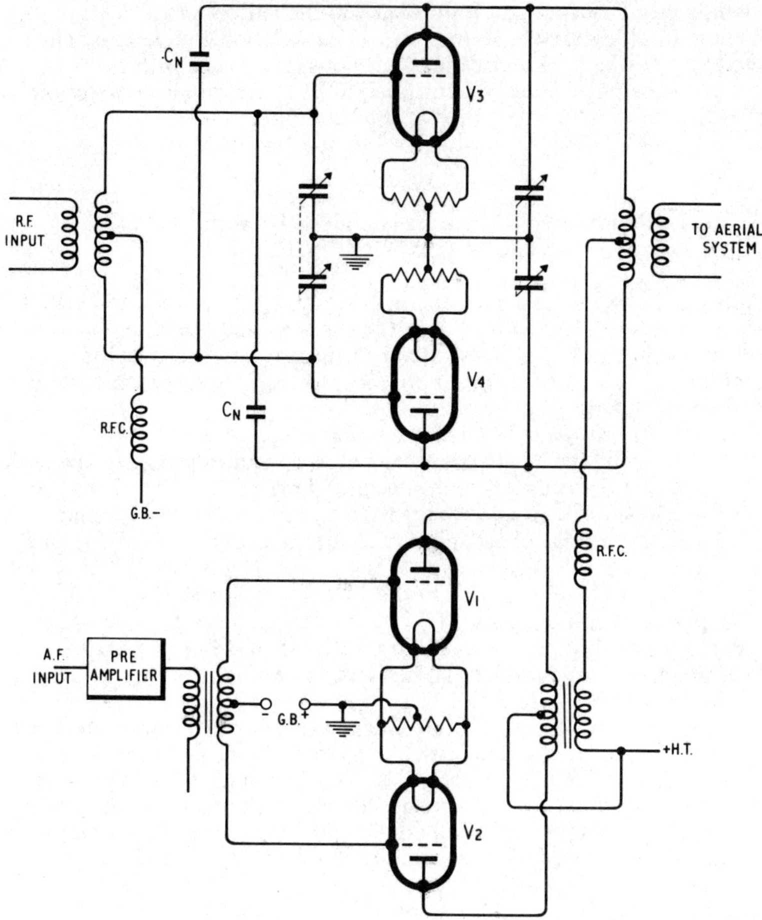

Fig. 6.5 Push-pull modulator and modulated amplifier

Because of the push-pull arrangement there is no steady polarisation of the modulation transformer from the primary winding. Additionally (not shown in the figure) the use of parallel feed choke-capacitance coupling between the transformer secondary and the modulated amplifier prevents the d.c. component of the modulated amplifier anode current from flowing in the modulation transformer secondary.

Considerable power at low impedance may be needed to drive the modulator stage. A cathode-follower stage is, therefore, sometimes used for the drive.

To reduce distortion, negative feedback is applied across the modulator stages. One way of doing this is to rectify part of the modulated r.f. output and feed it back in the correct phase. This is done in both high level and in low level modulation equipments. The phase shifts which take place in the various stages vary with frequency and the negative feedback circuit used must apply suitable correction over the audio-frequency range of the transmitter.

6.3. GRAPHICAL REPRESENTATION OF AMPLITUDE MODULATED WAVEFORM

A carrier $V_c \sin \omega t$, and a modulating waveform $V_a \sin pt$ have been shown in Figs. 6.1(a) and (b) with the resulting amplitude modulated waveform in Fig. 6.1(c). It is often helpful and easier to visualise the modulated wave in the form of phasors representing carrier and lower and upper side frequencies.

Fig. 6.6(a) shows phasors representing

1. the carrier (of length proportional to V_c and rotating at ω (= $2\pi f$). rad/s in the counter-clockwise direction),
2. the lower side frequency (of length proportional to $V_a/2$ and rotating at $(\omega - p)$ rad/s in the counter-clockwise direction), and
3. the upper side frequency (of length proportional to $V_a/2$ and rotating at $(\omega + p)$ rad/s in the counter-clockwise direction).

From Equation 6.2 phasor (i) is a sine function which is zero at $t = 0$ and rising positive; (ii) is maximum positive, and (iii) is maximum negative at the same instant. These facts determine the initial positions of the phasors in the figure.

To simplify matters the whole diagram (Fig. 6.6(a)) can be imagined to be rotated at angular velocity ω in a clockwise direction about 0. This yields Fig. 6.6(b) in which V_c is stationary and the $V_a/2$ phasors have angular velocities p in opposite directions as shown. Clearly the two $V_a/2$ phasors (now labelled OL and OU respectively) are always symmetrically placed with respect to the stationary V_c (OC).

Figs. 6.6(b) to (j) show successive positions of OL and OU at intervals of $\pi/4$ (i.e. 45°) during one cycle of modulating voltage. Figs. 6.6 (k) to (s) show the magnitudes of the resultant phasor at these instants (time here being measured vertically downwards) and the envelope.

In Fig. 6.7 the previous figure is drawn again with a horizontal time axis.

Remembering that a clockwise rotation was imposed on the diagram at the outset we can now, by removing this superimposed rotation, visualise how the resulting phasor in Figs. 6.6(b) to (j) is not only changing in magnitude but is also rotating at ω rad/s in a counter-clockwise direction. Applying this variation to Fig. 6.7 brings us back to the modulated waveform of Fig. 6.1(c) as would be expected.

The use of phasors will be very helpful in later work in investigating and understanding modulation problems.

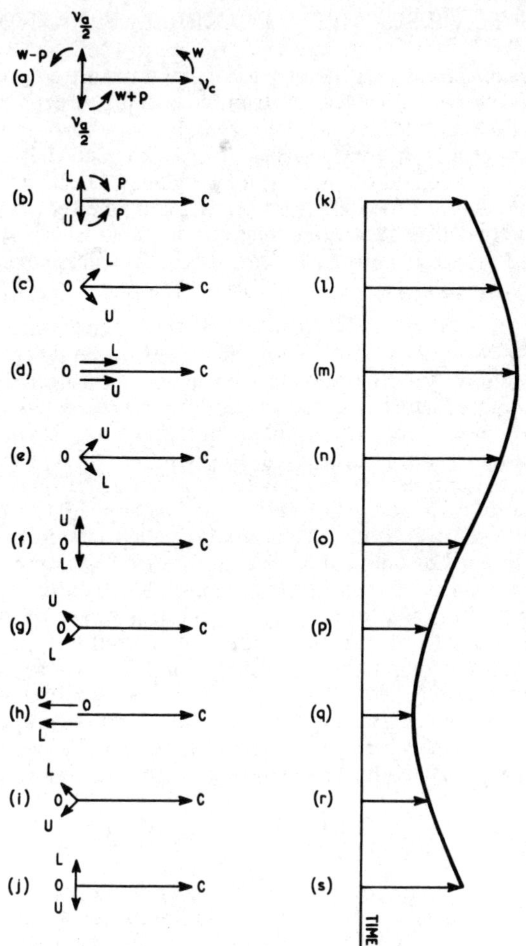

Fig. 6.6 Phasors representing carrier and side frequencies

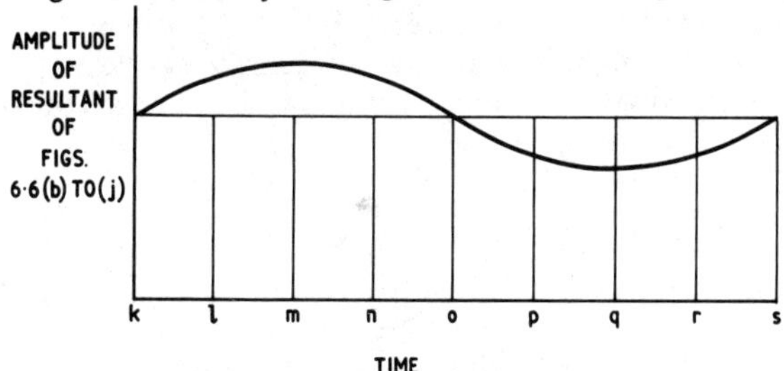

Fig. 6.7 The envelope of the resultant of the sum of the phasors

6.4. REDUCED OR SUPPRESSED CARRIER OPERATION

Using d.s.b. transmissions the major part of the transmitted power is in the carrier. The carrier, however, contributes nothing directly to the magnitude of the received intelligence signal, its only purpose is to enable the receiver detector to function satisfactorily. Provided that a signal of adequate amplitude and of the correct frequency can be made available at the receiver, the detection process can be effected without the carrier. Such an arrangement enables considerable saving of power to be effected at the transmitter. There are other possible advantages as we shall see shortly.

The locally generated substitute carrier at the receiver must have a frequency very close indeed to that of the carrier which it replaces. If the frequency stability at the transmitter and at the receiver is very good the locally generated substitute carrier need not be corrected and the carrier may be completely suppressed at the transmitter. If the overall frequency stability is not good enough to permit complete suppression of the carrier at the transmitter, a low level (pilot) carrier may be transmitted so that frequency correction may be applied at the receiver. This pilot carrier is often transmitted 26 dB down on its normal value.

With a suppressed or reduced carrier transmission, it is also better to eliminate one of the sidebands. If the remaining sideband is transmitted alone (with or without a pilot carrier), the result is *single sideband transmission (s.s.b.)*. In commercial practice it is usual to transmit two sidebands each conveying different information. This is *independent sideband transmission (i.s.b.)*. The sidebands are often 6 kHz wide and may each contain two 3-kHz speech channels, several telegraph channels, or some mixture of speech and telegraph channels so that each sideband is not more than 6-kHz wide. The different channels are separated in the receiver.

The advantages of single sideband or of independent sideband transmission may be summarised as follows:

1. If only one sideband (or two sidebands each conveying different information signals) is transmitted, the frequency spectrum occupied is only about one-half of that needed for double sideband transmission. About twice as many stations can, therefore, be accommodated in a given waveband.
2. The reduction in bandwidth reduces the receiver noise (and, other things being equal, increases the signal-to-noise ratio) in the same ratio as the bandwidth is reduced.
3. Considerable power is saved in not generating a carrier at high power.
4. Fading affects both carrier and sidebands. Therefore, if a steady carrier is introduced at the receiver the effects of fading are reduced.
5. With some types of detector the receiver output increases, up to a point, with the amplitude of the carrier. Hence a large locally

introduced carrier boosts the receiver output.

6. There is a reduction in cross-modulation on multiple channels and a corresponding reduction in the swing at the valve or transistor inputs.
7. Complete elimination of the carrier gives some secrecy because the transmission does not result in any output from an ordinary receiver. Reasonable reception, however, is usually possible on a receiver fitted with a beat frequency oscillator.

6.4.1. Sideband generation

There are several ways in which sidebands may be generated without the carrier signal appearing in the output. The ring modulator was introduced in *Radio and Line Transmission Volume 2,* in this series and now a ring bridge modulation circuit (Fig. 6.8) will be described. The audio-frequency input is applied to the first transistor which functions as a phase splitter and gives, at AB, a push-pull input to the ring bridge circuit. The potentiometers shown enable any small departures from symmetry to be corrected.

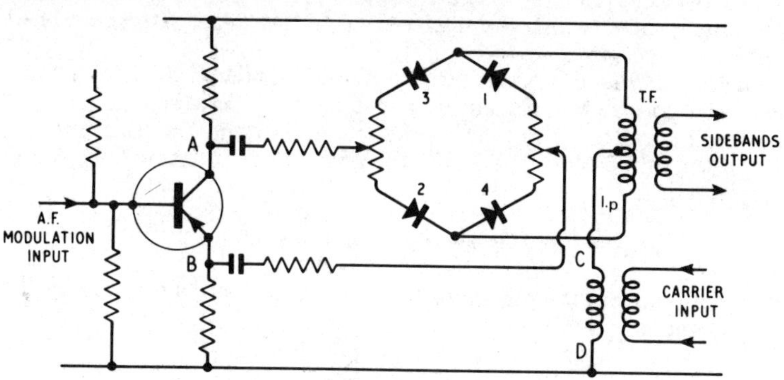

Fig. 6.8 Ring-bridge modulator circuit

The carrier input at *CD* is a signal whose amplitude is maintained at a substantially greater value than that of the modulating signal. The carrier frequency, at this stage, is often 100 kHz. The influence of the carrier voltage on the circuit is to switch the pairs of diodes in the bridge circuit on and off in accordance with the polarity of the voltage. Because of the carrier magnitude this action is independent of the modulating signal.

When the carrier voltage is positive at *C* with respect to *D* diodes 3 and 4 conduct but diodes 1 and 2 are cut off (Fig. 6.9(a)). For the duration of this half cycle of carrier voltage the current I_p circulates round the circuit shown and varies in magnitude in accoradance with the modulating voltage from the input transistor– for simplicity this latter voltage is shown as a single voltage in Fig. 6.9(a). Figs. 6.9(c) and (d) show the carrier and the modulating voltage (drawn as a sine wave).

So far period 1 has been considered in these figures. During this period the current I_p is of the form shown for the same period in Fig. 6.9(e).

During period 2 the carrier voltage polarity reverses cutting off diodes 3 and 4 and switching on diodes 1 and 2. The conducting circuit is now that shown in Fig. 6.9(b). The magnitude of the current is still controlled by the modulating voltage $E \sin pt$ and so is given by that shown for period 2 in Fig. 6.9(e). Thus with the switching of the diodes by the carrier voltage, the current round the circuit (and through the primary of the transformer *TF*) alternately increases and decreases and the magnitude of the change is determined by the modulating voltage $E \sin pt$.

The current changes are at the sideband frequencies (and certain, unwanted, higher frequencies). There is no component at the carrier frequency. The sideband output at the transformer secondary is taken to a filter circuit to remove the unwanted higher frequency modulation components after which the sideband signals are dealt with as outlined in the next section.

The circuit Fig. 6.8 may alternatively be considered by thinking of the carrier voltage as representing a switching device which changes the circuit admittance so that it conducts first in one direction and then in the other. Mathematically, the admittance function (Fig. 6.9(f)) is expressed:

$$y = \frac{4Y}{\pi}\left(\sin \omega t + \frac{1}{3}\sin 3\,\omega t + \frac{1}{5}\sin 5\,\omega t + \text{-----}\right)$$

The circuit current i_p is $yE \sin pt$ (i.e. the admittance multiplied by the voltage) and is:

$$i_p = \frac{4YE}{\pi}\sin pt\left(\sin \omega t + \frac{1}{3}\sin 3\,\omega t + \frac{1}{5}\sin 5\,\omega t + \text{----}\right)$$

Ignoring the second and higher terms within the bracket, and using only the first term we have:

$$i_p = \frac{4YE}{\pi}\sin \omega t \sin pt$$

$$= \frac{4YE}{\pi} \times \frac{1}{2}\left[\cos(\omega - p)t - \cos(\omega + p)t\right]$$

These are the upper and lower sidebands (and no carrier). The second and higher terms within the bracket, which have been ignored, yield unwanted modulation products. These are removed in the circuit filter which follows the ring modulator. Thus the resultant output does consist only of the upper and lower sideband terms.

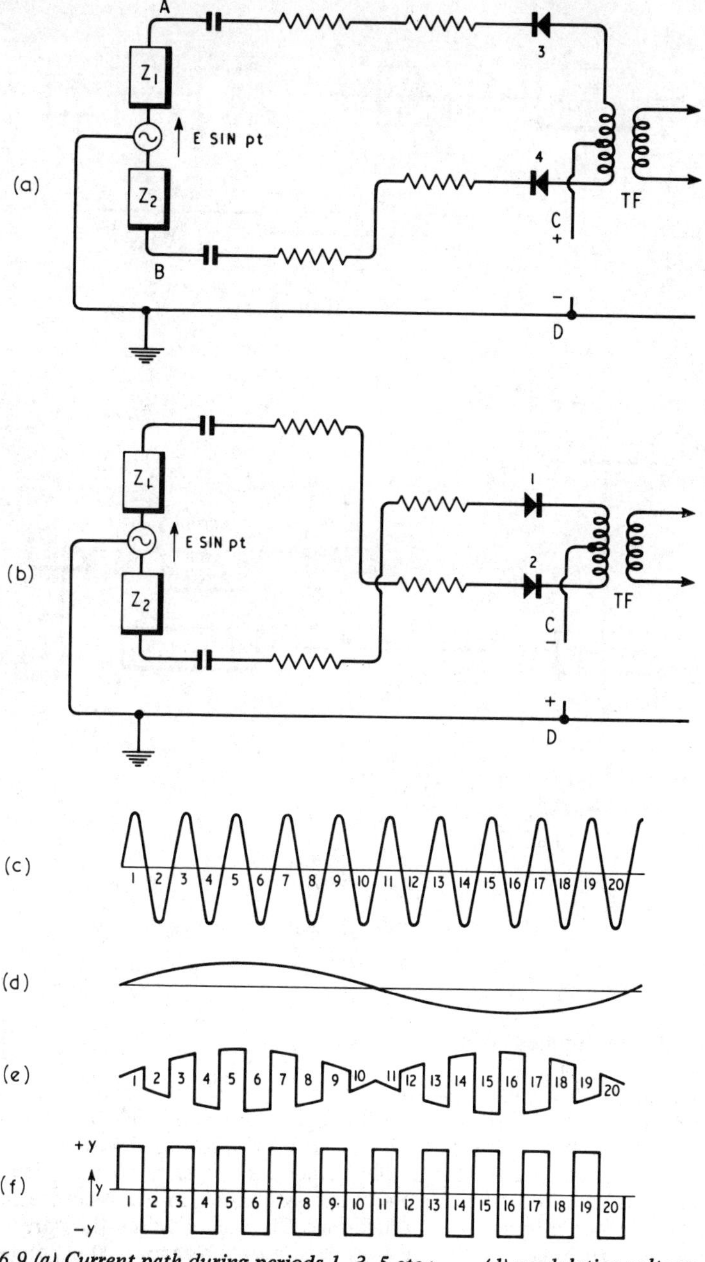

Fig. 6.9 (a) Current path during periods 1, 3, 5 etc.; (b) current path during periods 2, 4, 6 etc.; (c) carrier voltage; (d) modulating voltage; (e) circuit current; (f) admittance function

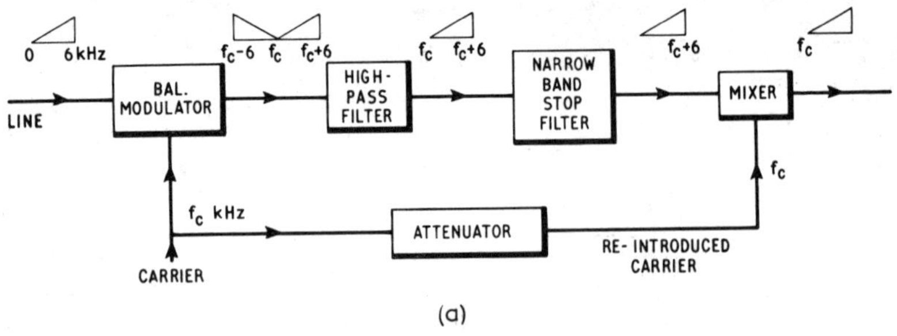

(a)

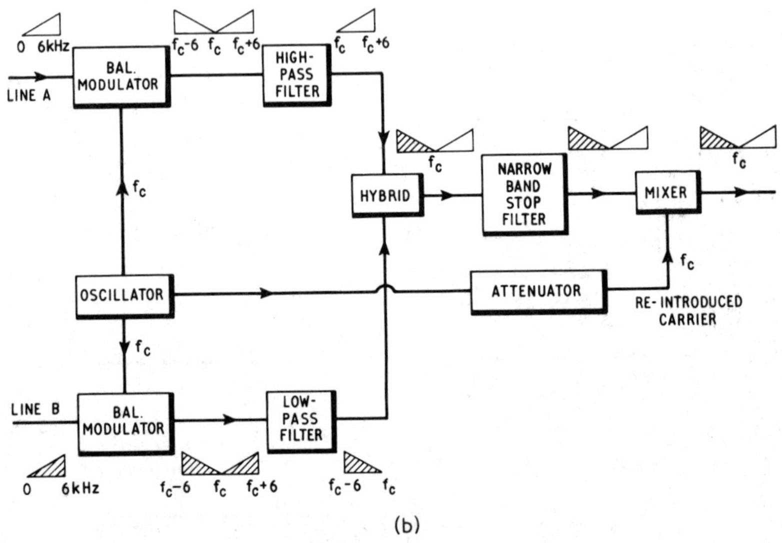

(b)

Fig. 6.10 Development of i.s.b. signal

6.4.2. Development of i.s.b. signal

In Chapter 10 are examples of i.s.b. transmitters. The general principle, however, is illustrated in Fig. 6.10(a). The information signal, shown as occupying a band of frequencies up to 6 kHz, is mixed with the carrier in the balanced modulator to give upper and lower sidebands with very little carrier in the output. One of the sidebands, in this instance the lower, is then eliminated in a high-pass filter after which the narrow bandstop filter completes the elimination of the carrier. This creates a single sideband signal with suppressed carrier.

Also shown in the figure is the line along which, with suitable attenuation, the carrier may be re-introduced at a known low level. We now have a single sideband signal with pilot carrier.

In Fig. 6.10(b) are shown two incoming information channels A and B. Signal A is treated in a manner generally similar to that shown in Fig. 6.10(a). Signal B has the upper sideband removed after modulation so that after combination in a hybrid there is a composite signal embodying, in two sidebands, the different information signals of lines A and B. The carrier is then eliminated and, if required, re-introduced at low level. This gives an independent sideband signal with either suppressed or pilot carrier.

QUESTIONS

1. Derive an expression for the output power of a double sideband amplitude-modulated transmitter in terms of the unmodulated carrier output power and depth of modulation.

The output power of a transmitter is 1 kW when modulated to a depth of 100%. If the depth of modulation is reduced to 50% what is the power in each sideband? *(C & G)*

2. Determine an expression for the relationship between the power in the sidebands and the carrier power for a double sideband amplitude-modulated transmitter. Use the expression to sketch a graph of the change in the ratio of sideband power to carrier power with increase in modulation factor from zero to unity.

3. An amplitude-modulated transmitter has an anode modulated class-C output stage in which an audio frequency sine wave of 3 kV peak value is developed across the secondary of the modulating transformer in series with the 5 kV h.t. supply. The stage has an anode efficiency of 75% and delivers 1·5 kW of carrier power into the tank circuit. Calculate:

(a) the depth of modulation,
(b) the mean anode current,
(c) the power supplied by the modulator,
(d) the total r.f. power delivered to the tank circuit.

State the assumptions you have made in these calculations. *(C & G)*

4. (a) Upon what factors does the anode dissipation capability of a modulated valve depend?
(b) With the aid of a schematic diagram explain the principles of operation of an anode modulated class-C r.f. amplifier stage.

5. (a) Why are class-B push-pull modulators in wide use for anode modulation?
   (b) Compare high level and low level modulation methods and their relative advantages and disadvantages.

6. (a) Why is trapezium modulation employed for certain types of transmissions and not for others? What are the advantages of its use?
   (b) Explain the terms: *sidebands, side frequencies, amplitude modulation, modulated amplifier, modulating amplifier.*

7. A carrier is 100% modulated by a single frequency sinusoidally varying voltage. Sketch phasors, in the correct relative phase, representing the carrier and side frequencies at $t = 0$ (when the carrier and the modulating voltage are both zero and rising positive) and at eight equally spaced intervals during one cycle of modulating voltage in order to determine the shape of the envelope of the resulting modulated voltage.

8. What are the advantages of using reduced or suppressed carrier transmissions compared with transmissions employing a full power carrier? When is it permissible to suppress the carrier completely and when must a reduced (pilot) carrier be transmitted?

9. Explain one method by which modulation may be effected to create sidebands without an accompanying carrier.

10. Draw a block diagram of a basic i.s.b. transmitter and explain in outline the method of operation. How are any unwanted unmodulation products removed?

7

Transmitter Output Stages

7.1. GENERAL REQUIREMENTS

The output stage of a transmitter comprises the r.f. power amplifier stage together with the coupling circuits by which power is transferred to the radiating aerial. The general requirements of the output stage may be listed as follows:

1. The valve(s) or transistor(s) should operate with the best efficiency consistent with freedom from distortion of the modulation envelope of the signal.
2. The aerial must be coupled to the power amplifier in a manner which avoids mismatch either between the output impedance of the active device and the feeder, or between the feeder and the aerial.
3. Any harmonics of the radiated frequencies which are products of the class-B or class-C biasing of the power amplifiers must be prevented from reaching the aerial and causing radiation which might interfere with harmonically related communication channels. This is especially important on h.f. where long range sky-wave transmission can occur on very low power.
4. Parasitic oscillations in power amplifiers must be prevented as these could give rise to spurious radiations causing interference with reception on other channels.
5. The bandwidth of the circuits used for coupling the output of the power amplifier to the aerial must be wide enough to accept all the side frequencies of the modulated signal without attenuation. If the sidebands are attenuated relative to the carrier, this is equivalent to a reduction in the depth of modulation. Further, if some side frequencies are attenuated more than others this is a form of attenuation distortion. Some transmitters use wideband power amplifier circuits which are designed to amplify and couple to the aerial all the operational channel frequencies without need of adjustment when one channel of working is exchanged for another (see Section 7.11).
6. Some form of safeguard for the output active device must be provided so as to limit or cut off the feed current if a fault occurs in the output circuit which would remove or diminish the loading of the device, or if the drive fails in such circuits as derive their bias by rectification of the drive voltage.

7.2. CAPACITIVE COUPLING TO THE AERIAL

Fig. 7.1 is the circuit of a single valve output stage capacity coupled to an aerial which is assumed to be long enough to have an inductive reactance at the frequency of transmission. The function of the components is as listed below.

C_1 is the coupling capacitor which insulates the grid of the power amplifier from the anode of the previous valve for the d.c. potential difference between them but acts as a coupling capacitor for the drive frequency. L_1 is an r.f. choke added in series with the grid leak R_1 so as to provide a sufficient impedance at r.f. in parallel with the input impedance of the valve. R_1 is the grid resistor across which a class-C bias is developed by the d.c. grid current. C_2 decouples the r.f. from the resistor R_1 so that r.f. power is not dissipated in R_1. Such an unn unnecessary loss would reduce the Q factor of the tuned load of the previous stage. L_2 is an r.f. choke through which the direct feed current reaches the valve anode and which directs the r.f. component of the valve current into the output circuit via C_3. C_3 is a d.c. blocking capacitor which insulates the aerial from the H.T. voltage of the valve anode. C_4 is an adjustable aerial coupling capacitor. L_3 is an aerial tuning variometer.

The combined action of C_4 and L_3 provide matching between the output impedance of the power amplifier and the aerial circuit and also secure resonance in the load.

An equivalent circuit of the aerial circuit is shown in Fig. 7.2.

R_0 is the effective output resistance of the valve under operating conditions. L is the inductance of the aerial conductor. R is the total resistance of the aerial. (This is a value in ohms which, when multiplied by the mean square of the aerial current, will give the total aerial power consumed. R represents both radiation and loss resistance.) C is the value of the distributed aerial capacitance to earth.

The branch of the circuit (Fig. 7.1) which includes the aerial has an inductive reactance given by: $j(\omega L_3 + \omega L - 1/\omega C)\ \Omega$, where ω is the angular frequency of aerial current. Let L' be the equivalent inductance of the aerial branch, then, $j\omega L' = j(\omega L_3 + \omega L - 1/\omega C)$ and $L' = (L_3 + L - 1/\omega^2 C)$ H.

With C_4 adjusted to bring the output circuit to resonance with the frequency of radiation, the valve is presented with a load of resistive value: $L'/C_4R\ \Omega$.

As both C_4 and L' can be varied, the load can be adjusted to match optimum load requirements of the valve. C_4 is first set to a minimum value and L_3 adjusted until resonance is obtained. Resonance is indicated by a dip in the reading of the feed milliammeter of the valve and by a maximum reading on the aerial ammeter, AA. C_4 is increased in steps, L_3 being altered at each new value of C_4 to produce resonance. As C_4 is made larger in value L_3 is made smaller and the dynamic impedance of the load becomes less. The settings of C_4 and L_3 which result in the best reading on the aerial ammeter are sought. The feed

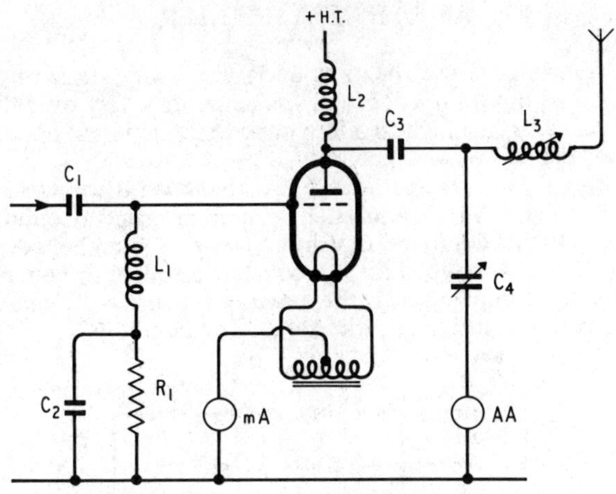

Fig. 7.1 Capacitance coupling to aerial

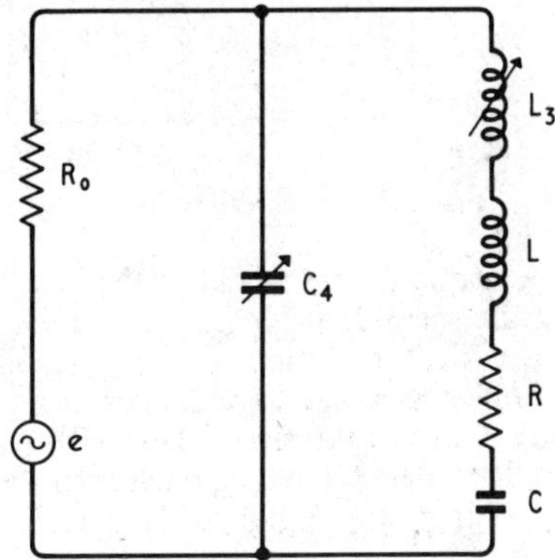

Fig. 7.2 Equivalent circuit of capacitance coupling

current of the power amplifier becomes larger as the power fed to the load increases. Care must be taken that the feed current does not become too large for the valve. An over-load relay may be provided to safeguard the valve when the feed current exceeds a safe limit. See Section 7.8.

The use of a capacitive coupling provides a means of reducing the radiation of harmonics of the required transmission frequency. C_4 has a reactance at the second-harmonic frequency which is only half of its value for the fundamental. The second harmonic and any higher order harmonics are, therefore, decoupled from the aerial by C_4.

7.3. USE OF A Π-FILTER

Some h.f. transmitters use the same aerial for a wide range of operating frequencies. The input impedance of the aerial may vary over the frequency range from a high to a low impedance and have either capacitive or inductive reactance.

A matching filter is, therefore, required between the power amplifier and the aerial. Fig. 7.3 is an equivalent diagram of such a coupling circuit. Components C_1, L and C_2 form a low-pass filter between the power amplifier and the load R_a. R_a, which is assumed to be resistive, may represent the impedance at the aerial terminals or the characteristic impedance of the feeder line to the aerial. The components of the filter

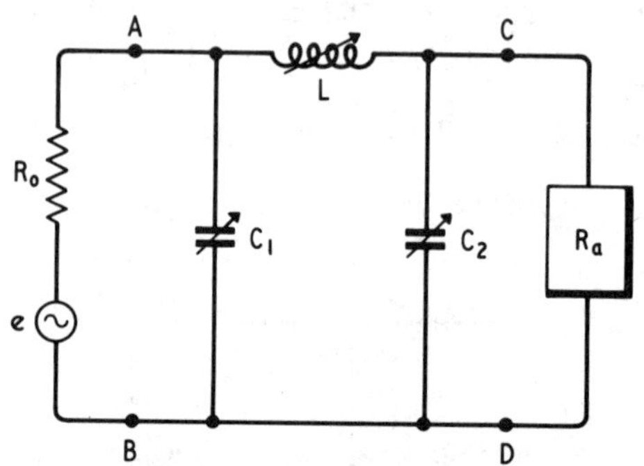

Fig. 7.3 A π matching filter

are chosen so that the frequency to be radiated lies within the pass band of the filter but the second and higher harmonics are in the attenuation band of the filter. For matching, R_0 the valve output resistance and R_a must be the image impedances of the filter. That is, the impedance looking in at the terminals *A-B*, with R_a connected, is equal to R_0, while the impedance looking back into the terminals *C-D*, with R_0 connected at *A-B*, is equal to R_a. The filter is redrawn in Fig. 7.4.

R_1 is the equivalent series resistance and C_1' is the equivalent series capacitance of the parallel arrangement of R_0 and C_1. Similarly, R_2 is the equivalent series resistance and C_2' is the equivalent series capacitance of the parallel arrangement of C_2 and R_a. It can be shown that

$$R_1 = R_0/(1 + \omega^2 C_1^2 R_0^2)$$

$$R_2 = R_a/(1 + \omega^2 C_2^2 R_a^2)$$

$$C_1' = C_1\ (1 + 1/\omega^2 C_1^2 R_0^2)$$

and,

$$C_2' = C_2\ (1 + 1/\omega^2 C_2^2 R_a^2)$$

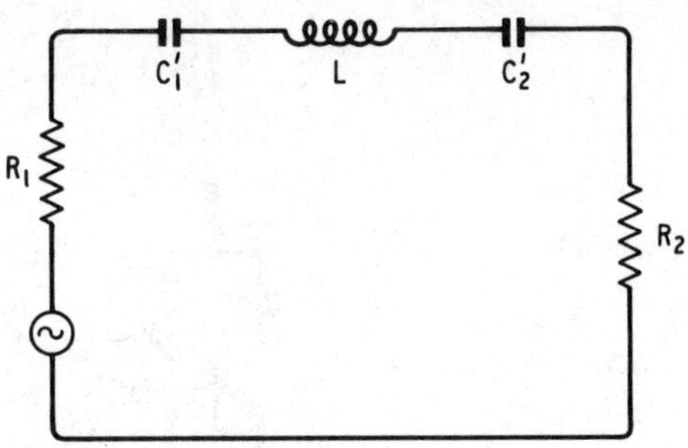

Fig 7.4 A series equivalent circuit for a π-filter

For matching, R_1 must equal R_2 and the reactance of L must equal the combined reactance of C_1' and C_2' in series so as to make the circuit resonant. These equalities can be obtained by adjustment of C_1, C_2 and L. The inductance of L is first set to a value suitable to the transmission frequency. The higher this frequency, the smaller the value of L required. C_2 is first set to minimum coupling (maximum capacitance) and C_1 adjusted until a dip in the feed current reading on a valve cathode current milliammeter shows that resonance has been obtained. Coupling is then increased in steps by successive reductions in the value of C_2, and C_1 is readjusted to restore resonance at each step. The process is continued until the final stage feed current reading reaches the maximum permitted value for the power amplifier. An aerial current meter in series with R_a now shows the best possible reading. The reactance of C_2 is small compared with the impedance R_a at the second harmonic of the radiation frequency.

It follows that the lower the radiation frequency, the larger is the value of coupling capacitance used. On the lowest frequency ranges it may be necessary to reduce the capacitive reactance of the coupling arm of the π-filter by including a small amount of inductive reactance in series with it. The added inductive reactance can be switched out on the higher frequency bands.

7.4. TRANSFORMER COUPLING

If an aerial presents a capacitive reactance and tuning requires the addition of inductive reactance in series with it, then matching over a limited frequency range may be secured by use of transformer coupling. Fig. 7.5 illustrates this. The tuned primary of the matching transformer selects the fundamental frequency component of the anode current for amplification. The gain at the higher harmonics is negligible if the Q of the primary is sufficiently high (say, over 10.) An electrostatic screen between primary and secondary guards against the capacitive

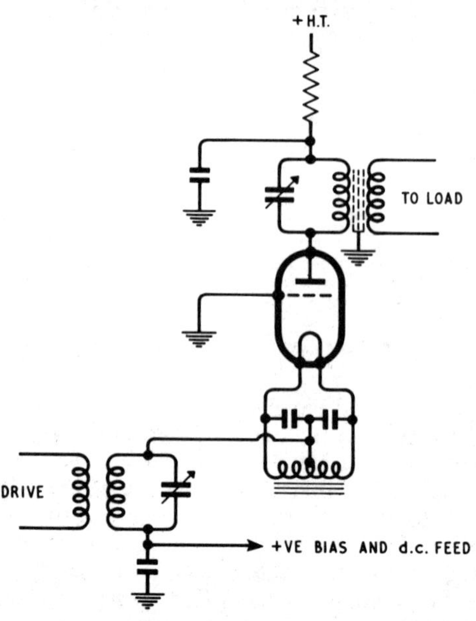

Fig. 7.5 A transformer coupled power amplifier

coupling of harmonics into the aerial or aerial feeder. This type of matching is suitable for either balanced or unbalanced feeders and may be used with either single ended or push-pull output stages.

The valve in Fig. 7.5 is shown as working in grounded-grid arrangement. This is advisable if a single triode valve is used in order to avoid feedback via the anode–grid capacitance of the valve.

7.5. Choice of Valves and Arrangements

Valves or transistors selected for the output stages of a transmitter must have the ability to dissipate without overheating, 25% or so of the power supplied to them. Silicon transistors are available which can dissipate 100 W or so. Two or more such devices in parallel can provide the output power of a small transmitter. Because transistors deteriorate rapidly if overheated, careful control of their power dissipation by effective ventilation is even more important than it is for valves.

Larger transmitters use valves in their output stages even though earlier stages may employ transistors. Tetrode valves capable of dissipating 6 kW or more are manufactured, but as the cooling of the screen grid presents some problems, water or vapour cooled triodes are usual in the output stages of very large transmitters. For example, a certain 250-kW broadcasting transmitter uses two vapour cooled triodes each with an anode rating of 125 kW as its output valves. They are connected in an earthed-grid push-pull circuit.

A choice may exist between one large valve and two or more smaller valves. The deciding factor may be cost. A single valve is cheaper than two or three smaller valves and takes up less space. However, the larger valve requires a higher H.T. voltage and the provision of this, especially in mobile equipment, may make the total outlay greater than that for the single valve choice. Where two or more valves operate in parallel, then despite the failure of one valve the transmitter may continue to operate albeit at reduced efficiency and output. (This assumes that the valves do not operate with their heaters in series as is sometimes the case.) A disadvantage of connecting valves in parallel is that their interelectrode capacitances are in parallel also—a feature which encourages parasitic oscillations (see Section 7.6).

Alternatively, if two valves are used, a push-pull connection can be used with the following advantages:

1. The valves do not have their capacitances in parallel.
2. They are less liable to parasitic oscillation.
3. Even harmonics of the drive frequency produced by the valves cancel in the output circuit so that the second harmonic which is the most troublesome is suppressed.
4. The opposite ends of the output transformer windings are balanced in their impedance to earth which facilitates coupling to a balanced line.
5. The push-pull circuit is easily neutralised (see Section 5.7).

Example 7.1

Show that the harmonic distortion produced by two similar valves connected in push-pull is less than if the same two valves are connected in parallel.

Equivalent circuits for the two alternative connections are given in Figs. 7.6(a) and (b). Let the equation for each valve current be given by:

$$i_a = A + B\,V_g + C\,V_g^2 + D\,V_g^3 + \ldots$$

where A, B, C and D are constants. Let the grid drive voltage for each valve be, $V_g \sin \omega t$.

In the parallel circuit both valves have their drive voltages connected with the same phase, but in the push-pull circuit the two valves have drive voltages which, though equal in amplitude, are opposite in phase. Thus, if one valve has a drive of $V_g \sin \omega t$, the other has a drive of $-V_g \sin \omega t$. Fig. 7.6(a) shows that for the parallel connection the total output current is the sum of the two valve currents. Fig. 7.6(b) shows that

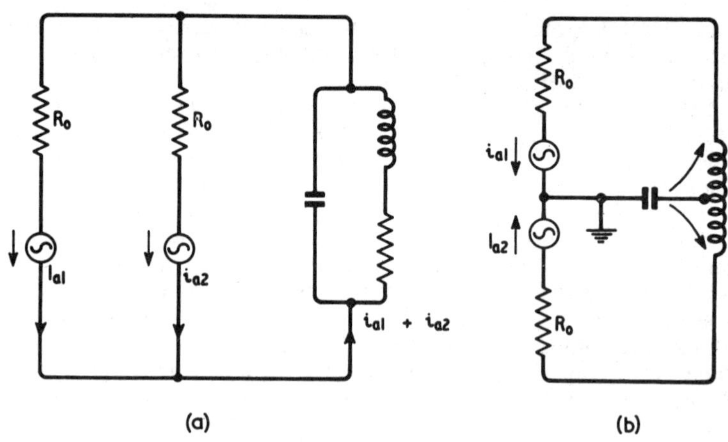

Fig. 7.6 Parallel and push-pull equivalent circuits

current is the sum of the two valve currents. Fig. 7.6 (b) shows that with the push-pull connection the two output currents feed in opposite directions through the two halves of the output transformer primary so that the output is proportional to the difference of the two valve currents.

Hence for the parallel connection the output is proportional to:

$$i_{a1} + i_{a2} = 2(A + B\, V_g \sin \omega t + C\, V_g^2 \sin^2 \omega t + D\, V_g^3 \sin^3 \omega t)$$

But for the parallel arrangement the output is proportional to $(i_{a1} - i_{a2})$ where,

$$i_{a1} = A + B\, V_g \sin \omega t + C\, V_g^2 \sin^2 \omega t + D\, V_g^3 \sin^3 \omega t$$

and

$$i_{a2} = A - B\, V_g \sin \omega t + C\, V_g^2 \sin^2 \omega t - D\, V_g^3 \sin^3 \omega t$$

so that, $(i_{a1} - i_{a2}) = 2B\, V_g \sin \omega t + 2D\, V_g^3 \sin^3 \omega t + \ldots$

In the above expression for the difference of the two valve currents, all components which are functions of the even powers of the grid voltage are reduced to zero assuming that the two valves have identical characteristics. The even powers of the grid voltage contain components at the even harmonics of the drive frequency, and the odd powers of the grid voltage contain components of the odd harmonics of the drive frequency. The lowest odd harmonic other than the fundamental is the third, and the amplitude of this is likely to be small compared with that of the second-harmonic output of the parallel circuit where the second-harmonic outputs of the two valves combine in phase. Thus the push-pull arrangement reduces the second harmonic to zero and contains less harmonic distortion in its output than the parallel circuit.

7.6. PARASITIC OSCILLATIONS

If an amplifier tends to sustain an oscillation and an output at a frequency other than the drive frequency, the oscillation is called parasitic. The parasitic oscillation removes power from the required output and, therefore, reduces the efficiency of the stage. In addition, the parasitic oscillation may cause spurious radiation which interferes with other services.

Fig. 7.7(a) and (b) show one way in which a v.h.f. parasitic oscillation may take place. The first diagram is of the nominal amplifying circuit. The second diagram shows only the features causing the parasitic

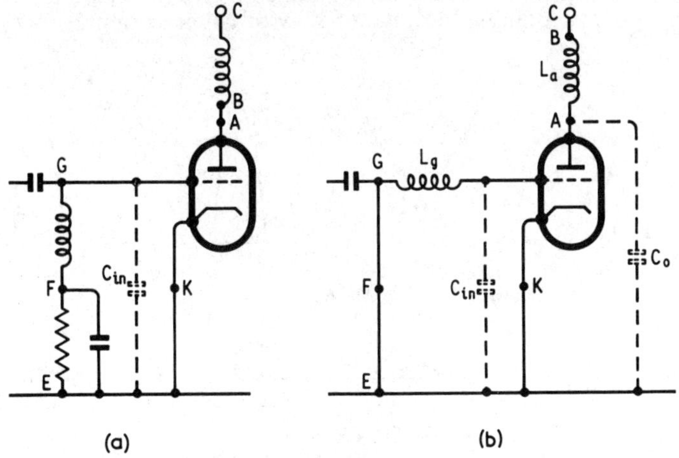

Fig. 7.7 Actual and parasite circuits

oscillation. The lettered points in the two diagrams correspond. In Fig. 7.7(b) the output capacitance of the valve is shown in parallel with L_a. The latter represents the inductance of the anode lead of the valve. The input capacitance C_{in} is effectively in parallel with the inductance L_g which represents the inductance of the grid lead of the valve. The r.f. chokes *GF* and *CB* of Fig. 7.7 (a) are effectively by-passed by their own winding capacitance and, therefore, are omitted from Fig. 7.7 (b). This second circuit is seen to be that of a tuned anode, tuned grid oscillator in which an oscillation may be sustained at a frequency determined mainly by the values of C_0, L_a, C_{in} and L_g.

The parasitic oscillations may be prevented by connecting resistors of the order of 100 Ω as near to the anode and grid pins of the valve as possible and in series with the valve leads. These resistances render the parasitic circuits non-oscillatory. Small chokes are connected in parallel with the parasitic oscillation stopper resistors to act as by-pass circuits both for the d.c. feed current to the anode and for the required output frequency.

7.7. AN INDICATION OF A RESONANT LOAD

Resonance in the tuned load of a class-B or a class-C amplifier may be detected by watching for a dip in the reading of the d.c. feed current which occurs when load is tuned through resonance. The reason for this dip in the feed current at resonance is explained with the aid of Figs. 7.8 (a) and (b).

The amplitude of the anode current pulse depends on how far positive the grid potential is driven and also on the instantaneous anode potential during the time of flow of anode current.

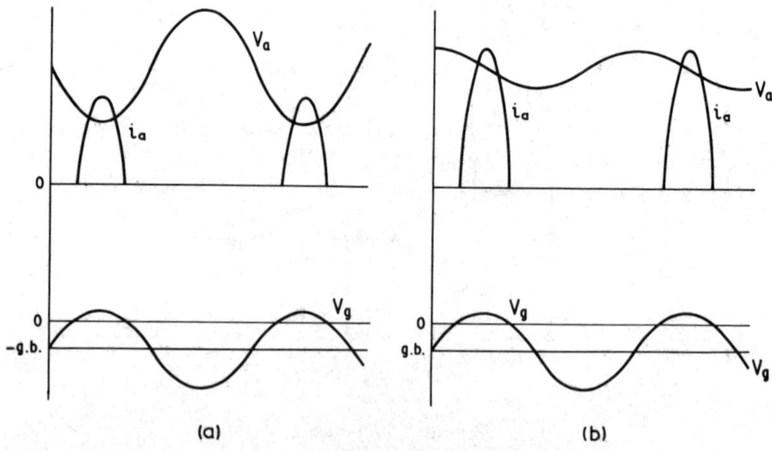

Fig. 7.8 The effect of detuning on anode current

In Fig. 7.8 (a) the load is assumed resonant and, therefore, resistive so that the anode potential is changing in anti-phase to the grid potential. Therefore, the anode current pulse peak occurs at the instant of minimum anode potential.

Fig. 7.8(b) shows conditions when the anode load is detuned The load impedance is smaller and has a phase angle of almost 90° since it is now reactive. With the load detuned, the anode voltage swing is smaller and the minimum anode voltage no longer occurs when the grid potential is maximum. The anode potential is thus higher while the valve is conducting and the peak current pulse is larger. The feed current meter for the stage shows the average valve current and the reading, therefore, rises when the load is detuned and dips to a minimum reading when the load is made resonant at the drive frequency.

If the load impedance is high as it is for a voltage amplifier, the anode voltage swing is large and the anode voltage v_a falls to a very low minimum at resonance which causes a very pronounced drop in the feed current reading. If the load impedance is made smaller by an increase of coupling between the aerial and the power amplifier, the dip in the feed current at resonance becomes less marked. When the average valve current for the resonant load conditions is as big as the anode rating of the valve permits, then the loading on the valve cannot be increased further.

7.8. SAFEGUARDS FOR POWER AMPLIFIERS

An r.f. short circuit across the load prevents any variation of the anode potential which thus remains fixed at the H.T. voltage. This high value of anode potential is present throughout the valve current pulse which is very large under these conditions. A power amplifier valve must, therefore, be protected against the high average current which flows if a fault occurs on the aerial feeder or aerial. Small valves may rely on the limiting effect of a cathode bias resistor which automatically increases

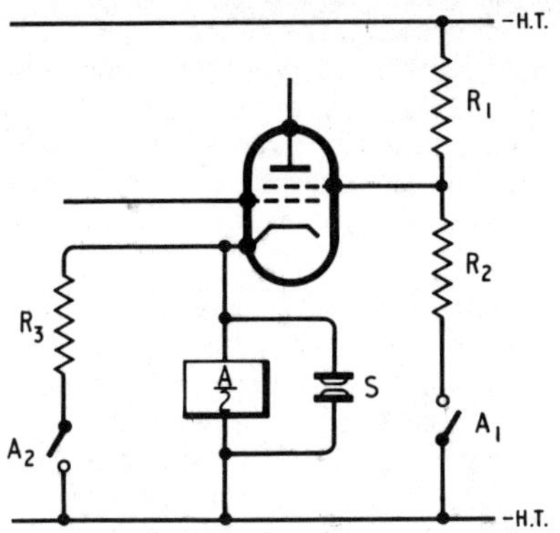

Fig. 7.9 A cathode relay circuit

the bias applied in direct proportion to the increase of anode current. Tetrode valves may use a cathode current relay which closes when a cathode current reaches a critical value. Fig. 7.9 shows one possible arrangement.

The relay A_2 will pull in if the cathode current reaches a value likely to overheat the valve anode. By closing contact A_1 the relay connects the resistor R_2 between screen grid and –H.T. The current which then flows through R_2 via R_1 increases the p.d. in R_1 so that the screen potential is reduced and the anode current held down to a safe level in this way. Relay contact A_2 opens to disconnect the shunt from the relay coil so that the whole of the now reduced cathode current flows through the relay coil and is able to hold the relay in until the fault which caused the relay to operate has been cleared. The relay can be reset by momentarily closing the switch contacts S.

An alternative safeguard is to use a photo-electric cell mounted near to the power amplifier. If infra-red radiation from the valve anode reaches a critical level, the cell admits an operating current through a relay which disconnects the H.T. from the valve. This device has the advantage that it will operate if the valve anode temperature rises excessively because of a general rise in the ambient temperature and not only because the valve current is too big.

Large vapour-cooled valves of a type used in high power broadcast transmitters may use thermal fuses. These are fitted to the outside of the anode structure and melt if the anode temperature reaches a dangerous level. The melting of the fuse releases a spring-loaded switch which disconnects the high voltage d.c. supply from the valve.

7.9. EFFICIENCY OF POWER TRANSFER

An increase of coupling between the tuned anode circuit of a power amplifier and its load results in a larger effective resistance in the tuned circuit caused by the transfer of a larger power from the tuned circuit to the load. There is consequently a reduction of the Q factor of the tuned primary circuit. The percentage of power transferred to the load can be assessed in terms of the Q factor of the primary circuit before and after loading is applied.

Let V_m = the peak swing of the anode potential,
ω = the angular frequency of the output,
Q_1 = the Q factor of the tuned primary with the secondary on open circuit,
Q_2 = the Q factor of the tuned primary when the load impedance is connected, and,
C = the capacitance of the tuned primary circuit.

Then the anode load impedance with the load disconnected is:

$$Q_1/\omega C \ \Omega$$

and the anode load impedance with the load connected is

$$Q_2/\omega C \ \Omega.$$

The r.f. power with the secondary on open circuit is

$$\tfrac{1}{2} V_m^2 \, \omega C/Q_1 \text{ W}$$

and the r.f. power with the secondary circuit complete is

$$\tfrac{1}{2} V_m^2 \, \omega C/Q_2 \text{ W}$$

The difference between these two powers is transferred to the aerial and is equal to

$$\tfrac{1}{2} V_m^2 \, \omega C(1/Q_2 - 1/Q_1) \text{ W}$$

This power expressed as a fraction of the total r.f. power output is

$$\frac{\tfrac{1}{2} V_m^2 \, \omega C(1/Q_2 - 1/Q_1)}{\tfrac{1}{2} V_m^2 \, \omega C/Q_2} \text{ or } Q_2 \left[\frac{Q_1 - Q_2}{Q_2 \, Q_1}\right]$$

$$= \left[\frac{Q_1 - Q_2}{Q_1}\right] \times 100\%$$

Example 7.2

The tank circuit of a power amplifier has a Q of 100 when the load is not coupled. The effective Q of the circuit falls to 12 when the load is coupled to the primary circuit. What power is transferred to the load and what is the d.c. power supplied to the valve if the valve efficiency is 70% and the valve produces 1 kW of r.f. power? State also the overall efficiency of transformation of d.c. to r.f. power.

The percentage of power transferred to the load is

$$(Q_1 - Q_2)/Q_1 \times 100\% = (100 - 12)/100 \times 100\% = 88\%$$

The load power is thus, 0·88 × 1 000 W = 880 W. As the valve is 70% efficient 1 000 W is 70% of the d.c. power supplied and the d.c. supply power is 1 000/70 × 100 = 1 430 W.

The overall efficiency = 880/1 430
= 0·615 or 61·5%

7.10. EFFECT OF Q FACTOR ON MODULATION FACTOR

If the output circuit of a transmitter has a too high Q factor, then the side frequencies are amplified less than the carrier frequency. This is equivalent to a reduction in the depth of modulation. Fig. 7.10 shows the tuned anode lode of an output valve which is transformer coupled to a resistive load R_L. Fig. 7.11 is an equivalent circuit in which the load has been transferred to the primary as an equivalent series resistance. R' is the total resistance of the tank circuit including both the primary resistance and the transferred load resistance. The effective Q of the tuned anode circuit is, therefore, equal to

$$\frac{1}{R'}\sqrt{\left(\frac{L}{C}\right)}$$

The impedance of the parallel load circuit is

$$Z_L = \frac{(R' + j\omega L).\ 1/j\omega C}{R' + j(\omega L - 1/\omega C)} \text{ ohms}$$

This may be written as:

$$Z_L = \frac{R'/j\omega C + L/C}{R'\ [1 + j\ .\ 1/R'\ (\omega_0 L\ .\ \omega/\omega_0\ - 1/\omega_0 C\ .\ \omega_0/\omega)]}$$

where $\omega = 2\ \pi \times$ the frequency applied

and $\omega_0 = 2\ \pi \times$ the resonant frequency.

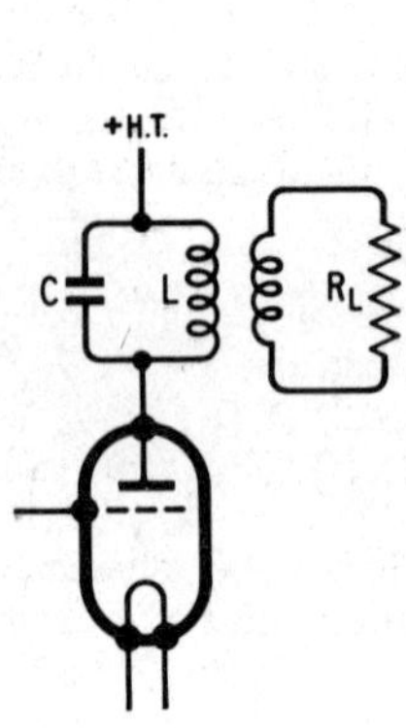

Fig. 1.10 A mutually coupled load

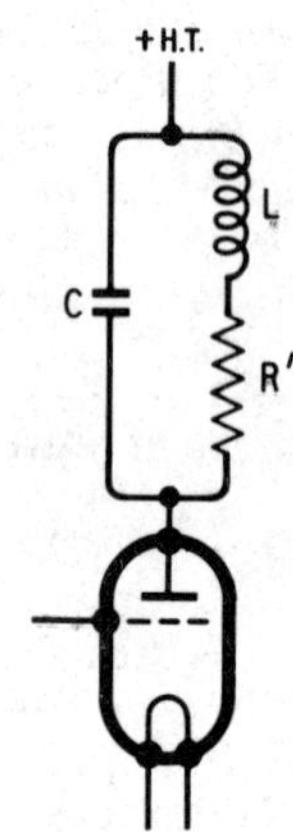

Fig. 7.11 Load transferred to the primary

Since $\omega_0 L/R' = 1/\omega_0 C \,.\, R' = Q$

$$Z_L = \frac{1/\mathrm{j}\omega C + L/CR'}{1 + \mathrm{j}Q\,(\omega/\omega_0 - \omega_0/\omega)}$$

If the reactance of C is small compared with the dynamic impedance L/CR', then the approximation may be made that

$$Z_L = \frac{L/CR'}{1 + \mathrm{j}Q\,(f/f_0 - f_0/f)}$$

(In this last expression angular frequency has been changed to frequency by division of all angular frequency expressions by $2\,\pi$.) The denominator bracket term $(f/f_0 - f_0/f)$ is equal to

$$\frac{f^2 - f_0^2}{f f_0} = \frac{(f + f_0)\,(f - f_0)}{f f_0}$$

If the frequency for which the impedance of the circuit is to be calculated is different from the resonance frequency by only a small percentage of the resonance frequency, then $(f + f_0)$ is approximately equal to $2f$ and

$$\frac{f^2 - f_0^2}{f f_0} \text{ may be written as, } 2\,\Delta f/f_0$$

where Δf is the value $(f - f_0)$.

With these changes

$$Z_L = \frac{L/CR'}{1 + \mathrm{j}\,2Q\,\Delta f/f_0}$$

and the modulus of this is

$$\frac{L/CR'}{\sqrt{(1 + x^2)}}$$

where

$$x = 2Q\,.\,\Delta f/f_0$$

The response of the amplifier to the carrier frequency is proportional to the dynamic impedance L/CR'. The relative response to the side frequencies is equal to the value $1/\sqrt{(1+x^2)}$ calculated for the frequencies involved. The relative response thus depends on the separation between the side frequencies and the carrier and on the Q factor of the load circuit.

The following example shows how the depth of modulation is affected when the relative response of the output circuit is markedly less than unity.

Example 7.3

A broadcast transmitter has an output circuit which has a Q of 10 when loaded. The depth of modulation is 70% with a modulating tone of 10 kHz. Find the effective depth of modulation of the radiated wave if the carrier frequency is (a) 200 kHz and (b) 1 500 kHz.

(a) The highest side frequency is 210 kHz and

$$x = 2Q\,\Delta f/f_0 = 2 \times 10 \times 10/200$$

$$= 1{\cdot}0$$

The relative response is thus

$$1/\sqrt{(1+1)} = 1/\sqrt{2} = 0{\cdot}707$$

The modulation factor of the radiated wave is, therefore

$$0{\cdot}7 \times 0{\cdot}707 = 0{\cdot}495$$

and the depth of modulation 49·5%.

(b) The highest side frequency is 1 510 kHz.

$$x = 2Q\,\Delta f/f_0 = 2 \times 10 \times 10/1\,500$$

$$= 0{\cdot}13\dot{3}$$

The relative response is thus

$$1/\sqrt{[1+(0{\cdot}13\dot{3})^2]} = 1/1{\cdot}01 = 0{\cdot}99$$

The modulation factor of the radiated wave is, therefore

$$0{\cdot}7 \times 0{\cdot}99 = 0{\cdot}693$$

and the depth of modulation 69·3%.

From the foregoing it is seen that the problem of sideband amplitude reduction is worse for a long wave than for an m.f. or h.f. transmitter.

So long as the impedance of the power amplifier load falls symmetrically to either side of the carrier frequency, distortion can be avoided. The reduction in depth of modulation at the higher modulation frequencies can be countered by one of two methods. One of these uses an equalising filter in the output circuit whose frequency response is complementary to that of the remainder of the output circuit. The other method, which is only suitable when the depth of modulation is low, modifies the response of the a.f. amplifiers prior to the modulation stage, so that the higher audio frequencies are given an appropriate amount of boost relative to the lower modulation frequencies.

7.11. WIDEBAND OUTPUT STAGES

Short wave transmitters are required to change frequency with a minimum loss of operating time. The frequency change can be made more rapidly if as many stages as possible are of sufficient bandwidth to cover the whole of the h.f. spectrum from 2 MHz to 27·5 MHz without need of tuning adjustment.

Fig. 7.12 shows the manner of connection of the valves in one type of broadband amplifier. The circuit is called a *distributed amplifier* or a *transmission line amplifier.* The inductors marked L_a together with the output capacitances of the valves marked C_0 form the elements of an artificial transmission line. The inductors L_g together with the valve input capacitances marked C_i form the elements of another artificial line. The line to which the anodes are connected is terminated at each end by its characteristic impedance Z_a. The value of Z_a is $\sqrt{(L_a/C_0)}$. Similarly, the grid transmission line is terminated by its characteristic impedance the value of which is $\sqrt{(L_g/C_i)}$. This is marked as Z_g in Fig. 7.11

When a drive voltage is applied at the input terminals 1 and 2, a travelling wave is set up on the input artificial line which reaches the valves distributed along the line with a phase delay which is progressively greater from valve to valve. Each valve contributes a travelling wave of current and voltage to the output transmission line. Half of each valve current at the signal frequency travels backwards to the terminating impedance Z_a' where its power is dissipated. The waves which travel in the opposite direction all reach the load impedance Z_a in phase with each other. This in-phase addition of the valve contributions to the load power is arranged by making the phase constant of the anode artificial line equal to that of the grid line. Thus

$$\omega\sqrt{(L_a C_0)} = \omega\sqrt{(L_g C_i)}$$

radians per section. The input voltage to valve 2 is thus delayed on the input to valve 1 by the time taken by the output from valve 1 to travel along the output line section to reach the anode of valve 2. Likewise the input to valve 3 is delayed on the input to valve 1 by the time taken

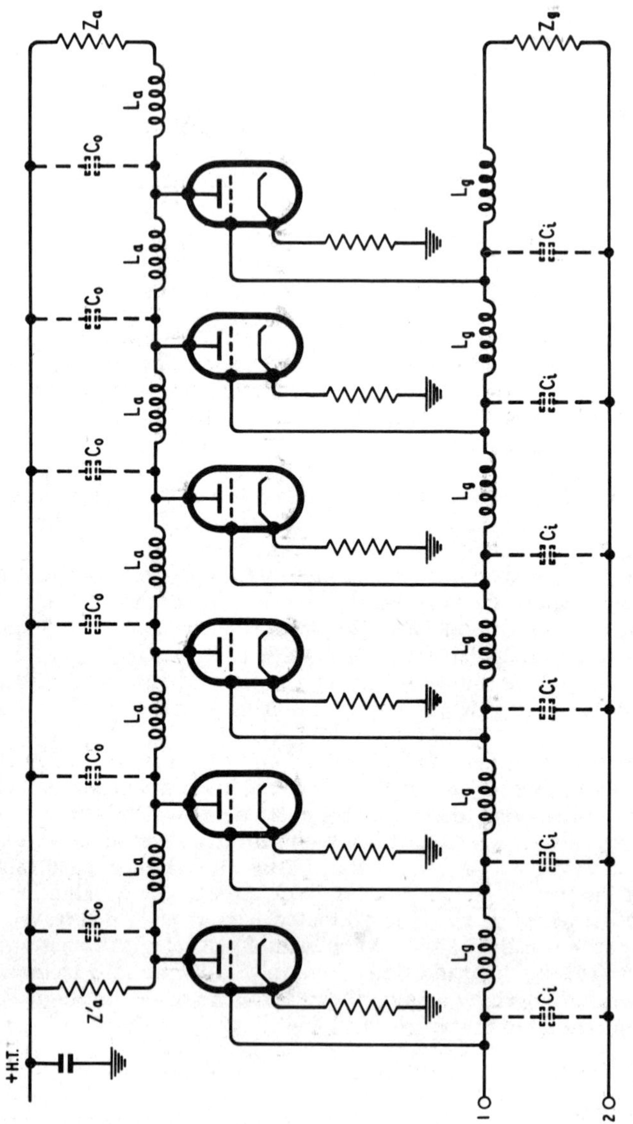

Fig. 7.12 A transmission line amplifier

by the output of valve 1 to travel along the output line as far as the anode of valve 3, and so on. Each artificial line behaves as a low-pass filter with a cut-off frequency given by:

$$f_c = 1/\pi\sqrt{(L_a C_o)} = 1/\pi\sqrt{(L_g C_i)} \text{ Hz}$$

The amplifier has a bandwidth which is approximately 0·8 of this frequency and can be designed to encompass the h.f. range necessary.

Because half the output power of each valve is wasted in the terminating impedance Z_a', the effective mutual conductance of each valve is halved and the voltage gain is given approximately by

$$A = n\, g_m\, Z_a/2$$

where n is the number of valves used. Because the grid transmission line is matched by the terminating impedance Z_g no reflection occurs and standing waves on the line are avoided.

The load impedance Z_a is the input impedance of a specially designed wideband matching transformer. The secondary of the transformer may connect to a twin wire transmission line of impedance of the order of 450 Ω. An h.f. wideband transformer is also needed to match the input feeder to the characteristic impedance of the artificial grid transmission line. The cut-off frequency of a wideband transformer depends on how small its leakage inductance and shunt capacitance are. The number of turns used is, therefore, very small and a ferrite core operates with a high maximum flux density. The core material must be able to retain its magnetic properties at a high working temperature. The core may be built up of discs of a nickel-zinc ferrite interspersed with aluminium cooling fins.

Wideband amplifier valves must operate on a linear part of their mutual characteristic as their load impedance is such that harmonics of the drive frequency may be amplified. Even harmonics produced by distortion can be balanced out by using two lines of valves operating in push-pull. Odd harmonics are not cancelled in this way so that valves must be selected which do not introduce third harmonic distortion. The characteristic impedance of the grid transmission line must be of a low value so as to avoid the distortion which would otherwise be caused by flow of grid current.

Overall negative feedback is not applied on account of the large phase difference which exists between input and output signals. Some negative feedback in individual stages improves linearity and reduces valve input capacitance. Cathode bias resistors without by-pass capacitors provide such negative feedback.

Wideband amplifiers make possible simultaneous transmissions on several wavebands. However, when this is done, the power available for each channel is reduced to W/n watts where W is the total power available and n is the number of channels for which drive is supplied.

Transmitters with an output power of 1 or 2 kW may use a wideband output stage. Higher power transmitters, although possibly using wideband amplifiers for several stages, must use a tuned output stage in order to reduce sufficiently the power on frequencies which are harmonics of the intended channel frequencies.

QUESTIONS

1. Make a list of six requirements which must be met by the coupling arrangements between a transmitter aerial and the output transistor or valve.

2. In the equivalent circuit of Fig. 7.2, C_4 is set to 200 pF and the value of L_3 is adjusted to tune the output circuit to 500 kHz. The Q value of the parallel resonant circuit is 10. What impedance is presented to the output device of the transmitter? If C_4 is increased to 400 pF and the value of L_3 adjusted to restore resonance without change in the loss resistance R, find the new impedance of the parallel circuit.

3. If in the circuit of Fig. 7.3, $R_0 = 2{\cdot}5$ kΩ, $C_1 = 120$ pF, $C_2 = 820$ pF, calculate the value of R_a assuming correct matching and tuning for a frequency of 15 MHz. (Note two answers are possible.)

4. Outline the factors affecting choice of transistors or valves and their circuit arrangement for the output stage of a transmitter.

5. Explain how parasitic oscillation may occur in a power amplifier circuit and how his may be prevented.

6. If the coupling between a power amplifier and its aerial circuit is increased, the dip in the feed current which occurs on tuning becomes less pronounced and the level of the d.c. feed current is higher. Explain the reason for this.

7. A power amplifier has a parallel tuned anode load which is coupled by mutual inductance to the aerial. The Q of the tuned primary circuit is 80 when the aerial is disconnected but falls to 10 with the aerial circuit completed and coupled. If the valve efficiency is 60% and the d.c. power supplied is 1 kW, calculate, the total r.f. power produced, the power coupled to the aerial, and the r.f. dissipated in the resistance of the tuned primary circuit.

8. It is required that the response to side frequencies of a power amplifier load circuit shall not fall more than 3 dB below the response to the carrier frequency for modulation frequencies up to 8 kHz. If the carrier frequency is 692 kHz, find the maximum Q factor permissible for the output circuit.

9. What factors govern the bandwidth necessary in the output stage of a transmitter and what ill effects result from too narrow a passband?

10. If the output capacitance C_0 of each valve in the wideband amplifier shown in Fig. 7.12 is 5 pF and the bandwidth required is such that the cut-off frequency of the artificial line is 26 MHz, find the values necessary for L_a and Z_a.

8

Frequency Modulation —Transmission

8.1. THREE TYPES OF MODULATION

A sinusoidal alternating quantity may vary in three aspects–amplitude, frequency and phase. When one of these aspects of a radio wave varies periodically at some frequency which represents information to be carried from transmitter to receiver, the wave is modulated.

In an amplitude modulation system (a.m.) it is the amplitude of a radiated wave which is varied at the modulating frequency and to a degree which is proportional to the amplitude of the modulating waveform.

With frequency modulation (f.m.) the frequency of the radiated wave is varied at the modulation frequency and to an extent dependent on the amplitude of the modulating waveform.

A system of phase modulation (p.m.) varies the phase of the radiated wave at the modulating frequency and by an angle which is proportional to the amplitude of the modulating waveform.

From the foregoing it will be seen that each type of modulation is named according to which aspect of the transmitted wave is proportional to the amplitude of the modulating function.

It is helpful to use phasors to compare the modulated wave with the unmodulated wave for each of the three types of modulation.

In each instance let the modulating function be $b = B \sin \omega_m t$ and the unmodulated carrier $a = A \sin \omega_c t$, with a frequency f_c Hz.

The phasor for the unmodulated wave is of length proportional to A and rotates with a constant angular velocity at f_c rev/s (see Fig. 8.1(a)). The phasor for the amplitude modulated wave rotates at constant angular velocity f_c rev/s but varies in length by mA, where m is the modulation factor, $m = B/A$ (see Fig. 8.1(b)).A complete cycle of change in the amplitude of the rotating phasor requires f_c/f_m revolutions of the carrier phasor.

The phasor representing f.m. varies in angular velocity at the modulation frequency but remains of constant length representing the constant amplitude of the transmitted wave (see Fig. 8.2). The f.m. wave may be written as: $a = A \sin (2\pi f_c t - \Delta f_c/f_m \cdot \cos \omega_m t)$ where f_m is the frequency of modulation and Δf_c is given by kB, where k is a constant which relates the amplitude of the modulating function with the change in

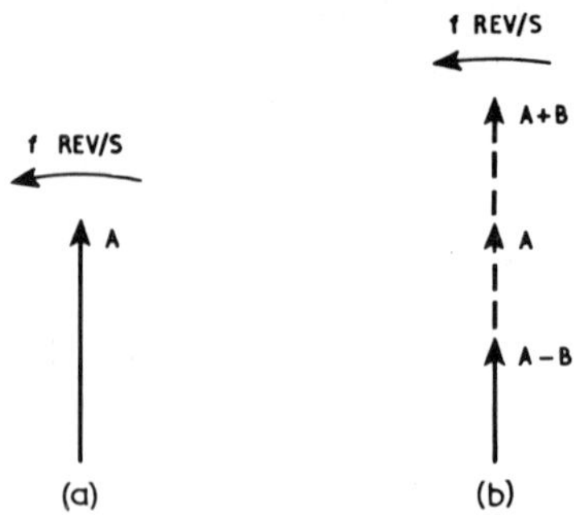

Fig. 8.1 Amplitude modulation shown by phasors

frequency caused. Note that when the phasor velocity increases it must take up an angle of phase lead on the phase of the unmodulated phasor. Similarly, when the phasor slows down, it must take up an angle of lag on the unmodulated phasor. The change of phase angle is proportional to:

1. the amplitude of the modulating function,
2. the period of the modulating function.

It is the second factor which makes f.m. a different concept from p.m. Fig. 8.3 represents phase modulation. The phasor for the unmodulated wave remains as in Fig. 8.1(a). The phasor for the p.m. wave alternately leads and lags on the phasor for the unmodulated wave and the angle of lead or lag is proportional to the amplitude of the modulating function. The p.m. wave can be written as: $a = A \sin(\omega_c t + \phi)$ where $\phi = k'B \sin \omega_m t$. Here k' is a constant which relates the amplitude of the modulating function to the change in phase caused. Note that when the phasor is taking up an angle of phase lead it must acquire a higher angular velocity in order to do this, and that this represents a higher frequency. Similarly, while the phasor is taking up an angle of phase lag, it must have a decreased angular velocity which represents a reduced frequency. Thus phase modulation involves a change of frequency. The change of frequency is proportional to:

1. the amplitude of the modulating function,
2. the frequency of the modulating function.

Factor (2) arises because the higher the modulating frequency, the shorter is the modulation period in which the change of phase angle must occur. It is this factor which differentiates p.m. from f.m. Although f.m. involves a variation of phase, and although p.m. involves a change of frequency, yet each type of modulation is named according to the aspect of the modulated wave which is changed in direct proportion to the amplitude of the modulating function.

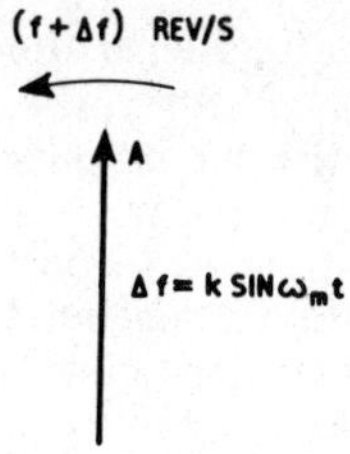

Fig. 8.2 Frequency modulation shown by a phasor

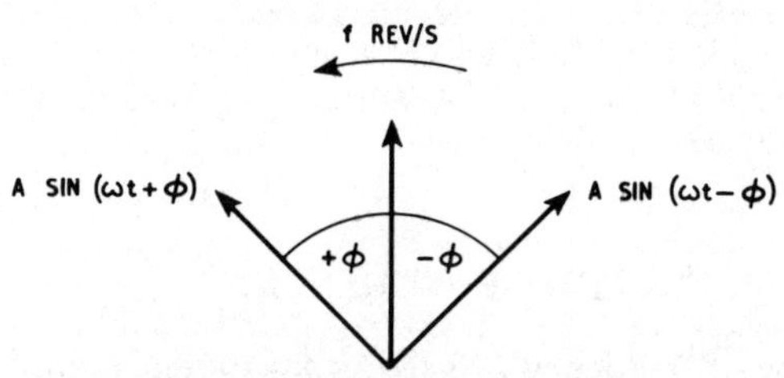

Fig. 8.3 Phase modulation shown by a phasor

The term *angle modulation* can be applied to both f.m. and p.m., as both types of modulation involve a variation of phase as distinct from the amplitude of a radio wave.

Phase modulation is of practical importance because it provides an indirect method of achieving f.m. which is suitable for the designs of compact transmitters used in mobile stations (see Section 8.8).

8.2. MODULATION INDEX AND RELATED TERMS

In any particular f.m. service there is a stipulated maximum variation of frequency above or below the mean frequency value. This is called the rated *system deviation.* It represents the largest possible amplitude of modulating function which can be accepted by the system. A larger frequency variation is over-modulation. The limit has to be set in order to avoid mutual interference between adjacent communication channels. In broadcasting, the maximum deviation is 75 kHz, whereas in mobile radio telephony service 15 kHz, 5 kHz or even 2·5 kHz may be the limiting deviation.

The deviation at any one time depends on the amplitude of the modulating signal. With a sinusoidal modulating waveform the ratio of the frequency deviation to the frequency of the modulating waveform is called the *modulation index.* This is represented as Δ in Section 8.3 and has a significant bearing on the bandwidth used by the emission. The index varies during transmission because both the magnitude and the frequency of the modulating signal alter. High values of Δ occur with high amplitude and low frequencies of modulation.

The difference between the maximum and minimum values of instantaneous frequency of an f.m. wave is called the *frequency swing.* With a symmetrical modulation the highest possible value of frequency swing is twice the rated system deviation. Half the frequency swing is the frequency deviation for the particular modulation index applicable at the time. In some applications, however, the deviations above and below the unmodulated carrier frequency are unequal. The term frequency swing is then of more importance.

The ratio of the system deviation to the highest modulating frequency is called the *deviation ratio.* For broadcasting services the deviation ratio is given by: 75 kHz/15 kHz = 5. The ratio for mobile radio-telephony transmitters may be the same, i.e. 15 kHz/3 kHz = 5, or less with reduced system deviations.

8.3. SIDE FREQUENCIES

If a carrier wave of frequency f_c is frequency modulated by a signal of frequency f_a, the frequencies which are radiated in addition to the carrier frequency are given by: $(f_c \pm f_a)$, $(f_c \pm 2f_a)$, $(f_c \pm 3f_a)$ and so on. These are called the first, second, third etc. order of side frequencies. Fortunately the higher order side frequencies are usually of a relatively small amplitude, but the presence of several orders of pairs of side frequencies is one reason why f.m. transmissions require a wide channel bandwidth. For this same reason f.m. systems generally make use of the v.h.f. and higher frequency wavebands. A practical way of assessing the requisite bandwidth is to assume it equal to $2 \times (f_d + 2f_a)$ where f_d is the system deviation and f_a is the highest modulation frequency used.

The f.m. transmitters of the band 2 broadcast service have a channel spacing of 2·2 MHz for two transmitters serving the same locality. This allows a generous channel separation band and the effective bandwidth of each channel is 200 kHz. Mobile radiotelephone transmitters may have a channel spacing of 50 kHz or less between carrier frequencies with an effective channel bandwidth of 36 kHz or less depending on the system deviation.

The amplitude of the side frequencies relative to the unmodulated carrier amplitude depends on the ratio:

$$\Delta = \frac{\text{FREQUENCY DEVIATION}}{\text{MODULATION FREQUENCY}}$$

As this ratio becomes larger, the number of significant side frequencies increases. For a deviation ratio of 5, side frequency pairs up to the 8th order have amplitudes greater than 1% of the amplitude of the of the unmodulated carrier. If the deviation ratio is less than unity, only the first and second order side frequencies need be considered.

As the instantaneous amplitude of the modulating signal changes, the side frequencies vary in number and amplitude. Nevertheless, the total power in the modulated wave remains constant. If the modulation index increases, power radiated on the carrier frequency diminishes while power radiated on the side frequencies increases. When an r.f. wave is amplitude modulated, the power of the carrier wave remains unchanged and the power provided by the modulating stage for the side frequencies is additional to the carrier power. If frequency modulation is used, the carrier wave power diminishes by an amount equal to the power in the side frequencies so that the total radiated power is unchanged by modulation.

Table 8.1 shows the amplitudes of the side frequencies relative to the amplitude of the unmodulated carrier for a range of values of Δ, from 1 to 5. If, therefore, the carrier amplitude is assumed to be equal to unity then these values are the amplitudes of the side frequencies.

Table 8.1

Δ	1	2	3	4	5
Carrier	0·7652	0·2239	0·2601	0·3971	0·1776
1st order	0·4401	0·5767	0·3391	0·0660	0·3276
2nd ,,	0·1149	0·3528	0·4861	0·3641	0·0466
3rd ,,	0·0196	0·1289	0·3091	0·4302	0·3648
4th ,,	–	0·0340	0·1320	0·2811	0·3912
5th ,,	–	–	0·0430	0·1321	0·2611
6th ,,	–	–	0·1140	0·0491	0·1310
7th ,,	–	–	–	0·0152	0·0534
8th ,,	–	–	–	–	0·0184

Assuming for example that the deviation ratio is 2, than the total of the mean squares of the amplitudes of the various frequency components present can be found as follows:

Mean square of the carrier is	$(0{\cdot}2239)^2/2 = 0{\cdot}02506$
For the two 1st order side frequencies, the mean square is	$2 \times (0{\cdot}5767)^2/2 = 0{\cdot}3326$
For the 2nd order side frequencies the mean square is	$2 \times (0{\cdot}3528)^2/2 = 0{\cdot}1245$
For the 3rd order side frequencies the mean square is	$2 \times (0{\cdot}1289)^2/2 = 0{\cdot}0166$
For the 4th order side frequencies the mean square is	$2 \times (0{\cdot}0340)^2/2 = 0{\cdot}0012$
Total of the mean squares:	$= 0{\cdot}5000$
The square root of this (the r.m.s.)	$= 0{\cdot}707$

This r.m.s. value is that of a carrier with an amplitude of 1·0. If the r.m.s. of all the side frequency and carrier amplitudes is calculated for any other value of Δ, it is found to be 0·707 in each instance on the assumption that the carrier amplitude when not modulated is unity.

8.4. THE REACTANCE VALVE

This section describes a circuit by which the frequency of an oscillator can be varied by a change in a voltage and by which an alternating voltage may alternately increase and decrease the frequency of an oscillator. The circuit and its associated valve is called a *reactance valve.* The essential characteristic of the reactance valve is that it has an alternating component of anode current which is in quadrature with the alternating component of anode voltage. The valve, therefore, acts as a reactive impedance to the alternating voltage across it. If, by variation of the mutual conductance of the valve, the alternating current through it is changed, without a corresponding change in the alternating voltage across the valve, then the effective reactance of the valve is also changed. A valve with variable mu characteristics can be used as a variable reactance which changes at the frequency of a control voltage applied to its grid and is thus a means of achieving frequency modulation of an oscillator.

Fig. 8.4 shows four different circuits of reactance valves. In each circuit a potential dividing network is connected across the applied alternating voltage V_a. This network in each instance provides an alternating component of grid voltage which is approximately 90° out of phase with V_a. The section of the potential dividing network which is between the anode and grid is in each circuit much greater in impedance than the section between the grid and cathode.

In circuits (b), (c) and (d), C' is merely a d.c. blocking capacitor to insulate the grid from the d.c. anode voltage. (This d.c. supply is

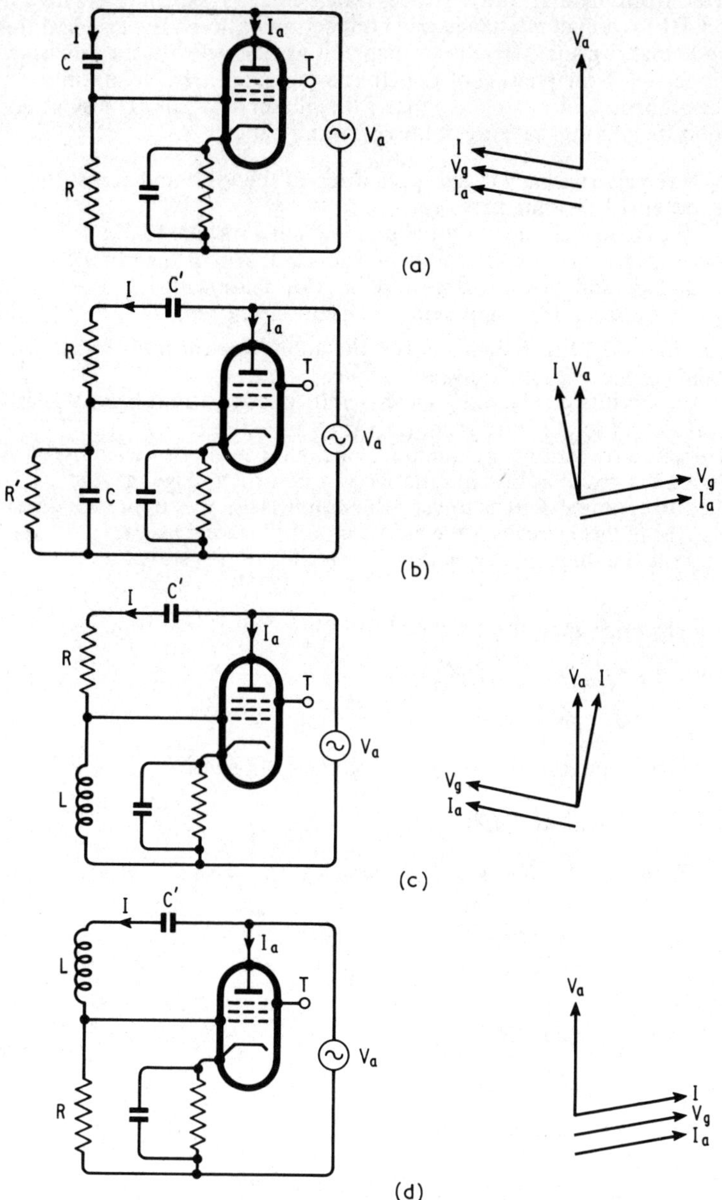

Fig. 8.4 Reactance valves with associated phasors

omitted from the diagrams). The reactance of C' is negligible. R' in Fig. 8.4(b) is a high resistance grid connection between the grid and the cathode bias resistor. The phasor diagrams associated with the circuit diagrams relate the phasing of the alternating quantities. No attempt has been made to draw the lengths of the phasors to scale. The symbols used on the phasors have the following significance:

V_a is the alternating voltage applied across the valve and across the potential dividing network;
I is the current in the potential dividing network due to V_a;
V_g is the grid voltage derived from the lower arm of the potential divider, and which is almost 90° out of phase with V_a;
I_a is the alternating component of the anode current.

In the circuits of Fig. 8.4(a) and (c), the anode current leads by almost 90° and the valve is the equivalent of a capacitor.

In the circuits of Fig. 8.4(b) and (d) the anode current lags by nearly 90° on V_a and so the valve is equivalent to an inductor.

The effective mutual conductance of the valve can be varied by changing the grid bias but, alternatively, a control voltage may be applied to terminal T so as to vary the potential of the suppressor grid.

The first of these circuits is the most convenient since it needs neither the d.c. blocking capacitor C' nor the grid resistor R'.

An algebraic analysis of the circuit in Fig. 8.4(a) is as follows:

$$V_g = [R/(R + 1/j\omega C)] \,.\, V_a \tag{8.1}$$

Also,

$$I_a = g_m V_g \tag{8.2}$$

Substituting equation 8.1 for V_g in equation 8.2 gives,

$$I_a = g_m R \; V_a/(R + 1/j\omega C)$$

Dividing by V_a gives the ratio I_a/V_a which is the valve admittance, as:

$$Y = I_a/V_a = g_m R/(R + 1/j\omega C)$$

Rationalising:

$$Y = \frac{g_m R \,(R - 1/j\omega C)}{R^2 + 1/\omega^2 C^2}$$

or separating real and unreal parts of the admittance,

$$Y = \frac{g_m R^2}{R^2 + 1/\omega^2 C^2} + \frac{j\, g_m R/\omega C}{R^2 + 1/\omega^2 C^2}$$

Multiplying the numerators and denominators by, $\omega^2 C^2$ gives,

$$Y = \frac{g_m \omega^2 C^2 R^2}{1 + \omega^2 C^2 R^2} + \frac{\mathrm{j}\, g_m \omega C R}{1 + \omega^2 C^2 R^2}$$

Since in practice $\omega^2 C^2 R^2$ is very small compared with unity, the conductance of the valve is approximately, $g_m \omega^2 C^2 R^2$ and its susceptance is approximately, $g_m \omega C R$.

The reactance valve may, therefore, be represented as a resistance of $1/g_m \omega^2 C^2 R^2$ Ω, in parallel with a capacitance of $g_m CR$ F. It is to be noted that the equivalent value of the capacitance is directly proportional to the mutual conductance of the valve. Hence any change in the mutual conductance will bring about a proportionate change in the effective capacitance.

A similar analysis shows that the circuit of Fig. 8.4(b) is equivalent to a resistance of $\omega^2 C^2 R^2 / g_m$ in parallel with an inductance of CR/g_m H.

The circuit of Fig. 8.4(c) is equivalent to a resistance of $R^2/g_m \omega^2 L^2$ in parallel with a capacitance of $g_m L/R$ F.

The circuit of Fig. 8.4(d) is equivalent to a resistance of $\omega^2 L^2/g_m R^2$ in parallel with an inductance of $L/g_m R$ H.

8.5. USE OF A REACTANCE VALVE FOR F.M.

Fig. 8.5 shows how a reactance valve may be connected in parallel with the tuned circuit of a Colpitts oscillator to produce frequency modulation.

V_1 is the reactance valve which varies the effective LC valve of the oscillator circuit L C_1, C_2, V_2 is connected as an inverted Colpitts

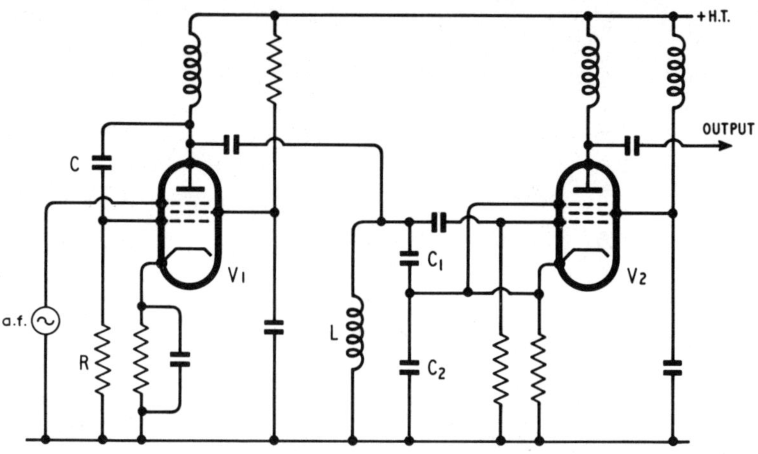

Fig. 8.5 An f.m. modulated oscillator

oscillator. If C' is the capacitance presented by the reactance valve, then the LC value of the oscillator circuit is given by:

$$L \times C + \left(\frac{C_1 C_2}{C_1 + C_2}\right)$$

The a.f. modulating signal is applied to the suppressor grid of the reactance valve and varies the effective mutual conductance of this valve. When the signal increases the potential of the suppressor grid more of the cathode current is collected by the anode and less by the screen grid. A reduction of the suppressor grid potential diverts more of the cathode current to the screen grid so that the anode current is reduced. The alternating current component of the valve which passes through the equivalent capacitance reactance is varied at the modulating signal frequency.

The effect on the frequency of oscillation of the Colpitts oscillator may be evaluated thus:

Let

f be the frequency before a modulating voltage is applied to the suppressor grid of V_1,
Δf be the change in frequency,
L be the oscillatory circuit inductance,
C_T be the total circuit capacitance before any modulating signal is applied, including the mean value of the capacitance presented by V_1, and
ΔC the change in capacitance presented by V_1.

Then $$f = 1/2\pi\sqrt{(LC_T)} \text{ Hz} \tag{8.3}$$

and $$f - \Delta f = 1/2\pi\sqrt{L\,(C_T + \Delta C)} \text{ Hz} \tag{8.4}$$

Dividing equation 8.4 by equation 8.3 gives,

$$1 - \Delta f/f = \sqrt{\left[(C_T + \Delta C)/C_T\right]}$$

Rearranging, we have,

$$\Delta f/f = 1 - \sqrt{(1 + \Delta C/C_T)}$$

This may be written as,

$$\Delta f/f = 1 - (1 + \Delta C/C_T)^{-1/2}$$

Expanding this by the binomial theorem gives:

$$\Delta f/f = 1 - \left[1 - \Delta C/2C_T + \frac{(\frac{1}{2})(\frac{3}{2})(\Delta C/C_T)^2}{\underline{|2}} \ldots\right]$$

Neglecting the third and later terms of this expansion, which are very small, leaves $\Delta f/f = \Delta C/2C_T$ where ΔC may be written as $\Delta g_m CR$ if Δg_m is the change in mutual conductance of the valve and C and R are the capacitance and resistance values in the reactance valve phase changing potential divider.

8.6. STAGES OF AN F.M. TRANSMITTER

The block diagram of Fig. 8.6 is that of an f.m. transmitter which makes use of a reactance valve circuit.

Apart from providing amplification of the small output voltage of the microphone, the a.f. stages usually include a filter to restrict the range of frequencies reaching the modulating stage. Some form of amplitude limiter is also required to ensure that the over-modulation can not occur and the system deviation exceeded.

The reactance valve varies the frequency of output at the a.f. signal frequencies. At this stage the frequency swing is only small compared with the deviation required at the final stage. For this reason several

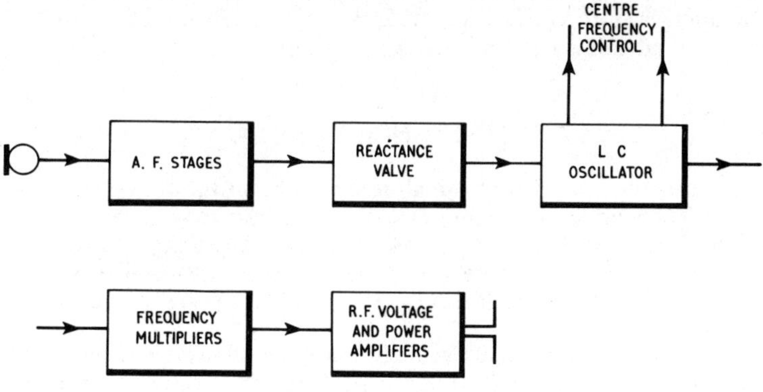

Fig. 8.6 A simplified block diagram of an f.m. transmitter

stages of frequency multiplication follow the oscillator. A broadcast transmitter might have a maximum deviation of the order of +4 kHz at the modulation stage and then use frequency multipliers to multiply both the carrier frequency and the frequency swing by a factor of 18.

When frequency modulation occurs it is essential that the centre or mean frequency about which deviation occurs, must remain constant. Unwanted momentary deviations in the frequency of the carrier or centre frequency are referred to as *scintillation.* Any shift in the centre frequency, momentary or otherwise will give rise to some distortion of the received signal. An f.m. transmitter, therefore, must have some means of holding the centre frequency stable with the order of stability associated with quartz crystal controlled oscillators ($\pm 20/10^6$ is the order of stability required). Some transmitters modulate the frequency of a crystal controlled master oscillator.

Fig. 8.6 assumes an *LC* oscillator with provision for comparing its centre frequency with that of a separate fixed frequency crystal controlled oscillator. The outputs of the *LC* oscillator and the crystal oscillator are fed to a comparator circuit. Any difference between the mean frequency of the *LC* oscillator and the quartz crystal oscillator frequency provides a drive voltage at the difference frequency to a small tuning motor. This turns the plate of a small trimming capacitor until the centre frequency of the *LC* oscillator is restored to its correct value and the drive to the motor thus removed.

Frequency multipliers shown by one block in Fig. 8.6 may multiply the output frequency of the modulated oscillator in three stages by a factor as high as 18. This multiplies both the carrier frequency and the deviation by this factor.

If the modulated oscillator is quartz crystal controlled then frequency multiplication by a factor of 24 is necessary.

The r.f. voltage and power amplifiers in f.m. transmitters are all class-C operated since their drive voltage is of constant amplitude. The operation of all r.f. stages with the high efficiency of class-C is a marked advantage of f.m. over a.m. systems.

8.7. A SUSCEPTANCE MODULATOR

A susceptance modulator is an alternative circuit for controlling current through an oscillatory circuit in quadrature with the p.d. across the circuit. It does not, however, like the reactance valve connect a phase splitting network directly across the oscillatory circuit, a procedure which reduces the Q factor of the oscillatory circuit.

Fig. 8.7 is a simplified circuit of one type of susceptance modulator. The voltage V_o is fed to the grid of the buffer amplifier *V1*. The phase changing networks L_1C_1 and L_2C_2 form its load impedance. The outputs V_{k2} and V_{k3} are of opposite phase as indicated in the phasor diagram Fig 8.8. The valves *V2* and *V3*, therefore, have r.f. drive voltages and r.f. anode currents which are of opposite phase. So long as *V2* and *V3* have equal values of mutual conductance their r.f. output currents are mutually cancelling and no r.f. current is fed to the oscillatory circuit.

The modulating voltage V_m applied to the grid of *V2* is in antiphase to that applied to the grid of *V3*. During one half cycle of modulation the mutual conductance of *V2* is increased by the modulating voltage while the mutual conductance of *V3* is reduced. As the phasor diagram shows, this causes a resultant current to be fed into the oscillatory circuit which leads by 90° on V_0. On the other half cycle of modulation, the mutual conductance of *V3* is increased and that of *V2* is reduced.

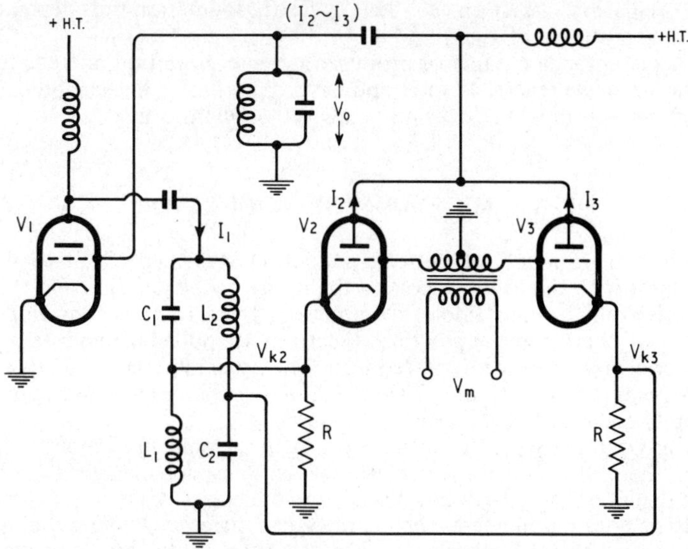

Fig. 8.7 A susceptance modulator

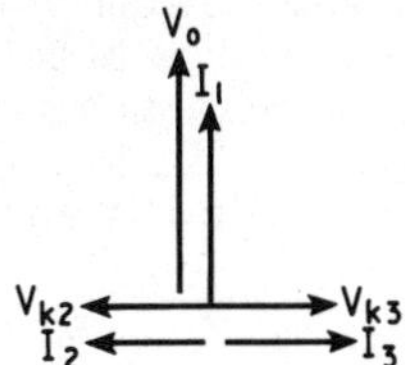

Fig. 8.8 Phasors for susceptance modulator

The resultant output r.f. current from the two valves now lags by 90° on V_O. The r.f. current fed by the two valves *V2* and *V3* through the oscillatory circuit is equivalent to positive or capacitive susceptance when it leads by 90° on the voltage V_0, and to negative or inductive susceptance when it lags by 90° on V_0.

Additional positive susceptance is equivalent to additional capacitance in parallel with the oscillatory circuit, and the effective *LC* value of the circuit is made larger. The frequency of oscillation falls. Additional inductive susceptance is equivalent to additional inductance in parallel with the oscillatory circuit which reduces the *LC* value so that the frequency of oscillation increases.

V2 and *V3* are variable mu valves and the extent to which their anode r.f. currents become unbalanced depends on the amplitude of the modulating voltage applied to their grids. The frequency deviation of the oscillatory circuit on each half cycle of modulation thus depends on the amplitude of V_m.

A susceptance modulator produces a greater deviation of frequency than the single reactance valve and has a more nearly linear modulation characteristic (Fig. 8.7 does not show the oscillator valve).

8.8. A SEMICONDUCTOR MODULATOR

A junction of p- and n-type semiconductors having a d.c. voltage applied across it with the negative pole to the p side of the junction and the positive pole to the n side of the junction, gives rise to a separation of opposite charges at the junction. Electrons are pulled across into p-type material and holes are displaced into the n-type material. The separated charges build up until they have a p.d. between them which is approximately equal to the applied d.c. voltage.

Between the separated charges is a volume depleted of mobile charges and, therefore, named the depletion layer. A capacitance exists across the depletion layer between the separated charges. An increase of applied voltage not only increases the quantity of separated charge but also draws the charges further apart. The depletion width is increased and the capacitance between the charges reduced. The depletion layer capacitance is approximately proportional to $V^{-1/2}$, where V is the applied voltage across the junction. Silicon semiconductor diodes which are designed to make use of this facility of control of capacitance by a voltage are called variable capacitance diodes or *varactors*. The capacitance may be of the order of 200 or 300 pF with 1V applied, reducing to half the capacitance with 10V applied.

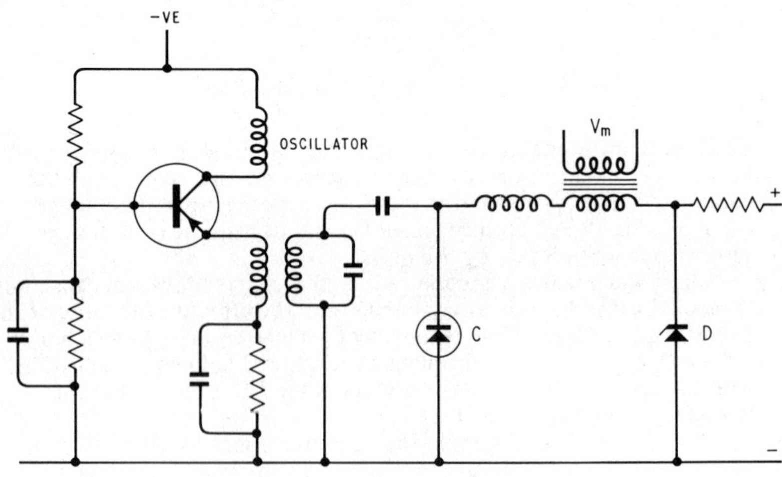

Fig. 8.9 A varactor modulator

Some low power narrow deviation f.m. transmitters use varactor diodes for modulation. A suitable circuit is shown in Fig. 8.9. The zener diode *D* stabilises the supply voltage so that the mean frequency of oscillation is not altered by variation of supply voltage. The capacitance *C* is varied at the modulation frequency by the a.f. voltage supplied by the modulating transformer. For simplicity, a simple feedback oscillator is shown but other *LC* oscillator circuits may be used.

8.9. INDIRECT F.M.

Mobile transmitters are often required to be switched rapidly from one channel to another and to have numerous communication channels available. These requirements, together with the need for a stable centre frequency, make necessary the use of a quartz crystal controlled oscillator. As direct modulation of the frequency of a quartz crystal is difficult, mobile transmitters often use a system of phase modulation. Audio correction is applied prior to frequency modulation so that although the process of modulation is one of p.m. the end product is f.m. (see Section 8.1). Fig. 8.10 shows the principle involved.

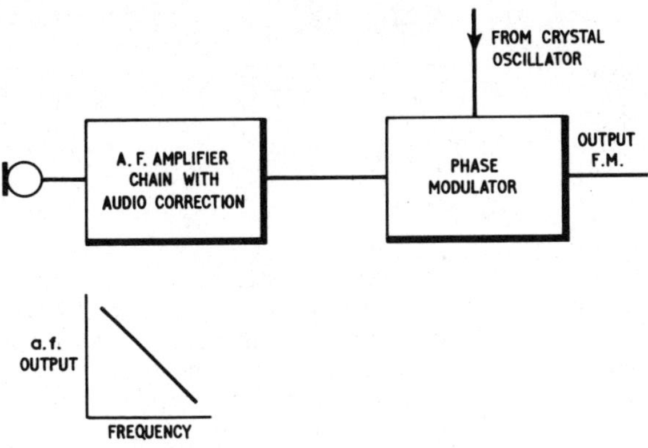

Fig. 8.10 Indirect f.m.

One circuit used for phase modulation is that of Fig. 8.11. The voltage from a master oscillator unit is applied at terminals 3 and 4. R_1C_1 and R_2C_2 are two phase changing networks connected in parallel across the oscillator voltage V_C. Both these networks have a phase angle of 45°. The reactance of C_1 is equal to the resistance of R_1 while the reactance of C_2 is equal to the resistance of R_2. The voltage V_1 which is coupled to the emitter of transistor *Tr1* lags by 90° on the current in R_1. The voltage V_2 which is coupled to the emitter of transistor *Tr2* is in phase with the current in R_2. V_1, therefore, lags on V_2 by 90°. Consequently

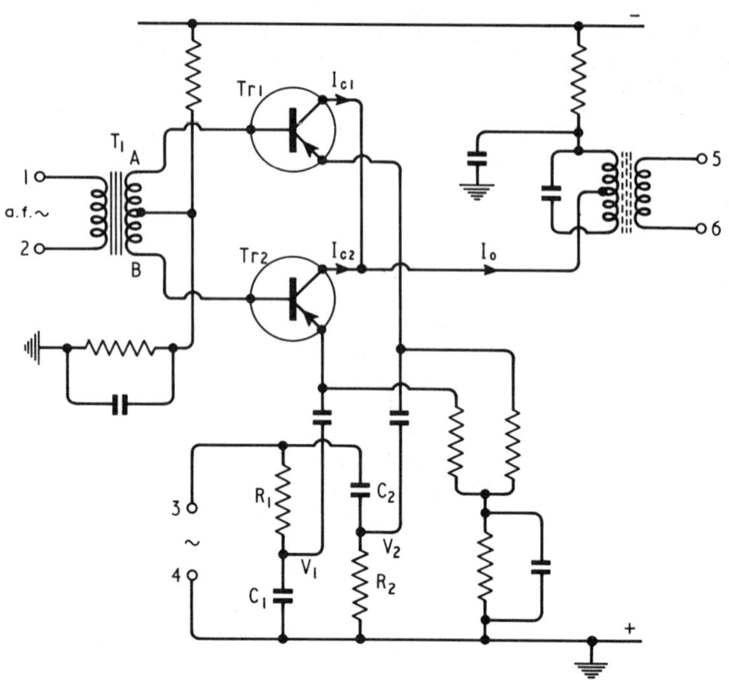

Fig. 8.11 A phase modulator

the emitter and collector current variations in the two transistors are 90° out of phase as shown by the phasor diagram Fig. 8.12.

The modulating voltage is applied to the primary of the transformer T_1 at terminals 1 and 2. As the centre tap of this transformer is decoupled to the battery line, the opposite ends of the secondary winding have potentials which vary in opposite phase at the modulation frequency. With no modulating voltage present, the two transistors have equal gains and their collector current changes at the carrier frequency are of equal amplitude. This condition is shown in Fig. 8.12(a). During the half cycle of modulation when end *A* of T_1 secondary swings negative in potential and end *B* positive, the amplitude of collector current at the carrier frequency in *Tr1* increases while that in *Tr2* decreases. The phasor diagram Fig. 8.12(b) shows these changes and also the way in which the phase of the resultant current at the carrier frequency I_o now lags in phase the carrier input voltage V_c. When the polarity of output voltage of the modulating transformer reverses, the collector current in *Tr2* at the carrier frequency increases and the collector current at this frequency in *Tr1* decreases. Phasors shown in Fig. 8.12(c) now apply and the resultant output current in the primary of transformer T_2 now lags in phase on the voltage V_C.

The graphs drawn in Fig. 8.13 relate the timing of voltage changes at A and B to the changing phase angle ϕ. The letters (a), (b) and (c) on the graph of the phase angle ϕ identify the points in time to which the phasors of Fig. 8.12 apply. There is one complete cycle of variation in the phase of the output current I_o for each cycle of modulation, so that phase modulation at the input frequency is achieved. The extent to which the gain factors of the transistors are altered by the modulating voltages applied in their base circuits is a function of the amplitude of the

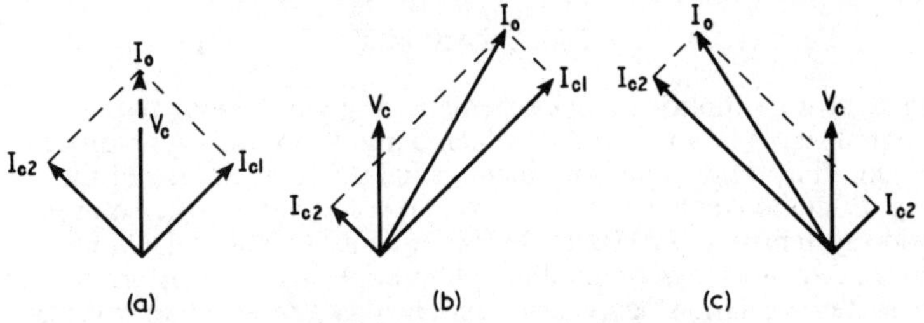

Fig. 8.12 Phasors for a phase modulator

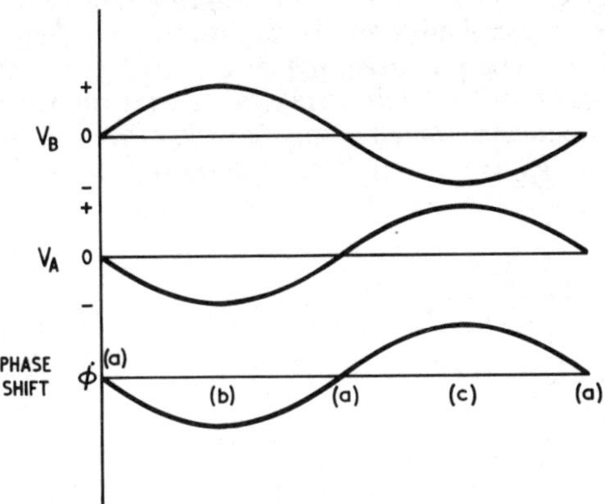

Fig. 8.13 Graphs relating a.f. and phase shift of output

modulating voltage. The greater the amplitude of the a.f., the greater is the difference caused in the collector currents Ic_1 and Ic_2. Therefore, the extent of phase variation of I_0 depends on the modulating amplitude. Thus the phase modulation index is a function of the input voltage at terminals 1 and 2.

As indicated in Fig. 8.10, the a.f. amplifying chain which precedes the phase modulating stage, provides an a.f. input in which the higher audio frequencies have been attenuated relative to the lower ones. The total effect is, therefore, of frequency modulation in which frequency deviation is proportional only to the amplitude of the modulating voltage and is independent of the frequency of the modulating voltage. Any changes in the amplitude of the resultant collector current I_0, are not important as their effects can be removed by an amplitude limiter at a later stage. .

A disadvantage of this indirect method of achieving frequency modulation is the smallness of the frequency swing produced. For mobile radiotelephone transmitters using system deviation values of ±15 kHz or less, the method is effective. Frequency multiplication by a factor as high as 24 may be necessary after the phase modulation process.

8.10. A P.M. TRANSMITTER

Fig. 8.14 is a simplified block schematic diagram showing the essential stages of a transmitter which obtains f.m. through a process of p.m. The audio amplifying chain includes a filter to restrict the range of audio frequencies to that which is adquate for the communication system–e.g. 200 Hz to 3 kHz for a mobile radiotelephone transmitter. A system of amplitude limitation is also necessary so that the system deviation figure is not exceeded as a result of an excessive sound amplitude entering the microphone. An audio *compressor* applies negative feedback over one or more a.f. stages when the amplitude of output exceeds a prescribed level. A simple audio corrector circuit is shown in Fig. 8.15. The potential divider consisting of R and C precedes an amplifier. C is chosen so that its reactance over the range of modulation frequencies is small compared with the resistance of R. The input voltage V_c to the amplifier is then inversely proportional to frequency.

$$V_c = \frac{1/\mathrm{j}\omega C}{R + 1/\mathrm{j}\omega C} \cdot V_{in}$$

or rearranging:

$$V_c = 1/(1 + \mathrm{j}\omega CR) \,.\, V_{in}$$

or if $\omega CR \gg 1$,

$$V_c = 1/\omega CR$$

$$= k/f \qquad \text{where } k \text{ is constant equal to } 1/2\,\pi CR\,.$$

The frequency generating unit may consist of a single crystal oscillator with a number of switched crystals to provide different carrier frequencies. An alternative method which is more economical of crystals employs two crystal oscillators and combines their outputs in a mixer stage. In the mixer, the two oscillator frequencies are inter-

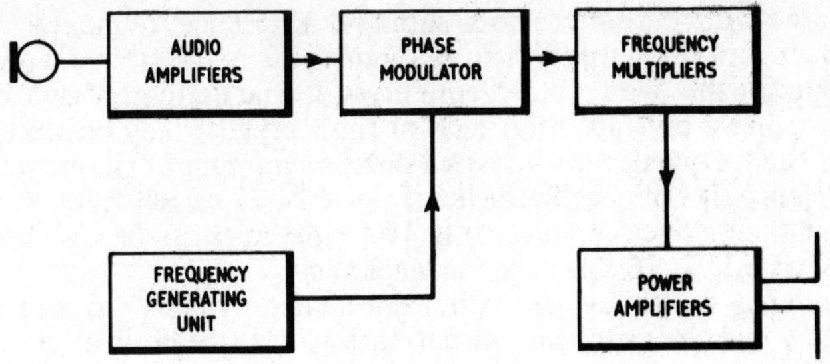

Fig. 8.14 Block diagram of a p.m. transmitter

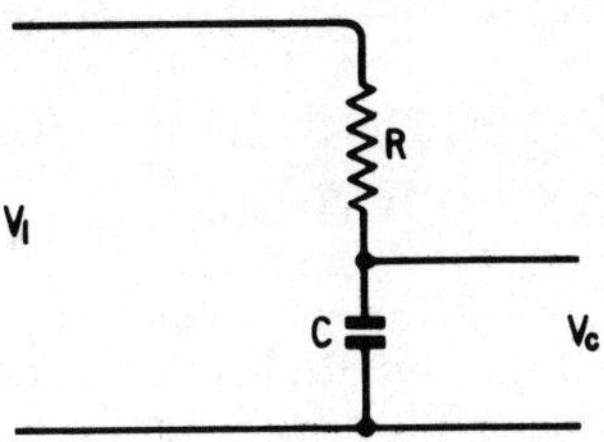

Fig. 8.15 An audio correction circuit

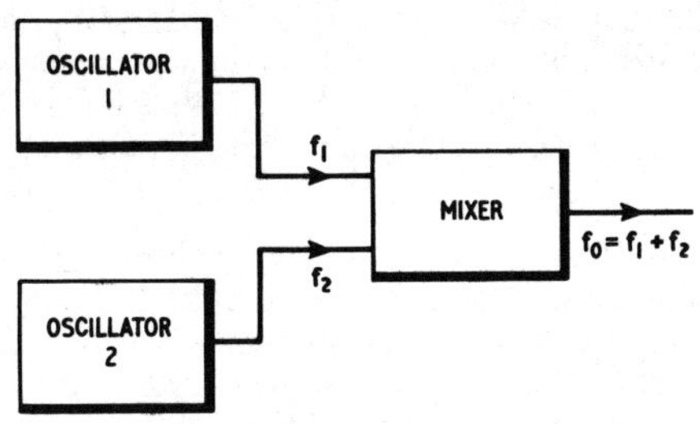

Fig. 8.16 A simple frequency generating unit

modulated which produces both sum and difference frequency components in the output. A filter circuit selects the frequency which is the sum of the two crystal frequencies. If one oscillator has x crystals and the other y crystals, then each of the x crystals may be mixed with each of the y crystals which gives a possible number of frequencies equal to xy. Hence, if each oscillator has 5 crystals, 25 carrier frequencies are available. Using one oscillator only 10 frequencies can be stabilised by 10 crystals. Fig. 8.16 shows the arrangement.

Frequency multipliers are either doublers or triplers. To multiply the frequency and the frequency deviation by 24, three doubler stages and one frequency tripler stage are needed. For a radiated centre frequency of 156 MHz, phase modulation is carried out with a centre frequency level of 6·5 MHz.

The power amplifiers use class-C bias and may operate in push-pull to feed a dipole aerial. The output power from a small transmitter of the type illustrated in Fig. 8.14 is of the order of 5 or 10 W. An f.m. broadcast transmitter on the other hand may radiate 30 kW.

QUESTIONS

1. Frequency modulation involves a change of phase. How is the change of phase affected if (a) the amplitude of the modulating voltage is doubled, (b) the frequency of the modulating voltage is doubled?

Phase modulation involves a change in frequency. How is the frequency variation affected if (c) the amplitude of the modulating voltage is doubled, (d) the frequency of the modulating voltage is doubled?

2. A carrier frequency of 93 MHz is frequency modulated with a modulation index of 2·0. If the modulation frequency is 7·5 kHz, state the maximum and minimum values of the instantaneous value of the radiated frequency and also the frequency swing.

3. An f.m. transmitter feeds the following frequency components into its aerial: 93,000 kHz, 93,002 kHz, 93,004 kHz, 93,006 kHz, 92,998 kHz, 92,996 kHz, 92,994 kHz. The power in the first four frequencies listed is respectively: 41 W, 13 W, 2 W, 0·3 W. State the frequency of modulation and the carrier power when no modulation is present.

4. The mutual conductance of a reactance valve varies by 0·2 mA/V for each volt change in its grid bias. The potential dividing network connected across the valve consists of a capacitor of 0·5 pF between anode and grid and a 2-kΩ resistor between grid and cathode. The reactance valve circuit is joined across the oscillatory circuit of a valve oscillator with a centre output frequency of 10 MHz. When no modulating voltage is applied to the grid of the reactance valve, the

total capacitance of the oscillatory circuit is 50 pF. Find the peak to peak swing of the frequency when an audio frequency voltage of 1 V peak value is applied to the grid of the reactance valve.

5. Draw the circuit of a reactance valve presenting inductive reactance and, with the aid of a phasor diagram. explain the principle of its action.

6. What is meant by the term *scintillation* and what precaution is taken in an f.m. transmitter to avoid it.

7. Explain briefly the principle of a capacitance diode and, with the aid of a circuit diagram, say how it may be used to modulate the frequency of an oscillator.

8. Why is a phase modulation process often preferred to a direct frequency modulation process for a mobile transmitter and how may the radiated signal be approximated to a true f.m. signal?

9. Draw the circuit diagram of a phase modulator and explain with the help of phasor diagrams the principle of its action.

10. The r.f. input frequency to a phase modulator is at a frequency of 8·8 MHz. The output has a maximum frequency deviation of 1·25 kHz. The modulator is followed by a frequency tripler and two stages of frequency doubling. What is the radiated carrier frequency and its maximum deviation?

9

Frequency Modulation—Reception

9.1. INTRODUCTION

A receiver for f.m. waves may be for broadcast reception in the band 87·5 MHz to 100 MHz, a part of a u.h.f. television receiver reproducing the sound information of the programme, or it may be a communications receiver operating in the v.h.f. or higher frequency bands.

Broadcast receivers usually have the benefit of a strong signal fields emitted by transmitters whose e.r.p. is of the order of 120 kW. Communication receivers must receive from transmitters whose output power may be only 20 W on full power and 0·5 W on low power. Such receivers are, therfore, more sensitive, more complex and more costly than a broadcast receiver. The latter must be manufactured at a competitive price for the mass market.

9.2. BLOCK DIAGRAM OF A V.H.F. BROADCAST RECEIVER

Fig. 9.1 shows the essential processes of an f.m. broadcast receiver.

The signal frequency amplifier improves the noise factor of the receiver and also both the image and adjacent channel selectivity. This stage also helps to reduce radiation from the receiver aerial arising from unwanted coupling to the local oscillator.

The frequency changer is a self-oscillating additive mixer producing an i.f. of 10·7 MHz.

The number of i.f. amplifiers which follow the frequency changer varies from one receiver to another. More stages are required if transistors are used than if the active devices are valves. The band width of the i.f. circuits is approximately 200 kHz.

The *ratio detector* is the stage at which demodulation occurs and the a.f. is regained (see Section 9·3 for details of the section of this stage). Unlike the Foster Seeley discriminator (described in Section 9·5), this type of detector performs the additional service of ensuring that variations of signal amplitude do not contribute to the output.

The a.f. voltage and power amplifier stages are not basically different from those used in a.m. receivers although the a.f. response should be wider in order to take advantage of the wider spectrum of modulating frequencies used at the transmitter. It should be noted that only a good quality loudspeaker system can produce the high fidelity response which an f.m. system is capable of providing.

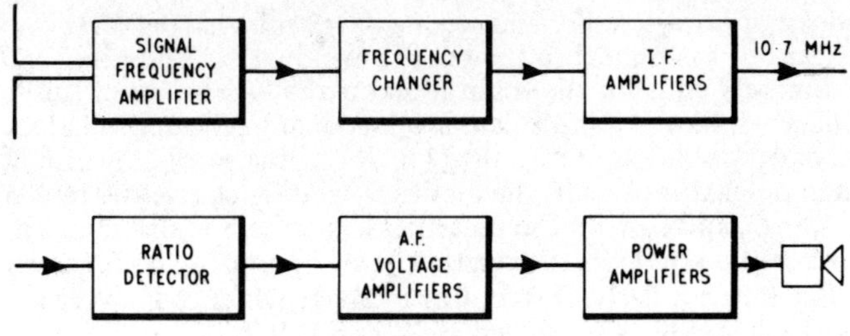

Fig. 9.1 An f.m. broadcast receiver

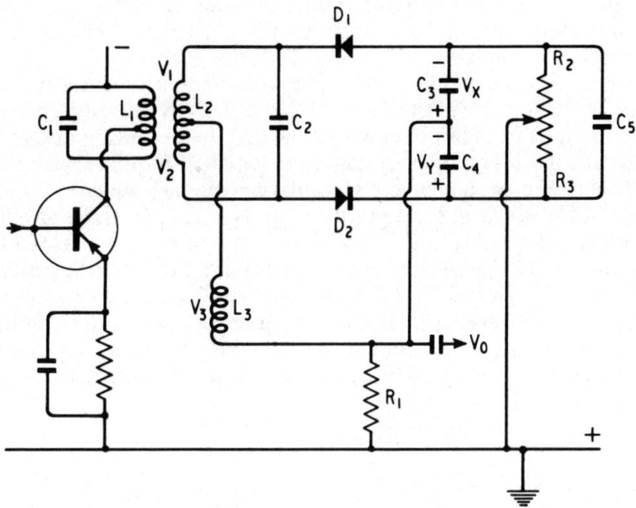

Fig. 9.2 A ratio detector

9.3. THE RATIO DETECTOR

The demodulation stage of an f.m. receiver has the general name of a *discriminator.* A discriminator is so called because it is able to discriminate between the mean carrier frequency, and frequencies which are either higher or lower than this centre frequency. The ratio detector is the type of discriminator found in broadcast receivers. One possible circuit for a ratio detector is shown in Fig. 9.2.

In this circuit L_1 and C_1 form the primary and L_2 and C_2 the secondary circuit of the final i.f. transformer of the receiver. The resonance frequency of both primary and secondary circuits is 10·7 MHz. L_2 is centre tapped to connect to the tertiary winding L_3. The e.m.f. in L_3 is approximately equal to the voltages across the two halves of the secondary winding. V_1 and V_2 are proportional to the e.m.f. induced into the secondary winding and to the Q factor of the secondary circuit.

When a signal is present, the diodes D_1 and D_2 charge the reservoir capacitors C_3 and C_4 with the polarities shown on the diagram. The voltage rectified by D_1 is the vector sum of V_1 and V_3 and is shown as V_x in Fig. 9.3. Similarly, D_2 rectifies a voltage which is the vector sum of V_2 and V_3. This voltage is shown as V_y in Fig. 9.3.

When the frequency received is modulated the values V_x and V_y must vary in proportion to the deviation and at the modulation frequency. The phasor diagrams of Fig. 9.3 help to explain how this is done. Fig. 9.3(a) applies when the radiated frequency is on its centre value and the i.f. has its mean value of 10·7 MHz. All the phasors are related to I_1 which represents the circulating current in the primary of the i.f. transformer. The secondary e.m.f. of mutual induction is shown as e_2 and lags by 90° on I_1. As the tertiary e.m.f. in L_3 is caused by the same alternating flux, V_3 is in phase with e_2. With the secondary circuit resonant at 10·7 MHz, the secondary current I_2 is in phase with e_2. Neglecting the resistance of the secondary coil, the voltages V_1 and V_2 are the e.m.f.s of self induction caused by I_2 in L_2 and these are 90° out of phase with I_2. V_1 and V_2 are shown as opposite in phase in the phasor diagrams because they act in opposite directions relative to the centre tap on L_2. In Fig. 9.3(a) V_x is equal to V_y and the two diodes conduct equally charging C_3 and C_4 to equal voltages. The potential at the junction of C_3 and C_4 is equal to $(V_x - V_y)/2$, as Fig. 9.4 and the accompanying algebra show: The current in the closed mesh comprising C_3, C_4, R_2 and R_3 is given by,

$$I = (V_y + V_x) / (R_2 + R_3)$$

or since R_2 is equal to R_3,

$$I = (V_y + V_x) / 2R_2 \cdot$$

The output voltage V_o is,

$$V_o = V_x - IR_2$$

$$= V_x - \frac{(V_y + V_x) \cdot R_2}{2R_2}$$

$$= (2V_x - V_y - V_x)/2$$

$$= (V_x - V_y)/2 \cdot$$

The output is thus zero when V_x is equal to V_y ·

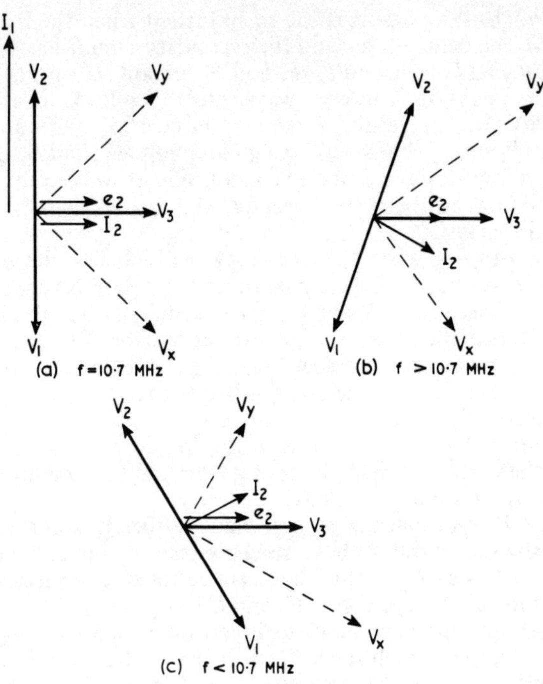

Fig. 9.3 Phasors for the ratio detector

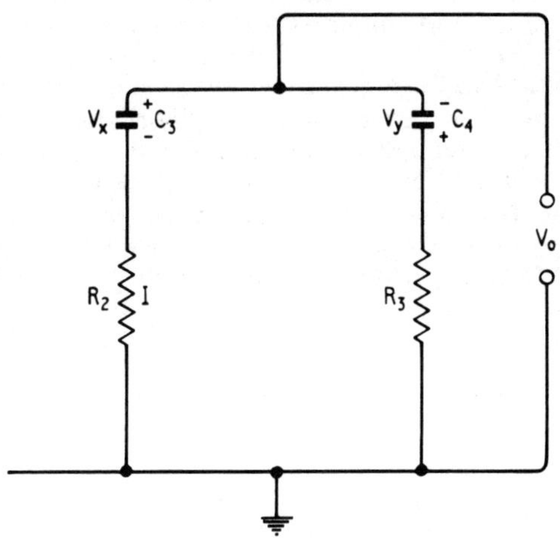

Fig. 9.4 Output circuit redrawn

In Fig. 9.3(b) the phasors relate to an instant when the frequency has swung above the centre value and the secondary circuit has an inductive reactance causing I_2 to lag on e_2. V_1 and V_2 remain 90° out of phase with I_2 so that the phase of V_1 moves away from V_3 while the phase of V_2 moves towards it. This results in the amplitude of V_y exceeding that of V_x and that diode D_2 charges C_4 to a greater voltage than D_1 charges C_3. A negative going output results. The amplitude of the output is proportional to the phase difference between I_2 and e_2 and, therefore, to the frequency deviation.

Fig. 9.3(c) applies when the frequency swings below the centre value and the reactance of the transformer secondary is capacitive. I_2 thus leads in phase on e_2. Again V_1 and V_2 remain at 90° to I_2 but now V_1 moves towards the phase of V_3 and V_2 away from the phase of V_3. The vector sum V_x is thus greater than V_y so that D_1 is now the diode with the bigger current. The charge on C_3 is bigger than that on C_4 so that the output polarity is positive.

The output voltage V_o alternates at the frequency of modulation applied to the carrier wave at the transmitter and the amplitude is proportional to the modulation index.

C_5 (Fig. 9.2) is a capacitor of large value, typically 16 or 25 μF, which with the resistors R_2 and R_3 has a time constant of the order of 0·1 s. The purpose of this *CR* combination is to suppress any variation in the total amplitude of voltage across C_3 and C_4. $(V_x + V_y)$ is, therefore, held constant and only the ratio of V_x to V_y varies at the modulation frequency, a feature which suggests the name of the ratio detector.
The self limiting action of the circuit depends on the long time constant given by $C_5 . (R_2 + R_3)$. Any momentary increase in the amplitude of the signal input increases the current through the diodes as they add charge to C_5. This increases the loading and damping of the i.f. transformer so that the gain of the last i.f. amplifier is momentarily reduced. The action may be aided by taking a lead from C_5 back to an earlier stage and providing a reverse bias for that stage whenever the output rises.
A reduction in signal amplitude causes C_5 to discharge slightly into C_3 and C_4. The diode currents are momentarily reduced and a slight increase in the gain of the last i.f. amplifier results. In these ways amplitude variations of the signal are smoothed out and the output depends only on the frequency deviation and not on any accidental changes in signal amplitude such as might be caused by interfering noise voltages.

9.4. A COMMUNICATIONS RECEIVER BLOCK DIAGRAM

Fig. 9.5 shows the basic requirements of an f.m. communications receiver. (For further details see Chapter 10).

The signal frequency amplifier provides the same advantages as in the broadcast receiver but there may be more stages in order to provide

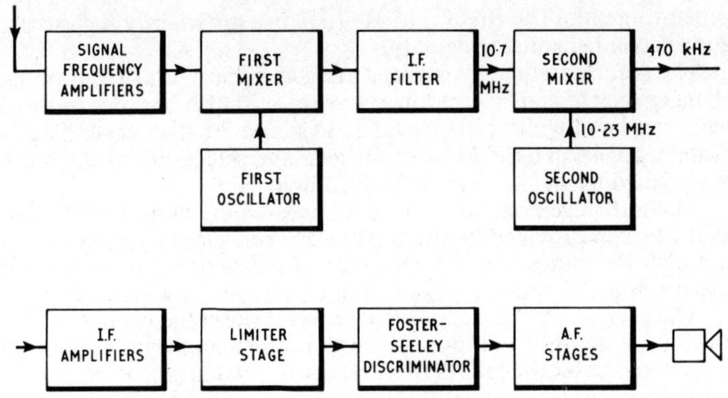

Fig. 9.5 A communications receiver

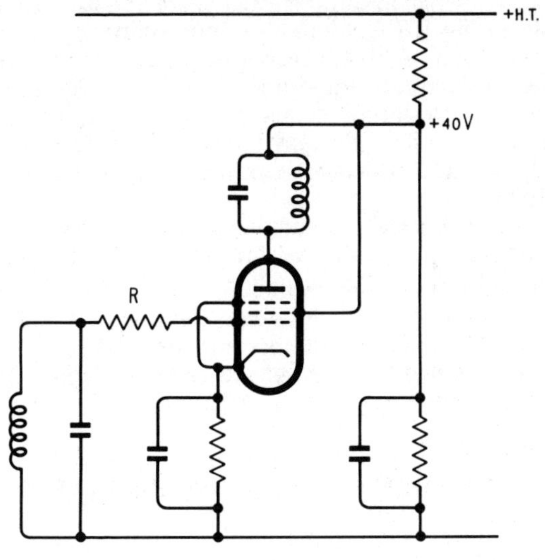

Fig. 9.6 A limiter circuit

greater sensitivity and selectivity. The first frequency changer uses a separate oscillator which is usually crystal controlled in order to provide a high order of frequency stability. To alter the frequency to be received, a channel switch is turned to select appropriate crystals in a frequency generating unit. The first i.f. of 10·7 MHz is sufficiently high to provide good second channel selectivity.

The second frequency changer using an oscillator of fixed frequency changes the frequency to a lower level of 470 kHz in order to provide a narrower band width of between 35 kHz and 8·5 kHz depending on the system deviation used. Most of the gain and selectivity of the receiver is obtained by the i.f. stages which follow.

The last stage prior to the discriminator is the limiter. It is essential that the gain provided before this stage is sufficient to make the signal input to the limiter big enough to secure efficient action of the limiter circuit. Fig. 9.6 shows a possible limiter circuit for a valve receiver.

The grid resistor R ensures that the grid potential cannot rise more than a fraction of a volt positive. Any rise of input signal amplitude merely increases the grid current flow and the p.d. across R. With the grid at a positive potential the grid-cathode space resistance is only one or two kilohms and is small compared with R. Thus, nearly all the signal voltage is dropped across R and only a small fraction of it appears between grid and cathode. When the grid swings in a negative direction, the anode current is cut off because of the low values of anode and screen potentials which give the mutual characteristic a short grid base. As the graphs in Fig. 9.7 illustrate, the anode current swings between the value it reaches for a slightly positive grid potential and zero value. If, however, the input voltage amplitude drops below that needed to cut the anode current off during the negative half cycle, then variations of output amplitude are possible. Any increase above this minimum amplitude causes little effective alteration in the limits of change of anode current.

The discriminator used in communications receivers is the Foster-Seeley discriminator, a circuit for which is shown in Fig. 9.8.

The audio amplifying stages which follow are designed to give adequate response to the speech frequencies essential to good quality radiotelephony. A bandwidth of approximately 3 kHz is adequate. The output may be coupled to a loudspeaker or to a telephone hand set comprising both a microphone and an earpiece.

9.5. THE FOSTER-SEELEY DISCRIMINATOR

The input circuit to the discriminator has some features in common with that used in the ratio detector but comparison of the two circuits shows that one diode is reversed so that both diodes have their anodes connected to the transformer secondary circuit. Further, the transformer has no tertiary winding (Fig. 9.8).

The resultant voltage to be rectified by each diode is obtained in a similar way to that used in the ratio detector. For diode D_1 the voltage to be rectified is the vector sum of V_1 and V_p where V_p is the voltage across the primary of the i.f. transformer. V_1 and V_p have 90° difference in phase so long as the frequency is the resonant frequency of the

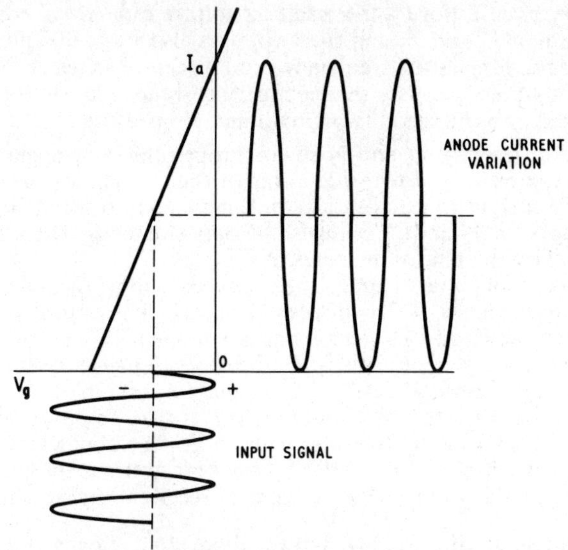

Fig. 9.7 Limitations of anode current changes

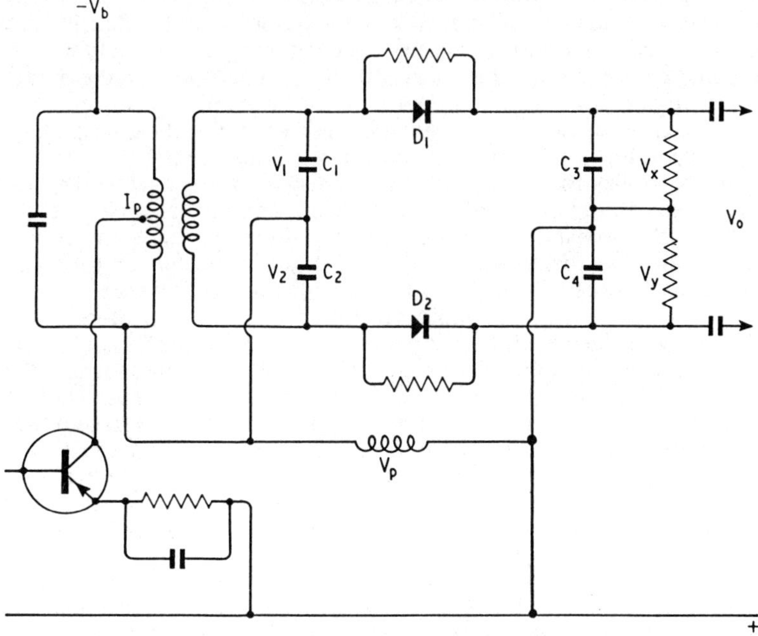

Fig. 9.8 A Foster-Seeley Discriminator

secondary circuit. For D_2 the resultant voltage to be rectified is the vector sum of V_2 and V_p and these voltages also are at 90° phase difference so long as the frequency is on its centre value. An upward swing of frequency causes the secondary current to lag on the e.m.f. introduced into the secondary circuit and the phasors for the reactive potential differences V_1 and V_2 swing through the same angle as the secondary current. A downward swing of the frequency causes the secondary current to lead on the e.m.f. induced into secondary winding. The phasors for V_1 and V_2 swing in the opposite sense. These changes are illustrated by the phasor diagrams of Fig. 9.9.

The rectified voltages appearing on the capacitors C_3 and C_4 are almost equal to peak values of the voltages V_x and V_y. The output voltage V_o is zero so long as V_x and V_y remain equal. On one half cycle of modulation V_x increases and V_y decreases while the reverse changes occur on the other half cycle of modulation. V_o is equal to the sum of the change in value of V_x and V_y. It will be noted that this gives an output which is double that achieved by the ratio detector for the same variation of reservoir capacitor voltage. Older types of receiver use thermionic diodes in the circuit shown but newer designs of receiver employ semiconductor diodes.

R_1 and R_2 are the d.c. loads of the diodes and allow V_x and V_y to vary at the modulation frequency. Additionally each diode is shunted by a high resistor to prevent the accumulation of charge on C_1 and C_2.

If the frequency swing is excessive, the output from the discriminator does not rise in proportion to the frequency swing. Beyond a limit which is determined by the bandwidth of the tuned circuits, the output begins to decrease because of the rise of the impedance of the circuits at frequencies outside their passband. The response of the discriminator is shown in Fig. 9.10. The frequencies f_1 and f_2 are at the limits of the frequency swing for which the output remains proportional to the deviation. The separation of f_1 and f_2 depends on the coupling used between the primary and secondary circuits of the last i.f. transformer and on the Q factors of the tuned circuits. For low distortion the difference between f_2 and f_1 must be greater than twice the system deviation. An increase of coupling factor between the primary and secondary of the last i.f. transformer increases the separation of f_1 and f_2 but reduces the slope of the charateristic and thus the efficiency. The frequency f_c is the centre or carrier frequency. The circuits must be carefully adjusted by trimming the inductance value of the transformer secondary so that f_c coincides with the resonance frequency of the circuits. When the circuits are correctly adjusted the d.c. potential difference between the cathodes of the two diodes is zero for an unmodulated signal fed into the receiver at the frequency to which the r.f. circuits are correctly tuned.

The Foster-Seeley discriminator does not provide amplitude limiting as does the ratio detector. The relative advantages of the Foster-Seeley discriminator are that it is more sensitive and less likely to produce distortion as its alignment for the best operating conditions is less critical.

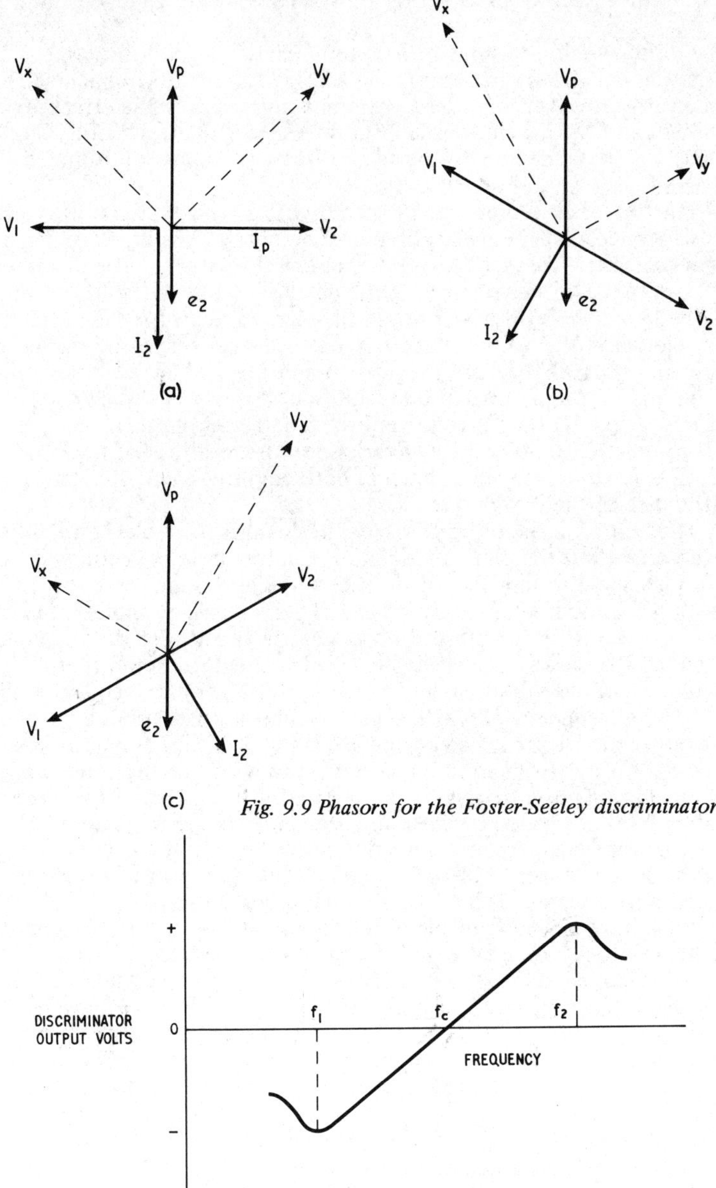

Fig. 9.9 Phasors for the Foster-Seeley discriminator

Fig. 9.10 Discriminator response graph

9.6. RESPONSE OF AN F.M. SYSTEM TO NOISE

The noise which interferes with communication signals is no one particular frequency but covers the whole of the radio-frequency spectrum. However, in order to examine more simply the effect of noise in an f.m. system, it is helpful to consider the result of adding a single noise frequency to the carrier frequency. This addition by phasors is shown in Fig. 9.11.

The phasor V_c represents the carrier voltage and V_n the noise voltage. It is assumed that the particular noise frequency considered is higher than the carrier frequency so that the direction of the noise phasor relative to the carrier phasor is anticlockwise. *OB* is the resultant voltage of the two and its tip passes round the circular locus for each cycle completed by the noise voltage relative to the carrier. If the carrier frequency is 100 MHz and the noise frequency 100·01 MHz, then the noise phasor completes 10 000 revolutions per second relative to the carrier phasor. During each second the length and phase of the resultant *OB* completes 10 000 cycles of variation. The resultant of the two voltages is, therefore, modulated in both amplitude and phase at the difference of the two frequencies.

The limiter in the receiver makes the changes in resultant amplitude of no effect but the phase modulation involves a variation of frequency to which the discriminator of the receiver can respond. As stated in Section 8.1 the frequency swing associated with phase modulation is proportional to the amount of phase change and also to the frequency at which the phase modulation occurs. The greater, therefore, the difference between the carrier frequency and the noise frequency, the higher the frequency at which their resultant is phase modulated and the larger the frequency swing involved. Fig. 9.12 is a graph to show how the output from an f.m. receiver varies with the difference between an interfering noise frequency and the carrier frequency. The maximum ordinate *OY* is limited by the noise level and also by the bandwidth of the receiver signal circuits. Yet an effective limit to the audible noise is set by the a.f. response of the receiver and by the limit of audio response of the human ear. This limit is assumed to be 15 kHz.

An a.m. receiver is capable of responding to the changes in amplitude of the resultant voltage *OB* and the variation of the amplitude can be assumed independent of frequency. An output from an a.m. receiver may thus be at the maximum level indicated by *OY* in Fig. 9.12, over the whole a.f. spectrum. A comparison of the area *OYZX* with the smaller area *OWX* reveals that the noise output from the f.m. receiver is much smaller than that from an a.m. receiver of similar sensitivity and a.f. response.

A quantitative comparison of the noise output power of an f.m. receiver with that of an a.m. receiver can be arrived at by integrating the noise output power over the audio spectrum for each type of receiver. For purposes of comparison it must be assumed that the maximum noise voltage which is present at the output of the discriminator is equal

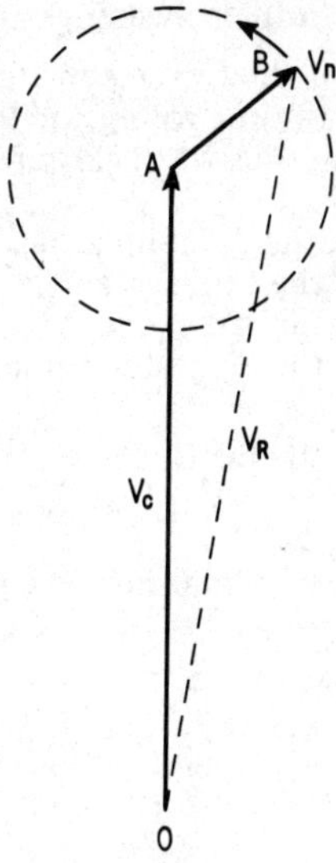

Fig. 9.11 Phasor addition of noise and carrier

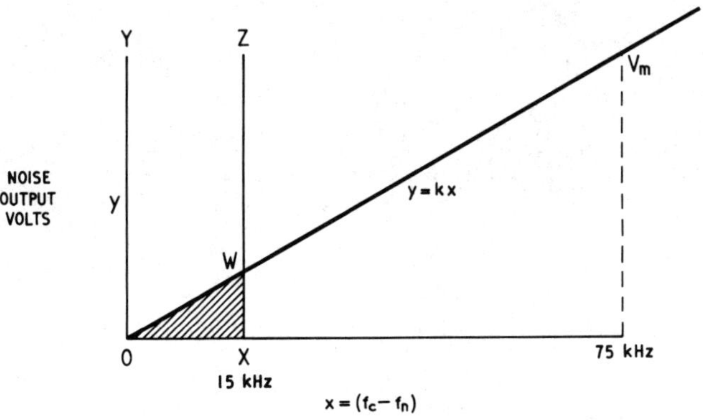

Fig. 9.12 Response of discriminator to noise

to the maximum noise output voltage of the detector of the a.m. receiver. The a.f. stages of the two receivers are taken to be the same in gain and bandwidth. Symbols used with their meanings are listed below:

δ is the system deviation of the f.m. receiver;
V_m is the maximum noise output voltage of the demodulator;
x is the frequency difference between carrier and the interfering noise frequency;
y is the noise output from the demodulator;
Δ is the deviation ratio of the f.m. system;
F is the maximum modulation frequency;
R is the load resistance of the demodulator across which the voltage y appears;
K is a constant of proportionality.

The equation to the noise output graph for the f.m. receiver is,

$$y = K\,x \tag{9.1}$$

When $x = \delta, y = V_m$

Therefore $V_m = K\delta$ and

$$K = V_m/\delta \tag{9.2}$$

Inserting equation 9.2 into equation 9.1

$$y = V_m x/\delta$$

The total noise output power from the discriminator,

is

$$P_f = \frac{1}{R}\int_o^F y^2 \,.\, \mathrm{d}x$$

$$= \frac{1}{R}\int_o^F V_m^2 x^2/\delta^2 \,.\, \mathrm{d}x$$

$$= \frac{1}{R}\left[V_m^2 \cdot x^3/3\delta^2\right]_o^F$$

$$= \frac{1}{R}\left[V_m^2 F^3/3\delta^2\right]$$

But by definition $\delta = \Delta F$, and $F = \delta/\Delta$

Thus $\quad P_f = V_m^2\delta/3\Delta^3 R$

The total noise output power from the detector of the a.m. receiver is,

$$P_a = \frac{1}{R}\int_o^F y^2 \cdot \mathrm{d}x$$

where $y = V_m$ throughout the audio range.

Thus
$$P_a = \frac{1}{R}\int_o^F V_m^2 \cdot \mathrm{d}x$$

$$= \frac{1}{R}\left[V_m^2 \, . \, x\right]_o^F$$

$$= V_m^2 F/R$$

$$= V_m^2\delta/\Delta R$$

The ratio P_a/P_f is then:

$$\frac{V_m^2\delta/\Delta R}{V_m^2\delta/3\Delta^3 R}$$

$$= 3\Delta^2$$

In decibels, the advantage in noise power of the f.m. compared with the a.m. receiver is given by: $10 \log 3\Delta^2$ dB. If, typically Δ has a value of 5, the improvement is

$$10 \log 75 = 19 \text{ dB}$$

9.7. PRE-EMPHASIS AND DE-EMPHASIS

As Fig. 9.12 shows, the noise frequencies which differ most from the carrier frequency cause the greatest noise output from the receiver. Such interfering noise frequencies occupy the same positions in the frequency spectrum as the side frequencies set up by the higher modulating frequencies. The signal to noise ratio at the output tends therefore to be worse for higher audio frequencies than for lower ones. A method of improving this situation is to emphasise the higher audio frequencies at the transmitter prior to modulation. This can be accomplished by including in the amplifier chain which precedes the modulator, a stage with a gain which increases with frequency for frequencies above approximately 3 kHz. A possible circuit for such a stage is shown in Fig. 9.13.

The values of L and R may vary from circuit to circuit but the degree of pre-emphasis is dependent upon the time constant of the load. In the U.K. this is 50 μs for broadcast transmitters. A simple calculation shows that the reactance of the inductance is approximately 5 kΩ at 3 kHz. Above this frequency the value of R is only a small fraction of the load impedance and the gain rises almost in direct proportion to the frequency within the limits of the audio range. That is, the gain rises at a rate of 6 dB per octave. The transistor which follows the stage at which pre-emphasis occurs is shown as an emitter follower. The high input impedance of this stage will not shunt excessively the frequency sensitive load of the first transistor. Side frequencies set up by modulating frequencies higher than 3 kHz are thus given proportionately higher amplitudes when modulation occurs after pre-emphasis.

At the receiver, a circuit which restores the correct relative amplitudes of the audio frequencies is inserted after the demodulation stage. This circuit attenuates the higher audio frequencies relative to the lower ones. A simple *CR* network as shown in Fig. 9.14 is sufficient.

The time constant of the de-emphasis circuit is equal to that of the pre-emphasis circuit used at the transmitter. In the process of attenuating the higher audio frequencies the circuit also attenuates the higher noise frequencies and thus provides a further advantage in signal-to-noise ratio for the f.m. receiver over an a.m. receiver. The theoretical maximum of this further improvement is 10 dB, but the presence of high impulse noise and the slight reduction of modulation index at the transmitter which is necessary on account of pre-emphasis, reduce the effective advantage to about 4·5 dB.

In addition to improving the signal-to-noise ratio the system of pre-emphasis and de-emphasis also improves the quality of the a.f. reproduction. This is because without pre-emphasis the modulation index tends to be small for modulation frequencies over 1 kHz and less distortion results on modulation and on demodulation if the modulation index is not very low.

9.8. ADVANTAGES OF A V.H.F./F.M. BROADCAST SERVICE

The wide frequency band occupied by a high quality f.m. broadcast service makes the use of the v.h.f. or higher frequency band essential. Some advantages arise because of the use of such high transmission frequencies. The range of a v.h.f. transmitter is generally determined by the heights of the positions of the transmitting and receiving aerials and interference from transmitters at distances of over 80 km or so is unlikely. A local service free from interference is thus available. Also the v.h.f.

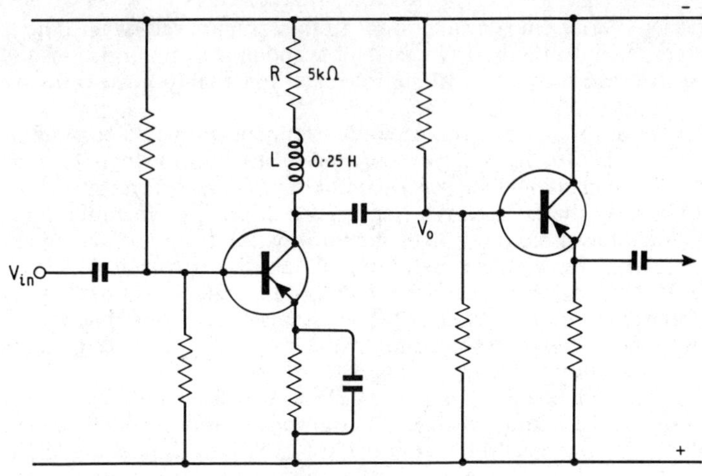

Fig. 9.13 A pre-emphasis circuit

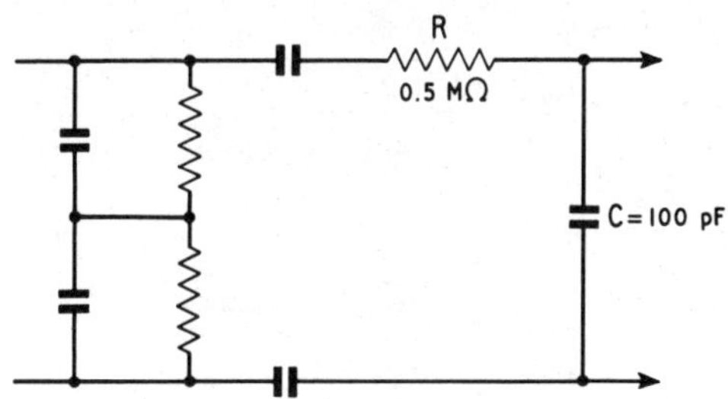

Fig. 9.14 A de-emphasis circuit

band has a sufficiently wide spectrum to accommodate the side frequencies set up by a broad base band of modulating frequencies. Audio frequencies up to the limit of the average listener's audio range may be reproduced at the receiver providing the receiver itself is adequately designed. A high fidelity broadcast service is, therefore, available.

Using f.m. rather than a.m. provides further advantages. The improved signal-to-noise ratio is marked provided that the receiver is correctly adjusted so that it does not respond to any amplitude modulation and provided that a good aerial system is provided. A carefully sited dipole with a screened down lead to the receiver is essential for the best results except in areas having a very strong signal from the transmitter. Without an efficient aerial, the ignition noises from cars may well swamp the required signal. In the best circumstances and in similar conditions the f.m. system can show a 24 dB improvement in signal-to-noise ratio over the a.m. system.

The ability of an f.m. receiver to discriminate against co-channel interference is a further relative advantage of the f.m. system. An a.m. receiver in an area in which signals are present from two transmitters of approximately the same carrier frequency will produce an audio output which is an unwanted mixture of the information from both channels. An f.m. receiver in similar conditions will take the signals of whichever carrier is the stronger in that area and reproduce these with negligible interference from the weaker of the two transmitter fields. The receiver is said to be captured by the stronger of the two stations and the effect is known as *capture effect.*

The phasor diagrams of Fig. 9.15 help to explain the effect. V_1 represents a side frequency from one transmitter and V_2 a side frequency from a second transmitter. In Fig. 9.15(a) it is assumed that V_1 has the larger amplitude. The resultant phasor R is modulated in both amplitude and phase at the difference of the two frequencies

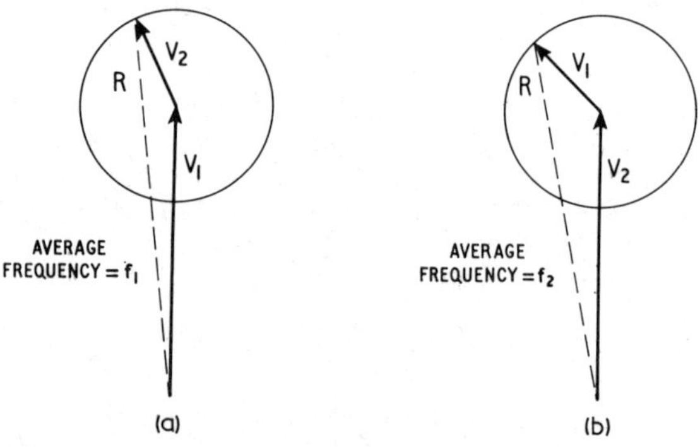

Fig. 9.15 Capture effect

but it has an average velocity equal to that of phasor V_1. In Fig. 9.15(b), V_2 is assumed to have the larger amplitude. The resultant is again modulated in amplitude and phase at the difference of the two side frequencies, but now has the average velocity of the phasor for V_2. It is seen, therefore, that when two frequencies are present together in the same circuit their resultant is changing in phase above and below the phase of an average frequency which is that of the one having the larger amplitude.

In a receiver which is susceptible to changes in the resultant amplitude an output is produced proportional to the large changes in amplitude of the resultant. (The resultant amplitude changes from ($V_1 + V_2$) to

$(V_1 - V_2)$). In a receiver which is only susceptible to changes of frequency the interference is less noticeable. The amplitude modulation of the resultant produces no output from an f.m. discriminator so long as the limiter is working effectively.

The phase modulation does involve a frequency variation to which the discriminator can respond but only in extreme conditions is this troublesome. If the two vectors are of equal amplitude and represent frequencies as far apart as 5 kHz, the phase modulation occurring is at an average rate of 45° in a quarter period of 50 μs. This represents an average rate of change of phase of 9×10^5 degrees per second or 2·5 kHz. The peak value of this, assuming a sinusoidal variation, is approximately 4 kHz. This change in frequency is small compared with a system deviation of 75 kHz, and so the interference output due to the frequency changes of R is small. The required output of the discriminator depends on the difference between the side frequencies of the wanted signal and on the related variations of amplitude of both side frequencies and the carrier.

Another advantage claimed for f.m. broadcasting is the greater dynamic range of sound amplitude which it is capable of reproducing. For an a.m. transmitter a practical range of modulation factor is from 0·05 to 0·9 which represents a 25 dB variation in level. Yet the range of sound amplitude from the quietest to the loudest output from an orchestra may be as large as 70 dB so that considerable compression of the audio level range is necessary. The modulation index of an f.m. wave can vary between 0·01 and 1·0 which is equivalent to a sound level variation of 100 times or 40 dB. Less compression of the audio level range is, therefore, necessary in the transmitting studios than would be required for a.m.

At the transmitter, f.m. allows greater efficiency in the use of amplifiers. Class-C operation of valves which come after the modulating stage is possible with f.m. whereas a waveform which is amplitude modulated requires class-B bias. Further, since the envelope power of an f.m. wave is constant, the transmitter output valves can operate with their optimum power output all the time. The same valves used for an a.m. output must have an average power output far less than their maximum in order to be able to increase their output power when the modulation envelope increases to its peak value.

The B.B.C. chose an f.m. system for their v.h.f. high quality broadcast service in preference to a.m. only after prolonged investigation. The better signal to noise ratio and the small response to co-channel interference of f.m. were among the major factors influencing their decision in favour of the f.m. system of modulation.

QUESTIONS

1. Draw a block diagram of a typical broadcast receiver for the reception of the v.h.f./f.m. broadcast programmes. State briefly the purpose of each stage shown.

2. Refer to Fig. 9.2 and Fig. 9.3. If V_1, V_2 and V_3 all have an amplitude of 300 mV, find by drawing a phasor diagram to scale (or by calculation) the values of V_y and V_x and the approximate output voltage V_0 when the phase angle of the circuit L_2C_2 is, (a) 0°, (b) +20°. Assume 100% efficiency for the diodes D_1 and D_2.

3. List the main stages to be found in a typical communications receiver for f.m. signals and state the purpose of each stage.

4. With the aid of a circuit diagram explain the action of a limiter stage in an f.m. receiver.

5. Assuming that in Fig. 9.8, $C_1 = C_2 = 400$ pF, and that the i.f. transformer secondary has a 3 dB bandwidth of 40 kHz centred on 470 kHz, find the effective resistance which the transformer secondary circuit must have.

6. Give a brief explanation of the principle of the action of the Foster-Seeley discriminator and state what advantages it has over a ratio detector.

7. Why is an f.m. receiver less responsive to noise interference than an a.m. receiver operating in the same frequency band?

8. If the deviation ratio of an f.m. system were increased from 1 to 5, what improvement in signal-to-noise ratio could be expected, other factors remaining unchanged?

9. Why is the signal-to-noise ratio in an f.m. system improved by pre-emphasis at the transmitter and corresponding de-emphasis of the higher audio frequencies at the receiver? If the gain of a pre-emphasis amplifier is 20 dB at 3 kHz, what gain has the amplifier at 12 kHz?

10. List the advantages which a v.h.f./f.m. broadcast service provides in entertainment value as compared with the same programme of broadcast service using an m.f./a.m. system.

10

Transmitters and Receivers

In the present context a transmitter is a piece of equipment in which is developed power at radio frequency suitably modulated by a signal characterising the information to be transmitted. From the transmitter the r.f. power is taken by transmission line (feeder) to the radiating system, i.e. the aerial or antenna system.

A receiver is a device which, connected to the receiving aerial, amplifies the signal received from the transmitter and converts it into the correct form for the interpretation of the information.

10.1. TRANSMITTERS–GENERAL

To convey information, a transmitter must be modulated in some way—perhaps by amplitude or frequency modulation to transmit speech, perhaps keyed for code transmission.

Modulation may be applied early in the transmitter (Fig. 10.1 (a)) before most of the amplification is applied. This is *low level modulation.* Or, modulation may be carried out in the last r.f. amplifying stage where the signal is at its greatest. This is *high level modulation.*

To effect the transmission it is necessary to radiate a relatively narrow band of frequencies centred on a carrier frequency, which, depending on the type of service, has a value of from some 15 kHz upwards to several gigahertz. For code transmission at slow speed, the width of the band of frequencies necessary to transmit the information is only one or two hundred hertz. For the rapid transmission of considerable detail, for instance a quickly changing picture, a bandwidth of several megahertz is needed.

The carrier frequency must conform to certain standards of accuracy and stability. These standards vary with the type of service, power radiated, frequency of transmission and other relevant matters. The frequency is determined by an oscillator, often crystal controlled, which must always provide an output frequency within the tolerance laid down for the particular type of transmission.

The final radiated carrier frequency may be the same as that of the oscillator; or it may be higher, in which case frequency multipliers (harmonic generators) are introduced; or it may be lower, necessitating the use, perhaps, of frequency dividers.

Frequency changing processes may be employed in a transmitter to shift the frequency by a given amount. Unless some special technique

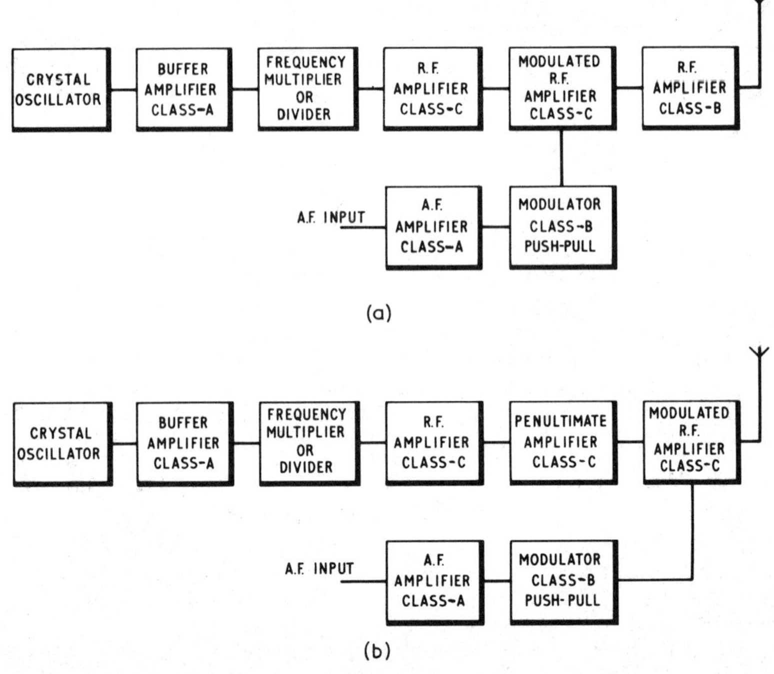

Fig. 10.1 Simple amplitude-modulation transmitters

is used (such as that employed in the transmitter described in Section 10.6.2) the resultant frequency accuracy and stability can be no better than that of the poorest oscillator in the system.

Amplitude modulation may not be effected before frequency multiplication (or division) is carried out (otherwise the sidebands are similarly multiplied or divided in frequency), but it may be, and is, applied before a frequency changing process is used (because this merely shifts all frequencies by the same amount).

An amplitude-modulated radio signal may be radiated in several forms, for example:

1. Two sidebands and full power carrier, as used in broadcast work.
2. A single sideband, which may convey more than one intelligence signal, or two sidebands, which may be different from one another and either or both of which may contain more than one intelligence signal, with either:
   (a) a full power carrier;
   (b) a much reduced (pilot) carrier, for automatic frequency control or automatic gain control purposes; or
   (c) no carrier at all (see Section 6.4.2).

10.1.1. Transmitter Frequency Control

Transmitters using the same carrier frequency for long periods have their frequency controlled by a single master oscillator designed for very high stability (a few parts in 10^8 maximum change over several months). Transmitters which need a range of carrier frequencies use *frequency synthesisers.* These units build up the required frequencies by using the harmonics or subharmonics of a single oscillator of the required stability. This avoids the high cost of providing a number of oscillators each with a high stability order (see Section 10.7.1).

There are two types of synthesiser:

1. The phase locked loop type. In this type a selected harmonic of a high stability source is fed to a phase comparator along with the output of a variable frequency oscillator. The phase comparator provides a d.c. voltage to control the frequency of the sub-standard oscillator by means of a capacitance diode. In this way the good long term stability of a crystal oscillator is linked with the good short term phase stability of a well designed v.f.o.
2. The Direct Synthesiser. In this type, frequency dividers, mixers and filters are used to build up a required frequency from harmonics of a high stability oscillator. Fig. 10.2 shows this principle. An unskilled operator can easily set the dials of the frequency synthesiser to select the required frequency accurately.

Automatic tuning of later stages of a transmitter may be used. Base-band information modulates a subcarrier (say 1 MHz). The band of frequencies

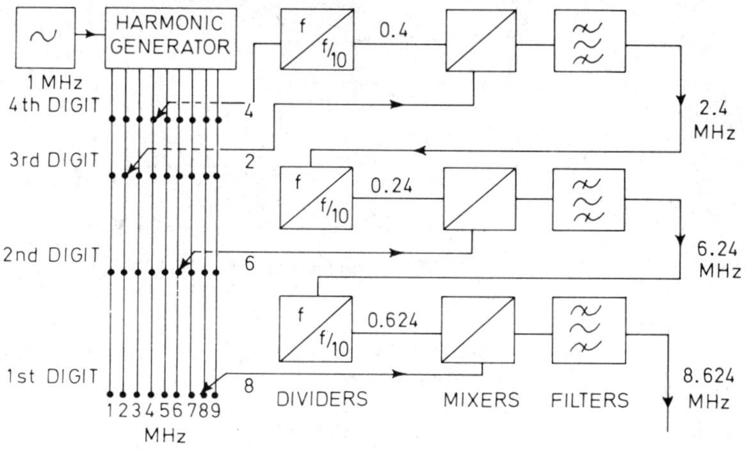

Fig. 10.2 A direct synthesiser

centred on this subcarrier frequency, is passed to the frequency changer where it is mixed with a signal, derived from the synthesiser, of such a frequency as to yield, after mixing, the required radiated frequency. Amplification follows:

this may be in a wideband amplifier, which covers the whole frequency spectrum of the transmitter without tuning, or in a tuned amplifier. The signal is now at the level needed to drive the final amplifier. For aerial powers greater than one or two kilowatts the final amplifier must be tuned. This is because wideband amplifier techniques cannot economically be applied to yield output powers greater than this. It can be arranged, however, that the tuning process is carried out automatically.

10.1.2. Telegraphy transmitters

For telegraphy transmission the transmitter must be operated in such a way as to cause the signals in the output of the receiver to be alternately 'on' and 'off'. The most elementary way of accomplishing this is to cause the telegraph key, or the keying relay, to switch the transmitter on and off. The two states are known respectively as *mark* and *space.* With low-power equipments this action may be effected by switching the supply voltage to one or more of the amplifying stages. The oscillator itself, and preferably also its following buffer amplifier, must not be switched or the radiated frequency will alter.

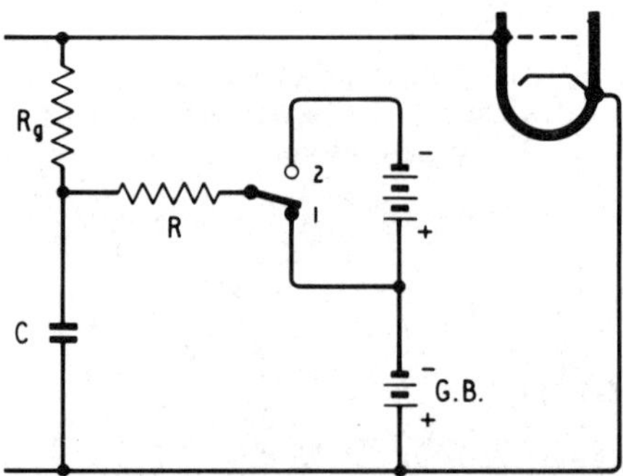

Fig. 10.3 Simple keying and filter circuit

Alternatively, keying may be carried out in the bias circuit(s) of one or two of the amplifiers–extra bias, sufficient to cut off the stage(s), being applied during the space periods (Fig. 10.3). Instead of switching extra voltage into circuit it may be arranged that the keying process, on space, removes a short circuit which otherwise exists across a bias resistor in series with the one which provides the normal bias. This, of course, also indirectly causes an increase in the bias voltage.

Rapid change in the shape of the radiated r.f. signal caused by on-off keying results in the generation and radiation of a wide band of frequencies. Such radiation of out-of-band frequencies, if permitted to take place, results in interference with other communication channels. The use of a filter in the keying circuit (a simple one is shown in Fig. 10.3) avoids the sudden changes in waveform. During mark periods the keying relay is in position 1, and the normal bias is applied; during space periods the keying relay is in position 2 and the additional bias is applied, thus cutting off the amplifier. Sudden changes are avoided because the capacitor *C*, in the filter circuit *CR*, cannot charge and discharge instantaneously. The time constant *CR* is chosen so that it is long enough to prevent undue out-of-band radiation but not so long as to prevent reasonably fast signalling.

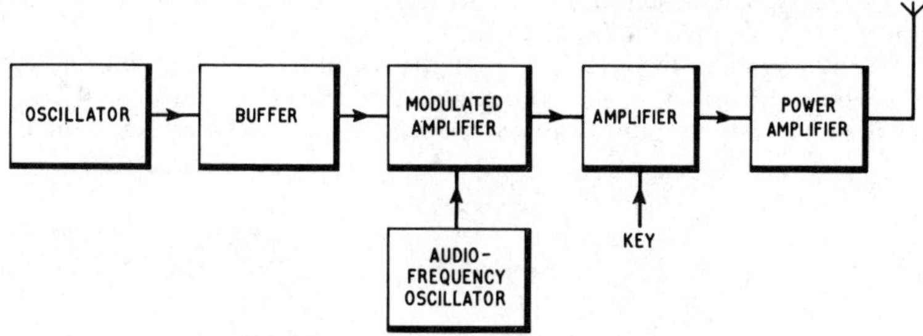

Fig. 10.4 Simple m.c.w. transmitter

For high power simple on-off keying cannot be used because of the large supply voltage surges created in the high voltage supply line. One method of avoiding this trouble is to employ a circuit which draws from the H.T. supply during the space periods about the same current as the keyed stage draws during the mark periods. This is *absorber keying*. It can readily be effected by including a valve, with resistive load, in parallel with the keyed valve. As the bias for the latter is increased to beyond cut-off value for space periods, the absorber valve bias is reduced and so draws current. At the end of each space period the keyed valve bias is reduced to normal, and it again draws anode current. Simultaneously the bias on the absorber valve is increased and it cuts off. The current drawn from the supply, therefore, remains substantially constant.

A keyed transmission of the types discussed above is known as a *continuous wave* (c.w.) transmission. For its audible reception the receiver must be equipped with a beat-frequency oscillator (see Section 4.7). If the transmitted carrier is interrupted or modulated at audible frequency the resulting type of transmission is called *interrupted continuous wave* (i.c.w.) or *modulated continuous wave* (m.c.w.), and the signal at the receiver is audible without the use of a beat-frequency oscillator. An m.c.w. or an i.c.w. transmission occupies a greater bandwidth than a c.w. transmission at the same speed.

Fig. 10.4 shows a block diagram of an m.c.w. transmitter. The audio-frequency oscillator provides the modulating signal. For c.w.

transmission the a.f. oscillator is omitted, or switched out of circuit, and the modulator functions as a straightforward amplifier.

Most high-power telegraph transmitters use not on-off keying but some form of *frequency shift keying* (f.s.k.) or *tone shift keying* (t.s.k.). Here the transmission is not interrupted during the space periods but is shifted to another frequency. Thus the transmitter is able to develop full power during the whole transmission.

A good frequency shift keying system must be capable of high-speed operation, have a stable and accurate centre frequency (so that adjacent transmission frequencies may be spaced as closely as possible to enable the maximum number of transmission to be accommodated in a given frequency spectrum), and must not radiate out of the band (so as not to interfere with other transmissions).

Many installations have arrangements for obtaining the required frequency shift by altering the frequency of the controlling oscillator by the necessary amount (a few hundred hertz). With a crystal controlled

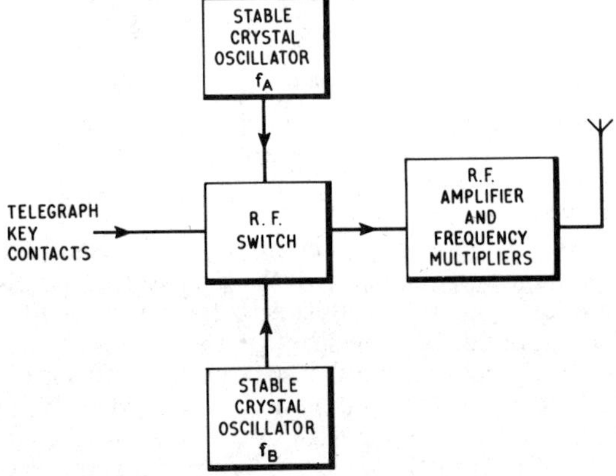

Fig. 10.5 Simple telegraph transmitter using two stable oscillators for frequency shift keying

oscillator the desired effect may be obtained by shunting the crystal by a small capacitor in series with a pair of contacts controlled by the telegraph key. When the key is pressed, the contacts close and bring the capacitor into circuit across the crystal thus slightly shifting the oscillatory frequency. The oscillator is followed by a buffer amplifier (to isolate the oscillatory stage from changes in the load imposed by the following stage), and a limiter to ensure that the output waveform is of constant amplitude. A varying amplitude results in the radiation of out-of-band signals.

Better frequency stability may be obtained if the oscillator frequency is not altered. The required frequency shift may be obtained by arranging that the manipulation of the key shifts the transmission from one stable oscillator to another (Fig. 10.5), so that with the key down the frequency is determined by one oscillator, with the key up the frequency is determined by the other oscillator. With this system it is difficult to avoid either a gap, or excessive overlap, in the transmission between mark and space. Either of these faults causes amplitude modulation and the radiation of frequencies outside the band.

With either of the last two methods the final radiated frequency may be obtained by the use of frequency multipliers following the crystal oscillator(s) and amplifiers. The frequency shift, of course, is multiplied in the same ratio as the frequency and this fact must be allowed for in determining the original shift.

Example 10.1

A 6-MHz crystal oscillator is being used in an f.s.k. transmitter in which the final frequency is to be 24 MHz. If the frequency difference between mark and space is to be 600 Hz, the shift in frequency at the oscillator when keyed must be 150 Hz. This is because the quadrupling circuits used to raise the crystal frequency of 6 MHz to the final radiated frequency of 24 MHz, also raise the shift frequency in the ratio four.

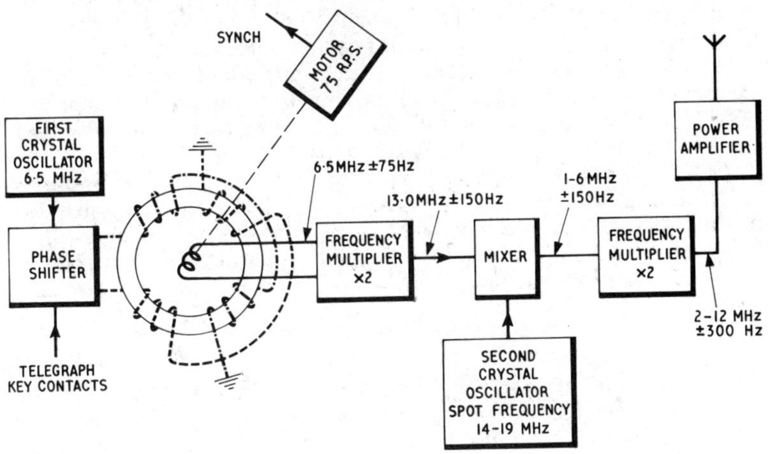

Fig. 10.6 Frequency shift keying using rotating frequency changer

Another method of obtaining the shift without changing the frequency of the controlling oscillator is employed in the G.E.C. transmitter which is shown in simplified block form in Fig. 10.6. This has some advantages in certain circumstances. The stable oscillator, labelled first crystal oscillator, operates at a frequency f_1 which is midway between the required mark and space frequencies. The oscillator output is supplied to two pairs of coils in such a way that a rotating field is created. The rotating field cuts a coil which itself is rotated at a speed f_2 equal to half the required shift frequency. The voltage induced in this coil is either at frequency $f_1 + f_2$ or at frequency $f_1 - f_2$ depending on whether the relative directions of the rotating field and the rotating coil are in the opposite or the same sense. The coil is mechanically rotated at constant speed, and always in the same direction. It is arranged that operation of the telegraph key changes the relative phases of the oscillator inputs (f_1) to the two pairs of coils, and so reverse the direction of the rotation of the rotating field. Thus the frequency of the mechanically driven coil output shifts, with keying, between $f_1 + f_2$ on mark and $f_1 - f_2$ on space. The frequency shift is $2f_2$.

In the block diagram of Fig. 10.6 the crystal oscillator frequency is shown as 6·5 MHz (this is f_1). The rotating coil introduces a change of 75 Hz (this is f_2). The total shift is thus 150 Hz. The motor which drives the coil is a small synchronous motor the drive for which is supplied by a low-power (a few watts) stable oscillator. (Such motors are available for speeds of rotation from about 1 500–15 000 rev/min.). Frequency doubling raises the mid-frequency to 13 MHz and the shift to 300 Hz.

The 13 MHz signal is mixed with the output of a second crystal controlled oscillator which has spot frequencies in the range 14–19 MHz. The difference frequency is selected to put the output centre frequencies in the range 1-6 MHz: the shift is unchanged at 300 Hz. The sum frequency output from the mixer is rejected in the tuned circuits.

Another frequency doubler brings the centre frequency to its final value, in the range 2–12 MHz, and the shift to the desired value of 600 Hz. Power amplification follows before the signal is transferred to the aerial system.

In order to keep out-of-band radiation to a minimum, it is important to synchronise the instant of reversal of the rotating field (by operation of the telegraph key relay) with the rotation of the pick-up coil. Provision is, therefore, made for a suitable synchronising signal.

10.2. PRIVACY EQUIPMENT

For major radio links and for confidential line circuits, privacy equipment is used. The recognised standard or privacy is that of dividing the speech band (250 Hz to 3 kHz) into five bands of 550 Hz each, and rearranging these, with frequency inversion in some cases, before transmission. At the receiver the bands are arranged into their original relationship.

At the transmitter the five bands are separated and each is modulated so that it occupies the range 100–100·55 kHz (Fig. 10.7). A filter excludes signals outside this range. Mixing each with a signal of a suitable frequency, and choosing the upper or lower sideband as appropriate in sharply selective circuits, enables each of the bands to be placed in any position relative to the others, either *erect* or *inverted,* and then recombined to occupy the speech frequency range 250 Hz to 3 kHz. In

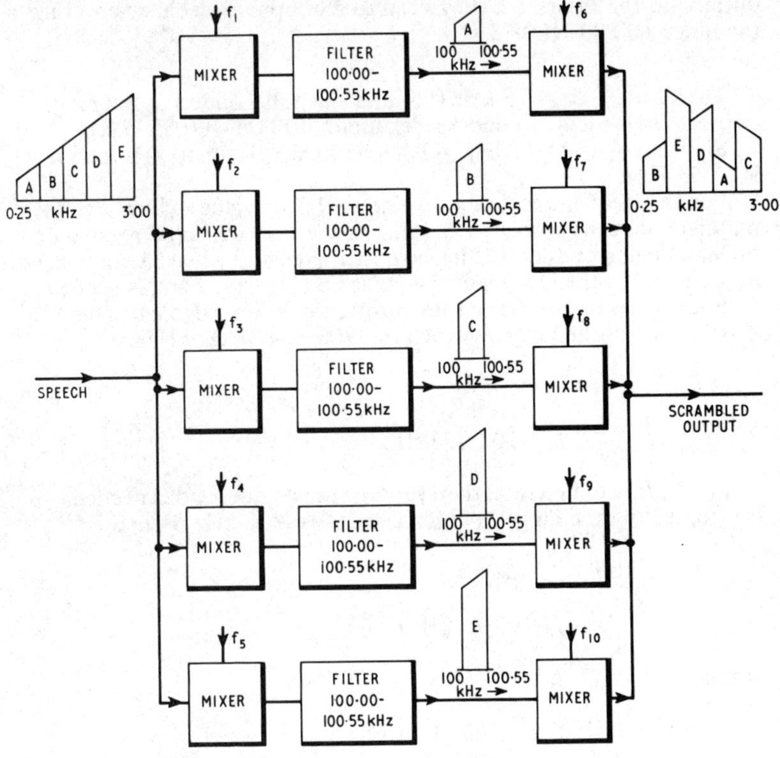

Fig. 10.7 Secrecy equipment

the figure, which oversimplifies the separation and recombination circuits, the band originallydesignated *B* is shown as translated erect to occupy the 250–800 Hz position—originally labelled *A*. Band *E* is inverted and transferred to the band *B* position, *D* is erect and moved to the band *C* position, *A* is inverted and moved to the band *D* position, and *C* is inverted and moved to the band *E* position.

At the receiver similar frequency changing and recombining arrangements reconstitute the signal in its original form. A quite unintelligible signal is obtained if a receiver is used which is not equipped with the proper unscrambling apparatus set up in accordance with the formula employed at the transmitter.

Example 10.2

To scramble the signal with the result shown in Fig. 10.7 the oscillator frequencies $f_1, f_2, \ldots f_5$ need to be such as to give erect outputs to the filters, i.e. they need to give upper sidebands covering the range 100·00–100·55 kHz.

Therefore, f_1 is 99·75 kHz (because this value added to the range 250–800 kHz yields an upper sideband of 100·00–100·55 kHz).

Similarly f_2 is 99·20 kHz, f_3 is 98·65 kHz, f_4 is 98·10 kHz, and f_5 is 97·55 kHz.

For the frequencies f_6 to f_{10} we must choose values which give the required frequency range (according to the position to be occupied by the band being considered) in the upper sideband when an erect band is needed, and in the lower sideband when an inverted band is needed.

For f_6: A is to be inverted, therefore, the lower sideband is needed. A is to occupy the range D which is 1·900–2·450 kHz. Hence

$$f_6 - 100 = 2{\cdot}450$$

$$f_6 = 102{\cdot}45 \text{ kHz}$$

For f_7: B is to be erect, therefore the upper sideband is needed. B is to occupy the range A which is 0·250–0·800 kHz. Hence

$$100 - f_7 = 0{\cdot}250$$

$$f_7 = 99{\cdot}75 \text{ kHz } (= f_1)$$

Similarly

$$f_8 = 103{\cdot}00 \text{ kHz}$$

$$f_9 = 99{\cdot}65 \text{ kHz } (= f_3)$$

$$f_{10} = 101{\cdot}35 \text{ kHz}$$

10.3. LINCOMPEX

Lincompex is the shortened form of *lin*ked *com*pressor and *ex*pandor. Line circuits use compression of speech amplitude ranges at the sending end and proportionate expansion at the receiving end in order to reduce cross-talk and improve signal-to-noise ratio. On radio links the expandor must be linked to the compressor by a method which remains effective despite fluctuation in the transmission path attenuation.

Lincompex provides this method in h.f. radio telephony links. The speech amplitude range is compressed at the transmitter before modulation.

Compensating expansion of the speech amplitude range occurs in the receiver after demodulation. The following advantages result:

1. Because the power in the side frequencies is almost constant, the transmitter operates with an average power approaching its peak possible power and output stages work with their maximum efficiency.
2. The signal-to-noise ratio is improved at the receiver since the expandor is an amplifier which offers minimum gain to low signal levels.

The link between the compressor at the transmitter and the expandor at the receiver is provided by an f.m. signal. The compressor is an amplifier the gain of which is controlled by a d.c. bias derived from its output. An increase in output level reduces the gain. The gain is increased by a fall in output level. The control applied to the compressor is also used to control the frequency of an l.f. oscillator. A deviation from a frequency of approximately 2800 Hz of 2 Hz for each dB change in signal level is caused. The a.f. band used for modulation is cut to 250–2700 Hz so that the speech band and the f.m. link control fall within a 3 kHz channel bandwidth. The entire band of speech plus f.m. control signal is used to amplitude modulate the h.f. carrier of the transmitter.

After demodulation at the receiver a filter separates the speech signal from the narrow-band f.m. link control frequencies. The f.m. signal is applied to a discriminator. The output of this is a direct voltage proportionate to the deviation from 2800 Hz present. An expandor is controlled by this d.c. voltage. In this way the increase in gain of the expandor is made to correspond with the decrease in gain of the compressor at the transmitter, and the range of a.f. amplitude is restored.

As the control signal is f.m. and not a.m., the degree of control is not a function of the received amplitude of signal and is not therefore hindered by changes in ionosphere attenuation so long as the signal path is satisfactory.

10.4. COOLING

Units of radio equipment must not be allowed to become too hot. Transistors are particularly susceptible to the effects of high temperature, and when necessary they are equipped with relatively extensive cooling areas: artificial cooling by air blast is additionally resorted to when necessary.

Valves, particularly the anodes, and sometimes the screens and the control grids, may tend to run very hot, and in high-power equipments require elaborate cooling arrangements. Valve filament heating contributes considerably to the total heat in a valved equipment.

Most components benefit from the availability of good cooling facilities even if these consist of no more than the provision of good ventilation.

10.4.1. Natural cooling

For receivers good ventilation is usually all that is needed (we are not here concerned with such matters as the precise control of temperature for, for instance, stable oscillator circuits).

For transmitters the needs vary with the power. Up to about 500 W, artificial cooling is usually not needed, but an adequate number of louvres and perforations must be included to give good natural ventilation.

10.4.2. Forced air cooling

For power between about half a kilowatt and two or three kilowatts the addition of an extractor fan is necessary to assist the natural convection. In particular, the system must be designed to give a good flow round the valve anodes.

For higher powers, up to some five or ten kilowatts, forced air cooling is used. To improve the effectiveness, valve anodes and other important areas are fitted with fins to increase the area exposed to the cooling air. Forced air cooling is less effective than water or vapour cooling but it costs less and is much simpler to instal and maintain. It has the advantage that the possibility of freezing does not exist and there are no leakage and similar difficulties. With cooling systems which use water, there is some electrical loss in the column of water: there is no similar loss with air-cooled systems. Also, unlike water and vapour cooling, forced air cooling extends to other items of equipment in the cooled cubicle. Unless care is taken there may be excessive noise developed in the air circulating system.

The air is drawn in from the transmitter room, or from outside, through filters into the cubicle which is to be cooled. It is directed so as to flow over the more important surfaces (such as valve anodes). The warmed air is exhausted either to the outside of the building or into the transmitter room to heat it. The cubicles need to be airtight and are equipped with filters at inlet and outlet. Ducts, often of galvanised sheet steel, may be used to lead the air to and from the cubicle. The use of ducts, however, tends to increase the noise level.

10.4.3. Water cooling

Valves dissipating upwards of about five kilowatts may have, in addition to forced air cooling, water cooling. The water is circulated round the anode jackets, the transmitter test load, the cooler and the water storage tank. The valve anodes are at a potential of several kilovolts, the cooler and the storage tank are at earth potential. So that the I^2R losses shall be as small as possible the water must have a high resistivity and be made to follow a lengthy path. Hence pure distilled water is used in an arrangement of plastic or similar pipes of considerable length. For compactness the pipe is formed into a coil. The water must not include solids or anything which might cause deposits in the anode jackets, or elsewhere, and so create hot spots.

Sometimes water cooling in high-power equipments is avoided by using several relatively small valves in some parallel arrangement, and cooling by air blast. The adoption of this scheme tends to lead to high values of input and output capacitance, so that a parallel push-pull circuit is usually better than a single-ended parallel one.

10.4.4. Vapour cooling

Vapour (steam) instead of water may be used for cooling. Water is circulated at such a rate round the anodes that the water is just caused to boil. The anode temperature is thus maintained constant at about 100°C. Considerable heat is extracted from the anodes in changing the state of the water from liquid to steam, without the temperature rising. The steam is taken to a condenser, cooled, and recirculated.

Advantages of vapour cooling compared with water may be:

1. Less water is used.
2. The pipe, therefore, may have a narrower bore so that the column resistance is greater and the losses smaller.
3. Less space is needed for the cooling system.
4. Operation is virtually silent, no large fans or pumps are needed.
5. The flow of water automatically adjusts itself to the value demanded by the anode heat dissipation. The system is thus relatively safe.
6. The cooler is small and needs little attention.

The relative permissible valve anode dissipation values for the three cooling systems forced air, water and vapour are roughly in the ratios 1:1. 5:2.

10.5. POWER SUPPLIES FOR TRANSMITTERS

Power supplies are needed in transmitters for H.T. voltages for valve anodes and screens, for grid bias voltages and for filament heating. Particularly stable supplies are often needed for any transistor circuits. Ancillary apparatus like fans and blowers need mains supplies.

Automatic voltage regulators are often included.

10.5.1. Filament supplies

Filaments are heated at low voltage (between about 6 V and 40 V, increasing with valve size), and with current values ranging between about 1 A and 500 A. Tungsten filaments when cold have a resistance which is only one-fourteenth of the hot value. They are, therefore, switched on gradually using series resistors which are cut out in turn, or by means of a tapped transformer. The transient current is never allowed to exceed about one and a half times the normal working value. The voltage may be stabilised; this is because any material departure from the correct value is bad, in particular quite small overvoltages reduced the working life of the filament considerably.

Heating may be by a.c. or d.c. The use of a.c. is more economical but there is the possibility of modulation of the anode current at the supply frequency unless adequate precautions are taken. For a.c. a step-down transformer is used to obtain the right voltage. A step-down transformer and metal rectifier arrangement is often used to provide d.c., motor generators are also often used. When they are the field rheostat may be employed to give the required gradual increase in the applied voltage as the filament warms up.

D.C. heated filaments give a greater emission from the negative end than from the positive end. Provision is, therefore, included for reversing the supply polarity at intervals.

10.5.2. Grid bias supplies

For low-power stages self bias (a resistor and capacitor in parallel in the cathode lead) may be used. For higher powers the loss incurred by the use of this method is too great and other types of supply are employed.

Cumulative grid biasing is sometimes used, often in conjunction with fixed bias. It is suitable for use only in unmodulated amplifiers because it causes too much distortion in the presence of modulation.

For large installations the power may be derived from motor generator equipment—the ripple must be kept to a low value.

The smoothed output of a rectifier, supplied by the a.c. mains, is very commonly used for all but low-power transmitter stages. A potential divider may be included across the rectifier output to enable the grid bias supplies for several stages to be drawn from one source. Such an arrangement must not be used when one of the stages, for example the final class-C stage, draws a current which fluctuates considerably, otherwise the common impedance of the potential divider couples the fluctuating voltage into the other stages.

10.5.3. Anode supplies

Motor generators have been in wide use for providing high-voltage supplies for high-power transmitters. Static rectifier equipments, however, have a number of advantages and are now generally preferred. They are silent, need little attention and are relatively cheap. They are very efficient, and the fact that there are no rotating parts constitutes a manifest advantage.

For low-power transmitters and for low-power stages of high-power transmitters, selenium, germanium or silicon rectifiers are often used, as are mercury-vapour rectifiers. Semiconductor power rectifiers have a higher efficiency and smaller heat dissipation than thermionic valves or dry-plate rectifiers. They are also very compact (the permissible current density in germanium is some 100 A/cm^2 compared with 0·25 A/cm^2 for selenium).

Mercury-vapour rectifiers are also used for medium powers.

For higher powers older equipments use mercury-arc rectifiers, and some even motor generator sets. Newer equipments use excitrons—these have a mercury pool cathode, a single anode and two grids supported by disc seals. For starting, an auxiliary anode is lifted from the cathode pool in order to draw an arc and thus create a copious emission of electrons.

The rectifiers are used with three-phase supplies, they are forced air cooled and the output is smoothed. The d.c. output voltage may be controlled by altering the firing instant, using for this purpose a negative bias voltage at the control grid. When this bias is removed, and a positive voltage substituted, current flows between cathode and anode.

10.6. HIGH-FREQUENCY TRANSMITTERS

10.6.1. H.F. transmitter with pilot carrier for i.s.b. operation

A typical i.s.b. transmitter with pilot carrier provides two 6-kHz channels, one to either side of the reduced carrier, Each of the two 6-kHz channels (designated *A* and *B* in what follows) is able to accommodate two 3-kHz speech channels, or combinations of speech and telegraph channels totalling not more than 6-kHz bandwidth. The transmitter to be described, one of those at the British Post Office station at Rugby for point to point working, was designed and built before the time when master frequency sources and synthesisers had been brought into wide use in transmitting and receiving stations. There are two main units the *drive*, and the *final mixer-amplifier* unit.

At the input to the drive are the two signals *A* and *B* each of which may have a maximum frequency spread from 100 Hz to 6 kHz. The narrow band between zero and 100 Hz is kept free of signals so that after modulation the pilot carrier will be free from adjacent frequencies which might distort the action of the automatic frequency correction circuits at the receiver. A 50-Hz filter may be included to eliminate mains frequency hum.

The signals in the two channels are amplified in the line amplifiers (Fig. 10.8) and limited in amplitude so that overloading cannot occur in subsequent stages—with consequent reduction in intelligibility.

The two channels are separately translated to the vicinity of 100 kHz

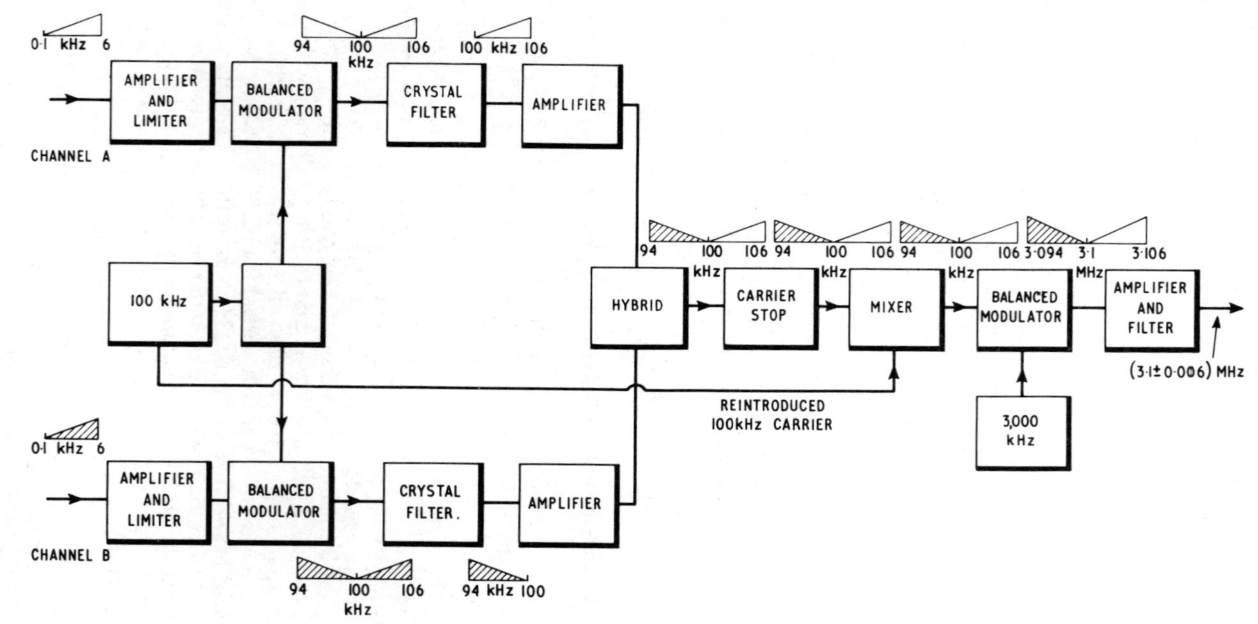

Fig. 10.8 I.S.B. transmitter–drive unit

by mixing with a precisely maintained input of 100 kHz which is accurate to a few parts in 10^8 per month.

Each line, *A* and *B*, now has a 100-kHz carrier and two 6-kHz sidebands, as shown diagrammatically in the figure, at the output of each balanced modulator. Either of these signals (channel *A* or channel *B*) can be passed through the remaining amplifying stages of the transmitter to energise the aerial as a normal double sideband transmission of perhaps 30 kW peak envelope power (p.e.p.).

For i.s.b. operation the 100-kHz carrier and one of the sidebands (the lower sideband for one of the channels, the upper sideband for the other) must be removed from each of the channel signals so that one upper and one lower sideband remain and may be combined to form one signal, centred on 100 kHz and occupying a total bandwidth of 12 kHz. In this example the lower sideband from channel *A* and the upper sideband from channel *B* are each removed in crystal filters (one high-pass and one low-pass) which cut off respectively just below and just above 100 kHz. The two remaining sidebands, the upper from channel *A* and the lower from channel *B*, are combined in a hybrid. After this the 100-kHz carrier, which at this stage is of unknown and uncertain amplitude following the filtering processes, is removed in a very narrow bandstop crystal filter. Because a low level carrier, of accurately known amplitude, is required for ultimate transmission (to operate the receiver automatic frequency and gain controls) a known level of 100-kHz signal is now reintroduced. At this stage we have the low level carrier flanked by the channel *A* information in the upper

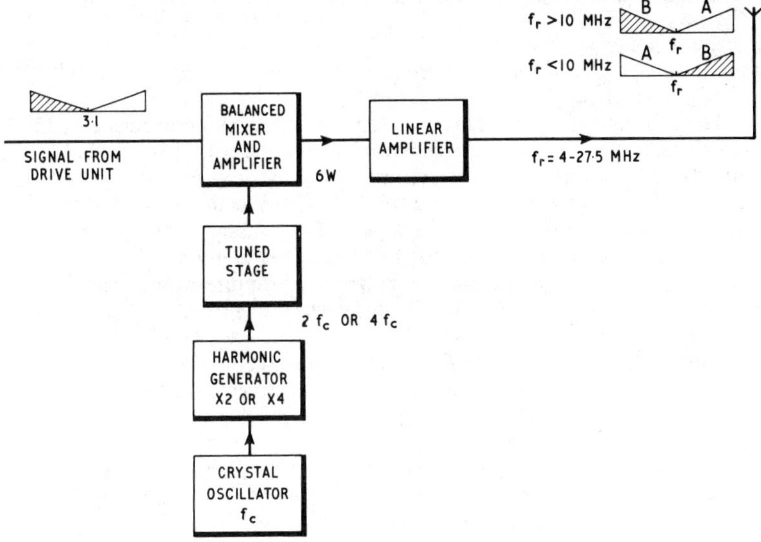

Fig. 10.9 I.S.B. transmitter–output unit

sideband (100·1–106 kHz) and by channel *B* information in the lower sideband (94–99·9 kHz).

The composite signal is now taken to a further balanced modulator and mixed with a 3 000 kHz input, derived from the same standard as drives the 100 kHz supply, to give the signal which constitutes the output of the drive unit, with a midband frequency of 3 100 kHz and the signals of channel *A* and channel *B* to either side. The other frequencies are filtered out. The peak sideband level is maintained at an appropriate value, perhaps 0·25 W.

Following the drive unit the signal is translated, in a balanced mixer, to the final radiated frequency and amplified to the required level—30 kW p.e.p. in this instance. This is shown in the block diagram of Fig. 10.9.

The oscillator input to the balanced mixer is derived from a crystal oscillator and harmonic generator which multiplies by either two or four. Selection of the appropriate crystal frequency and multiplying factor enables a number of spot frequencies in the h.f. band to be made available. The entire frequency changing operation requires little more than one minute. Because of the frequency range involved, a separate mixer unit is employed for each of the spot frequencies provided, perhaps half a dozen. The signal from the drive unit is fed to each, the desired mixer unit being selected by applying the H.T. supply and switching in the output circuit.

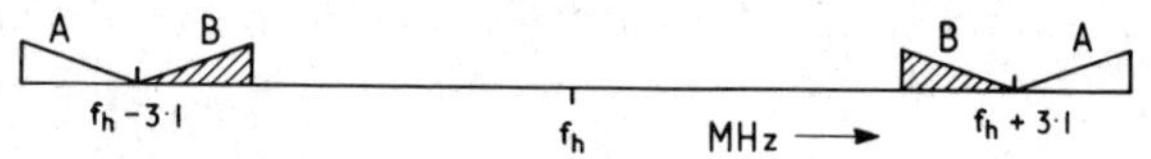

Fig. 10.10 Mixer output signals

Many unwanted frequencies are generated in mixing processes and these signals must be eliminated. Those which are not close to the wanted radiated frequency are removed or greatly attenuated in the tuned circuits between mixer and output. Those which are close to the radiated frequency are not so removed but many of them are severely attenuated in the balanced mixer. Others, deriving from unwanted multiples of the crystal frequency, are minimised considerably by the tuned stages between the harmonic generator and the mixer-amplifier.

At the balanced mixer are frequencies $f_h = 2f_c$ or $4f_c$ from the harmonic generator together with the composite signal centred on 3·1 MHz from the drive unit. Before filtering, therefore, the frequencies in the mixer are as shown in Fig. 10.10.

The mixer output and subsequent stages are tuned to the band centred on $f_h - 3{\cdot}1$ when the required frequency f_r is below 10 MHz, and to $f_h + 3{\cdot}1$ when the required frequency is above 10 MHz. Thus below 10 MHz the information contained in channel *A* constitutes the lower sideband and that of channel *B* the upper. For frequencies above

10 MHz channel A is the upper sideband and channel B is the lower. The arrangements at the distant receiver must take account of this. The signal at frequency f_h is removed in the tuned circuits along with the rejected sideband.

Example 10.3

The available crystal frequencies lie between 3·4 MHz and 7·0 MHz. Thus for a radiated carrier of:

27·5 MHz $f_h = (27{\cdot}5 - 3{\cdot}1)$ MHz = 24·4 MHz

$f_c = 24{\cdot}4/4$ MHz = 6·1 MHz

12·1 MHz $f_h = (12{\cdot}1 - 3{\cdot}1)$ MHz = 9·0 MHz

$f_c = 9{\cdot}0/2$ MHz = 4·5 MHz

8·7 MHz $f_h = (8{\cdot}7 + 3{\cdot}1)$ MHz = 11·8 MHz

$f_c = 11{\cdot}8/2$ MHz = 5·9 MHz

4·5 MHz $f_h = (4{\cdot}5 + 3{\cdot}1)$ MHz = 7·8 MHz

$f_c = 7{\cdot}8/2$ MHz = 3·9 MHz

The selected sideband, containing channels A and B and the pilot carrier f_r, is passed to a linear amplifier to raise the power to the level required for radiation. In the particular transmitter being described there are five stages tuned, by remote control, by continuously variable inductors (of copper tubing) and capacitors. All the valves, and the components in the cubicles, are air cooled.

The last four valve stages are single ended: this reduces the possibility of parasitic oscillation. All the valves are tetrodes except in the two final stages each of which has a single triode in the earthed grid configuration (thus avoiding the need to neutralise and reducing the necessary number of circuit adjustments).

Negative feedback is applied over the last four stages from the final anode and reduces third-order intermodulation products to better than –36 dB with respect to either tone at all power levels.

The cathodes of the last three stages are heated by d.c. power. This helps to maintain a good transmitted signal-to-noise ratio (better than 60 dB),

The high-voltage supply is at 8 kV from mercury arc rectifiers.

Frequency can be changed quickly by press button remote control to any of the six spot values. Operation of the control switches off the high-voltage supply and starts the frequency change mechanism which is controlled by cam-operated switches fitted to the tuning motors. (A rather complicated process compared with that for the more modern equipment to be described in the next section). The high-voltage supply is restored when the operation is completed.

10.6.2. H.F. transmitter using wideband amplifier

Quick frequency changing, and the possibility of an increase in the number of frequencies available, is facilitated by a reduction in the number of adjustments necessary in the frequency changing process. Such reduction not only saves time but may reduce the required technical capability in the operating personnel. The use of wideband amplifiers constitutes a considerable contribution to the achievement of both speed and simplicity of operation.

Fig. 10.11 shows the block diagram of a transmitter for the h.f. range (about 4–27·5 MHz). The main features are:

1. Any frequency, in steps of 0·1 kHz, may be selected by dialling.
2. Only one stage, the final one, is tuned. All the others are wideband so that the mechanical complexities of remote frequency control over many tuned circuits are mostly eliminated, and the problems of frequency changing much reduced.
3. The fundamental frequency control is by master oscillator with fine and coarse frequency controls to compensate for drift caused by crystal ageing. This drift is less than 5 parts in 10^8 per month. The basic output frequency is 1 MHz.
4. The decade selected final frequency is derived from a frequency synthesiser.

Example 10.4

To tune the transmitter to a frequency of 12 376·4 kHz the operator, faced by five dials, sets the left-hand one at 12 (thus obtaining the twelfth harmonic of 1 MHz), the next at 3 (thus obtaining 3 × 100 kHz), the next at 7 (thus obtaining 7 × 10 kHz), the next at 6 (thus obtaining 6 × 1 kHz) and the right-hand one at 4 (thus obtaining 4 × 0·1 kHz).

From left to right the dials read the desired frequency; 12 376·4.

Modulation, to the full 12 kHz total sideband figure, can readily be applied to the synthesiser output and there is a very low level of spurious components—well below the figure permitted by international regulations.

The information signal is developed initially on a carrier of probably 100 kHz and raised to the frequency level of 3·1 MHz (as for the transmitter of the last section).

To bring the final radiated frequency into the required range of 4–27·5 MHz an additional, accurate, frequency determining signal is

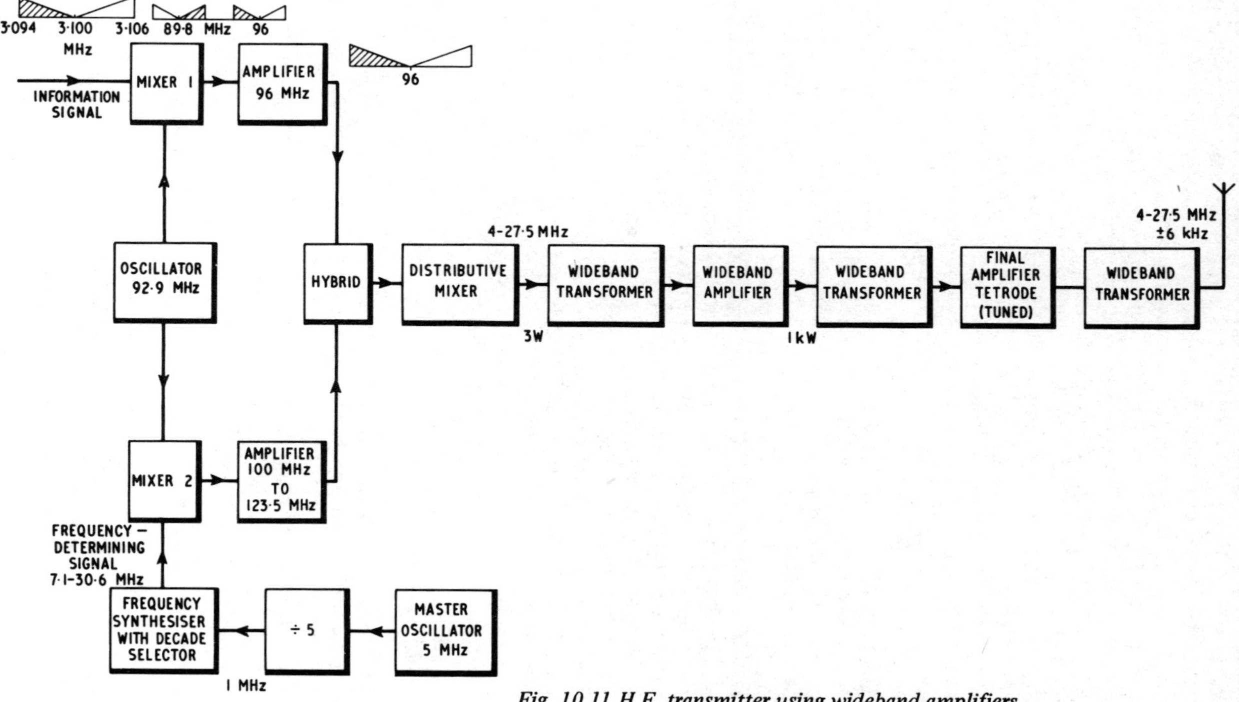

Fig. 10.11 H.F. transmitter using wideband amplifiers

required covering the range (4 + 3·1) MHz to (27·5 + 3·1) MHz; i.e. covering the range 7·1 – 30·6 MHz. This is the frequency range which is obtained from the synthesiser by setting the five dials. It is a simple matter to arrange that when the selectors are set to any desired final radiated frequency, the synthesiser frequency is in fact 3·1 MHz above this value so that the desired result shall be obtained.

The need for an extensive range of tuning adjustments at each frequency change is avoided by the use of the 92·9-MHz oscillator shown in the figure. The output of this oscillator is mixed (a) in Mixer 1 with the 3·1-MHz carrier and sidebands to give a signal based on 96 MHz, the signal based on 89·8 MHz being rejected in the filter, and it is mixed (b) in Mixer 2 with the frequency determining signal to yield an output in the range 100·0–123·5 MHz. The difference frequency range, 85·8–62·3 MHz, is rejected in the filter.

The 92·9-MHz oscillator, being applied to both the 3·1-MHz and the frequency determining signals, need not be particularly accurate or stable—but it must be good enough to enable the amplifier and rejection circuits to accept and reject the appropriate frequencies as described above.

Both the mixer for (a) above and the 96-MHz amplifier are narrow band, but they must have good gain, stability and linearity. The mixer for (b) must be of very good linearity otherwise a harmonic of the frequency determining signal, beating with the 92·9-MHz oscillator signal, appears in the output within the passband of the 100·0–123·5-MHz amplifier and remains as a spurious signal.

The two chosen signals, (a) centred on 96 MHz, and (b) a single frequency in the range 100·0–123·5 MHz, are combined in a hybrid and mixed in the distributed mixer. The output from the mixer is centred on the selected frequency in the range 4·0–27·5 MHz.

The power at this point is about 3 W.

Example 10.5

The wideband unit is to provide an output at 24 MHz to drive a high-power automatically tuned transmitter which is to radiate at the same frequency.

The frequency determining signal must be at a frequency of

$$(24{\cdot}0 + 3{\cdot}1)\ \text{MHz} = 27{\cdot}1\ \text{MHz}$$

The output from Mixer 1, as always, is at 96 MHz; that of Mixer 2 is at

$$(92{\cdot}9 + 27{\cdot}1)\ \text{MHz} = 120{\cdot}0\ \text{MHz}$$

From the distributed mixer the output is at

$$(120 - 96)\ \text{MHz} = 24\ \text{MHz, as required.}$$

The output of a wideband mixer, such as described, may be used to drive a wideband transmitter, which needs no tuning, of one or two kilowatts, or may be used as the drive for a higher power transmitter of perhaps several tens of kilowatts. Such a high-power transmitter requires to be tuned, manually or automatically, probably only in the last stage. Wideband transmitters of higher power than one or two kilowatts can be made but the arrangement is not usually economical.

In our example, which is a transmitter by the Marconi Co., the wideband mixer output is taken via a wideband transformer to a wideband amplifier consisting of five tetrodes with distributed load and thence, the power level now being about 1 kW, via another wideband transformer to the final amplifier. This uses a single-ended tetrode, but equally an earthed grid triode could be employed. Amplification in this final stage raises the power to the range 10–30 kW p.e.p. as required for the particular service.

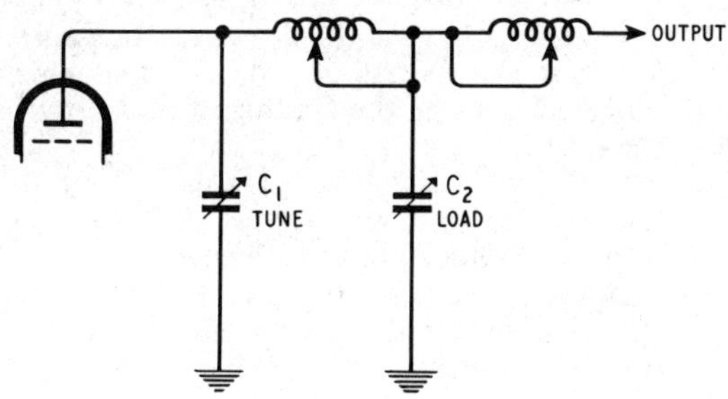

Fig. 10.12 Tuning and loading circuit

The final amplifier is the only one which is tuned and the number of adjustments is thus relatively small. The tuning may be done manually or automatically. For the latter, the controls may be pre-set and driven to their proper positions by control motors, or, and this is what is done in the transmitter being described, automatic following of the final stage input frequency, as set up on the synthesiser, may be arranged. The frequency following method is more flexible and rapid than the pre-set arrangement.

The adjustments are in two stages:

1. Coarse tuning, in which the correct range is selected and the approximate setting of the tuning and loading capacitors determined.
2. Fine tuning, in which the exact setting of the tuning and loading capacitors is determined.

Fig. 10.12 shows schematically the output arrangement. C_1 is the tuning capacitor, C_2 determines the loading. Both must be correctly adjusted, and an advantage of this automatic following arrangement is that any changes in feeder impedance which take place, and these are particularly prevalent in bad weather, meet with an immediate automatic response in the setting of the loading capacitor. With pre-set circuits, of course, no such correction takes place.

For accuracy of fine tuning the determination is not by the common method of tuning for minimum anode current—the indication is not sufficiently sharp. Instead a discriminator is used to detect when the r.f. anode voltage is exactly in antiphase to the grid voltage—this is the point of tune. At other settings the voltages are not in antiphase, an error voltage is developed in the discriminator and an amplified version of this voltage is used to drive a motor which corrects the tuning by driving C_1 in the right direction until no error remains.

The correct loading is indicated, depending on circumstances, either by the ratio of anode to grid r.f. voltage (this is an accurate indication when there is no grid current), or by the ratio of anode r.f. voltage to direct cathode current (this is an accurate indication when the loading is high). Thus by comparing the anode r.f. voltage with the grid r.f. voltage when the loading is small, and with the direct cathode current when the loading is greater, an error signal may be obtained which automatically causes C_2 to be adjusted to bring the loading to the correct value for the prevailing circumstances.

The output may be (unbalanced) to 50-Ω coaxial cable for aerial selection and then matched to a, say, balanced 600-Ω aerial input impedance by means of a wideband transformer.

Cooling is by forced air. Sensing devices are placed in the flow path so as to remove the H.T. supply if the flow falls below the safe value.

10.6.3. High-frequency broadcast transmitter

Unlike high-frequency transmitters for point to point working, broadcast transmitters cannot rely on sophisticated equipment at the receiving end and must, therefore, employ higher power. A full power carrier is transmitted. The Marconi transmitter, which will now be described, is one of those at the B.B.C. station at Ludlow and it develops about 250 kW in the carrier. It operates on any frequency between 4 and 18 MHz.

The driver unit employs wideband untuned amplifiers preceding a single-ended tetrode which is manually tuned (Fig. 10.13). This part of the equipment is like the corresponding unit described in Section 10.6.2. The tetrode output of 7·5 kW is insufficient to drive the final r.f. amplifier, so an intermediate stage, the penultimate r.f. amplifier, is interposed. This stage uses two vapour-cooled, earthed grid triodes in push-pull, and is matched to the tetrode output by a ferrite-cored wideband transformer.

Coupling from the penultimate to the final r.f. stage is by a π-network which uses different inductors to cover the tuning range and has vacuum capacitors for tuning.

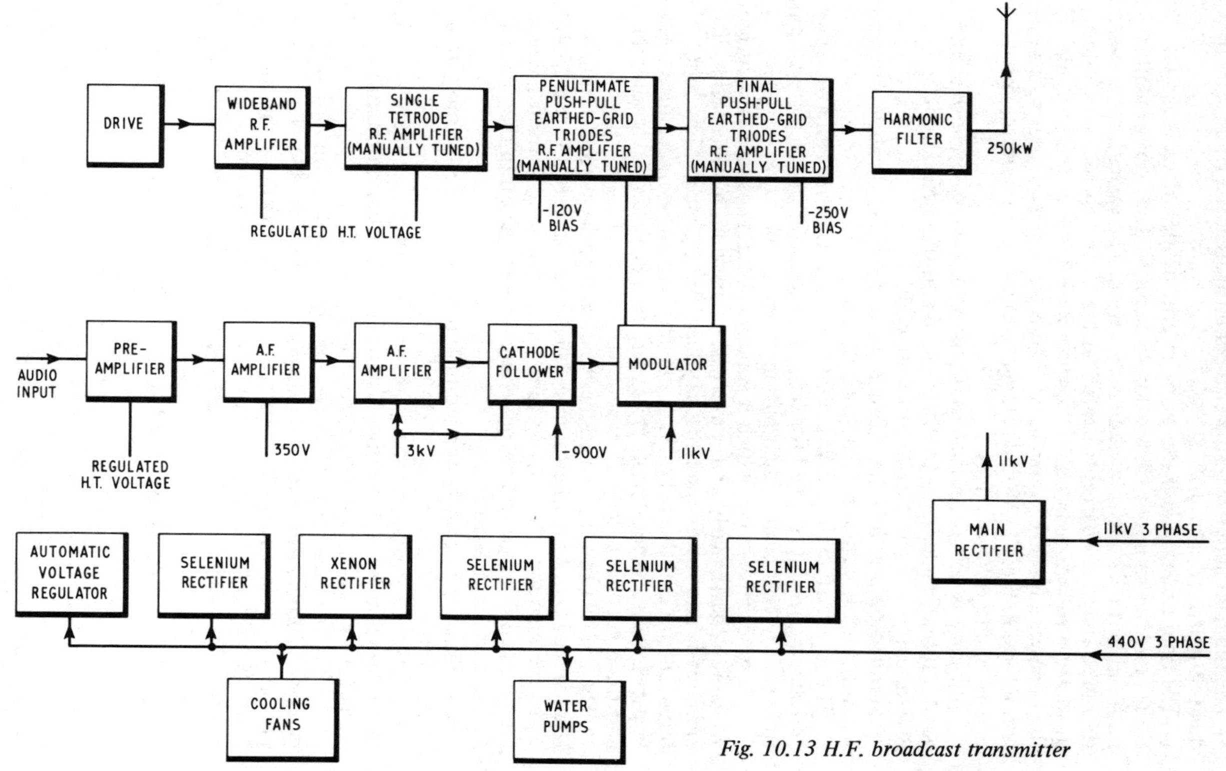

Fig. 10.13 H.F. broadcast transmitter

The filament supplies to the two penultimate stage triodes are in quadrature to reduce noise.

The final r.f. stage also uses earthed grid triodes. Each is vapour cooled and has an anode dissipation of 125 kW. The arrangement is push-pull with class-B amplitude modulation. The valves have two-section thoriated tungsten filaments heated with a.c., and these also operate in quadrature to reduce noise. The output impedance is 330 Ω to match the open-wire feeders to the aerial. There is a low-pass filter to attenuate harmonics, in the 40–240 MHz range, which could interfere with television and other services.

The stage is not neutralised as this is not necessary because of the good internal shielding of the valves used and the inclusion of low-inductance capacitors to earth the grids.

Modulation is at high level and is suitable for high-quality broadcast work or for trapezium modulation.

The input, at about 1 mW in 600-Ω lines, is taken via an attenuator and shaping circuit (needed for trapezium modulation) to a pre-amplifier followed by two push-pull amplifying stages.

The modulator valve, which is vapour cooled, runs into grid current, which is of particularly high value when trapezium modulation is employed. To provide the necessary high input power at low input impedance, a cathode follower stage, employing low impedance triodes in parallel (also vapour cooled) is interposed between the modulator and the preceding push-pull stage.

The modulator uses two triodes in class-B push-pull and supplies 180-kW audio output when sinusoidally modulated 100%, and 275 kW when modulated 95% by trapezium signals. The modulator filaments are supplied in phase so that noise components cancel in the primary of the modulation transformer.

The modulator output is applied in full to the final r.f. amplifier and in part (50%) to the penultimate r.f. amplifier.

Negative feedback—normally at some 16 dB below the output level of the modulation transformer—is applied from the modulation transformer to the first audio-frequency amplifier.

As has been mentioned, all the high-power valves (modulator, cathode follower and final amplifier) are vapour cooled. There is protection again excessive temperature rise provided by thermal fuses, fitted to the anode structure, which are linked to a spring-loaded tripswitch in the d.s. high-voltage supply. The cubicles are air cooled.

10.7. LOW- AND MEDIUM-FREQUENCY TRANSMITTERS

Transmitters for l.f. and m.f. use are, broadly speaking, of higher power than those for h.f. use because of the inefficiency and lack of directivity of l.f. and m.f. aerial systems.

In order to make the best use of the limited frequency spectrum, frequency control must be accurate and stable.

The ability to change frequency quickly is seldom needed.

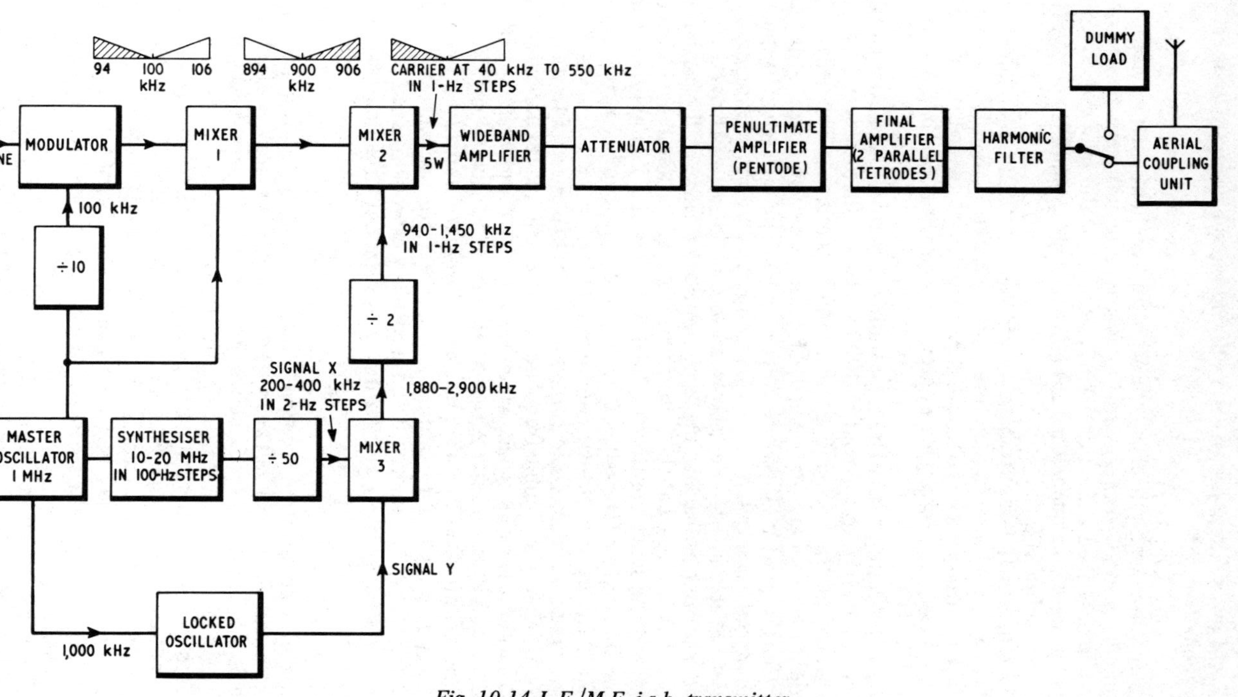

Fig. 10.14 L.F./M.F. i.s.b. transmitter

10.7.1. Low/medium-frequency i.s.b. transmitter using synthesiser frequency control

An interesting transmitter, by the Marconi Company, derives its frequency control from a master oscillator and synthesiser system of the type described in Section 10.6.2. Because of the smaller frequency range, however, frequency selection needs to be available in steps of 1 Hz instead of 100 Hz.

The information input, up to a total bandwidth of 12 kHz, modulates a 100-kHz signal derived from the 1-MHz master oscillator. The result, after the usual processes, is a 100-kHz carrier with two sidebands, each up to 6-kHz wide, conveying whatever information is to be transmitted (Fig. 10.14).

In Mixer 1 this is translated, by subtraction from the master oscillator output of 1 MHz, to a carrier of 900 kHz (with sidebands reversed because of subtraction) and then passed on to Mixer 2. The other modulation products are rejected. The other input to Mixer 2 is variable between 940 and 1 450 kHz in steps of 1 Hz. The difference output from Mixer 2 is selected in the tuned circuits, and it is centred on a carrier variable in 1-Hz steps between 40 and 550 kHz. The two sidebands are now back in the same positions relative to the carrier which they occupied at the outset. After amplification in a wideband amplifier, the signal constitutes the drive to the main amplifying stages to the transmitter. These could be designed to operate over any desired range of frequencies between 40 and 550 kHz.

Referring back to Mixer 2, the 940–1 450-kHz input is formed in two parts (originally at double frequency, 1 880–2 900 kHz) in Mixer 3. The first part consists of a signal, which we will call *X* and which is variable in steps of 2 kHz between 200 and 400 kHz. The second part, which we will call *Y*, provides spot frequencies at 200-kHz intervals between 1 600 and 2 600 kHz, By adding appropriate frequency values from the two parts *X* and *Y*, any required frequencies can be obtained in the range 1 880–2 900 kHz at 2-Hz intervals. Division by 2 yields the required Mixer 2 output.

Signal *X* comes from the synthesiser. This can be set, using decade dials, to any frequency between 10 and 20 MHz in steps of 100 Hz. The output is divided by 50 so that the frequency range is 200 to 400 kHz variable in steps of 2 Hz: this is signal *X*.

Signal *Y* is obtained from crystal oscillators which provide at will any frequency between 1 600 and 2 600 kHz in steps of 200 kHz. These signals are compared, for frequency accuracy, with signals derived from the master oscillator, and any necessary (small) corrections are made automatically by means of error signals generated in a comparator.

The output from the drive unit (that is the output from Mixer 2), variable in 1-Hz steps between 40 and 550 kHz, is at about 5 W. It can be used to drive an amplifier covering any desired range within this band. The particular transmitter being discussed covers the frequency range 40 to 160 kHz. It employs a linear amplifier delivering 100 kW p.e.p.

There are two stages of amplification, each operating without grid current, with resulting improvement in linearity. The first stage consists of a high-gain pentode fed from the drive unit through a 75-Ω coaxial, adjustable attenuator and wideband transformer. The anode circuit consists of a π-network, similar to that of Fig. 10.12, with a variable capacitor for tuning and a number of fixed capacitors for loading adjustments. The latter capacitors are monolithic, high temperature. and high current types of ceramic. To improve the already good linearity, 3 dB of r.f. negative feedback is included using a capacitance potentiometer across anode and earth to derive the feedback voltage. The pentode is a 1 500-W type with air cooling. The usual safeguards are included to prevent damage to the equipment should the air pressure fail.

The load resistor of the pentode is paralleled by a wideband ferrite transformer with a tapped secondary which constitutes the input to the final stage. This uses two tetrodes in parallel, and the cathodes are earthed. The secondary of the ferrite transformer is tapped so that the grid inputs to the two tetrodes may be separately adjusted to give the same cathode current in each. The anode circuit is a π-network with the main inductor of 1-in. diameter copper tube—not the usual litz wire conductors. Apart from the slight increase in loss, which widens the bandwidth, the results of the change from litz wire to tube are greater rigidity (with reduction of unwanted phase modulation), reduction in cost, and a simplification in construction with more easily provided tapping points. The tuning capacitance consists of fixed oil immersed mica capacitors, linked as required, in parallel with a motor-driven, nitrogen-filled, variable capacitor. 12 dB of negative feedback is applied, again using a capacitance potentiometer as the means of obtaining the feedback voltage. The large input capacitance of the two high-power tetrodes upsets the phase of the feedback loop unless corrected. In this amplifier the capacitance is tuned out with a shunt r.f. choke between grid and cathode. Air cooling is used in this stage also.

Normal tune indications cannot be used when the feedback is on. For both the amplifier stages, therefore, phase discriminators are used to give a precise indication of the correct tune position. This is additionally helpful because the circuits tend to have flat tuning responses, particularly at the lower frequencies.

In this frequency range a wide variety of aerial types exist. Provision is made, therefore, to match the amplifier output to any aerial system. The coupling unit, which contains inductors and capacitors which may be connected into any required configuration, is included for this purpose.

The amplifier output can be switched to the dummy load so that the transmitter may be tested on full load without the presence of an aerial.

To set or to change the frequency of transmission it is necessary to adjust the synthesiser decade dials and to tune the two final stages.

Frequency changing is seldom necessary for transmitters in this frequency range; but it may be done, however, in about half an hour.

10.7.2. Medium-frequency broadcast transmitter

The transmitter described in this section is by the Marconi Co. and it operates in the frequency range 525–1 605 kHz with the final power output of 50 kW. The frequency stability is ±1 part in 10^6 per day over a temperature range of ±15°C.

In the drive unit (Fig. 10.15) the frequency control is by crystal. Two crystals are used, one working, one standby. They are mounted in a thermostatically controlled oven. Each crystal has a pre-set capacitor in series to enable it to be pulled in to the exact frequency. A tri-tet circuit is used and this is followed by first an amplifier and then a frequency multiplier to raise the frequency to the desired radiation frequency. A power amplifier follows and provides 5 W to drive the main amplifier stages.

The first main r.f. amplifier (Fig. 10.16) uses two tetrodes in parallel, and it is coupled to the second stage by means of a capacitance potentiometer. This stage uses a single triode, operates in class C and is fully anode modulated. It is coupled to the final stage by a π-network. The final r.f. stage also uses a single triode. It is coupled to the output feeder by variable inductive coupling which provides a simple means of adjusting the power level.

In the modulation circuits the a.f. signals from the line are amplified in a three-stage pre-amplifier to bring the level up to 5 W. The remaining amplifiers in the modulation stages are push-pull. In the first stage two tetrodes are used in class A. The second stage uses a similar circuit and valves, but has two tetrodes in parallel in each half of the push-pull circuit. The final stage is a class-B modulator and needs a low impedance drive. This is provided by the cathode follower shown which has three paralleled tetrodes in each half of its push-pull circuit. The class-B modulator itself employs two triodes in push-pull, and it modulates fully the second and final r.f. amplifiers. The use of negative feedback to the first a.f. amplifier, together with a high-performance modulation transformer, results in high-quality performance.

Two similar transmitters may readily be operated in parallel in order to double the output power.

The power supplies are derived from three-phase mains. The main h.t. (at 8.5 kV) is obtained from a three-phase full-wave rectifier system using mercury pool excitrons. The other power supplies use selenium rectifiers. Filament heating is by alternating current. There is automatic voltage regulation for the filament and other supplies which need stability within ±1%.

Cooling is by forced air, this drawn in by fan at the top of the enclosure through air filters.

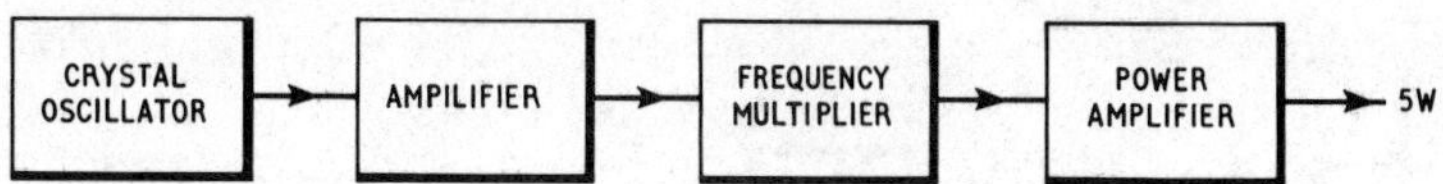

Fig. 10.15 M.F. broadcast transmitter–drive unit

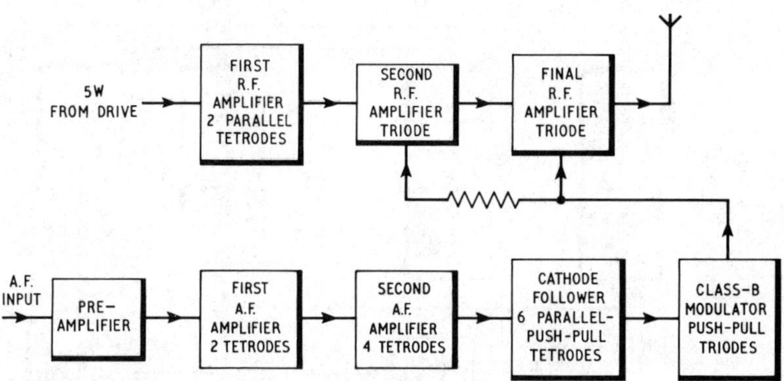

Fig. 10.16 M.F. broadcast transmitter–final stages and modulator

10.8. RECEIVERS

The basic principles of radio receivers have been covered in earlier work. The following is a brief description of an h.f. receiver for i.s.b. reception.

10.8.1. High-frequency receiver for i.s.b. reception with dual diversity

A block diagram, based on an interesting receiver by the Marconi Company, is shown in Fig. 10.17. The receiver is fully transistorised and provides for the reception of i.s.b. signals with pilot carrier in the h.f. band. The signal amplification and mixing circuits are duplicated to provide dual diversity reception.

The frequency range of 1·5 to 29·5 MHz is covered in four switched ranges. Two sidebands, each totalling 6 kHz, can be accepted.

Overall control of the frequency of internally generated signals is by a high stability master oscillator with main outputs at 1 MHz and 0·1 MHz. There is a synthesiser unit. A free-running oscillator is used to provide frequencies between the 0·1-MHz steps. Its accuracy may be checked at 10-kHz intervals by signals derived from the master oscillator.

The accuracy of frequency setting is within 200 Hz throughout the receiver tuning range.

For each of the dual diversity receivers the aerial signal is taken to

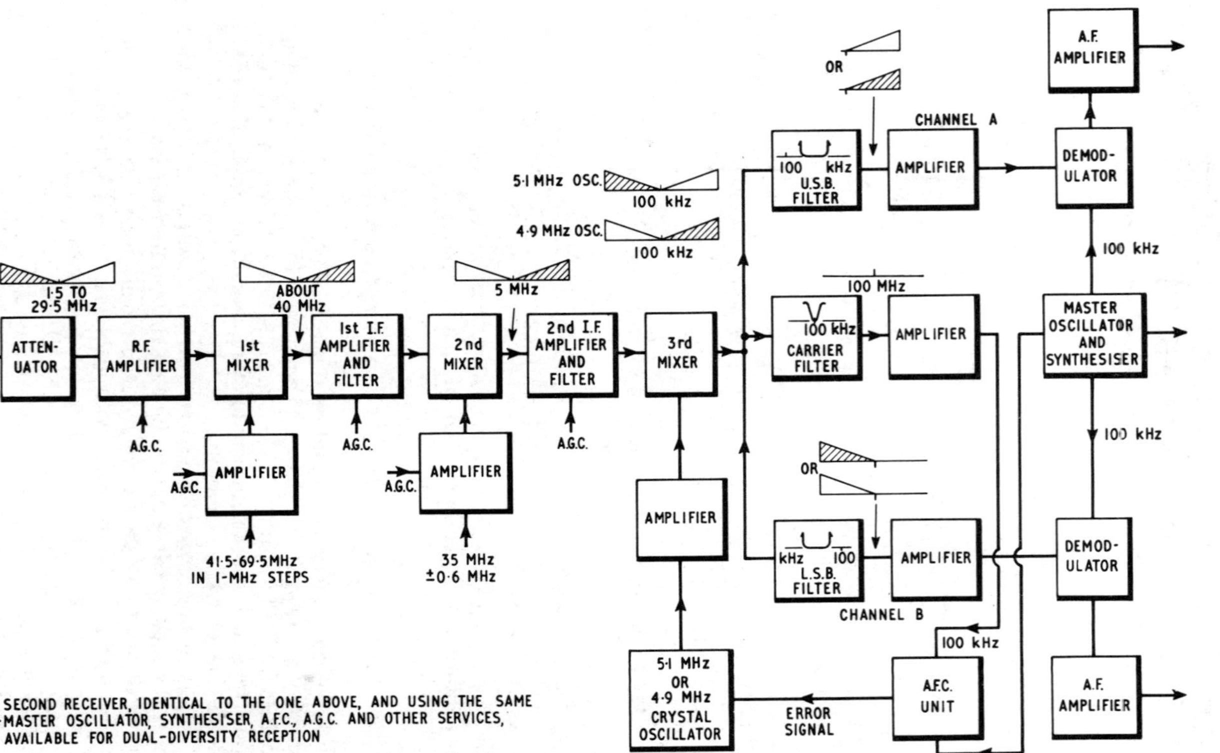

Fig. 10.17 H.F. i.s.b. receiver

an attenuator, which is variable in 10-dB steps to a maximum of 30 dB, so that unduly strong signals may be reduced to such a level as not to cause overloading. The attenuator leads to the s.f. amplifier via a bandpass coupled circuit which attenuates by over 30 dB all signals which are 10% or more off tune.

The s.f. amplifier is a cascode. The noise figure lies between 4 and 7 dB. This amplifier, and all the i.f. and mixer stages up to the third mixer, uses field effect transistors with their attendant advantages (Section 3.1.1).

The first intermediate frequency is high, 40 MHz, and this puts the image frequency at such a high value as to render it easy to eliminate.

The first local oscillator thus covers the range 41·5–69·5 MHz. It is variable in 1-MHz steps, a feature which requires that the bandwidth of the first i.f. amplifier is greater than 1-MHz.

After amplification at 40 MHz, the signal is converted to the second i.f. of 5 MHz. The second oscillator thus needs to be variable between 35·5 and 36·5 MHz. In the second i.f. amplifier the passband is limited, in a crystal filter circuit, to the value needed to pass the information signals—12 kHz.

To provide the final i.f. of 100 kHz the third, crystal-controlled, local oscillator operates at 5·1 MHz. This may be switched at will to 4·9 MHz so that the placings of the two sidebands with respect to the carrier may be exchanged.

The signal, now based on 100 kHz, fans out to three filters. One accepts the bands of frequency (3 kHz or 6 kHz wide at choice) immediately above 100 kHz: this is channel *A*. The second similarly accepts the bands immediately below 100 kHz: this is channel *B*. The third accepts the 100 kHz carrier alone: this is the pilot carrier. These three signals are individually amplified.

The signals in channels *A* and *B*, now at the level of 0·5 V, are separately taken to balanced demodulator circuits for mixing with 100-kHz signals from the master oscillator, to yield a.f. outputs which are then amplified as necessary. The master oscillator signals are injected at the level of 5 V. The high ratio of carrier voltage to signal voltage (10: 1) is used in order to ensure very small distortion in the detection process.

For automatic frequency control, one of the 100-kHz outputs from the master oscillator is used as a reference signal and compared with the 100-kHz pilot carrier from the narrow bandpass filter. The resulting error signal is fed to a voltage-controlled reactance circuit across the crystal-controlled third local oscillator (on 5·1 or 4·9 MHz) and reduces any mistune to a value which is never greater than 2 Hz.

The automatic gain control may be fed from the sideband amplifiers, or from the carrier amplifier. One of three time constants may be selected (0·1 s, 5 s, or 40 s). There are two a.g.c. lines, one for the f.e.t. amplifier stages, the other to control the level of voltage supplied by the local oscillators to the first two mixers. The system can reduce a 90 dB change in input signal to an audio output change of 6 dB.

10.9. V.H.F. COMMUNICATION. V.H.F. RECEIVERS

Important features of v.h.f. communication in general and reception in particular are dealt with in Chapters 3 and 9. Some of the salient points can now be reconsidered.

V.h.f. transmissions are valuable for short range communications up to about 50 miles.

Because of the high frequency, fixed stations are able to have, in a small space, aerial systems which give good directivity if required. Mobile stations can have very simple yet reasonably effective aerial systems which may be accommodated easily even on taxis, small boats and motor cycles. Transmissions needing wide bandwidth, like television broadcasting, frequency-modulated sound broadcasting, and multi-channel telephone links can readily be accommodated in the frequency spectrum available.

Frequency modulation or amplitude modulation may be employed. With high deviation values, and correspondingly wide bandwidth, the signal-to-noise ratio and quality at the receiver is much better with f.m. than with a.m. (within the service area). With narrow bandwidth there is little to choose between the two systems on this score.

For low power, amplitude-modulation transmitters are somewhat cheaper than those for frequency modulation. For high power, frequency-modulated transmitters are cheaper. Receivers cost much the same whichever system is used.

All v.h.f. receivers, except the most simple, are necessarily of the superheterodyne type. This is because the possible gain per stage at the signal frequency, with reasonably good quality, is relatively small.

The selectivity of the input tuned circuits must be sufficient to eliminate second channel signals: the action of a double or triple superheterodyne circuit, with high first intermediate frequency, simplifies image channel rejection. If considerable gain is essential, this fact alone may demand the use of a double superheterodyne circuit in order to maintain stability.

10.9.1. V.H.F. receiver for amplitude-modulated signals

Fig. 10.18 shows the block diagram of a straightforward double superheterodyne receiver for the reception of a.m. signals. Valves or transistors may be used throughout.

There are one or two stages of signal frequency amplification in order to bring the signal up to the required level for the first frequency changer, and to provide adequate rejection of unwanted signals at the image and intermediate frequencies.

The gain must not be so great as to give a signal amplitude large enough to cause cross-modulation or blocking. The provision of manual and automatic gain control is a considerable aid to this end.

Separate crystal oscillator circuits, providing as many spot frequencies as are needed, are used as the source of local signals for mixing purposes.

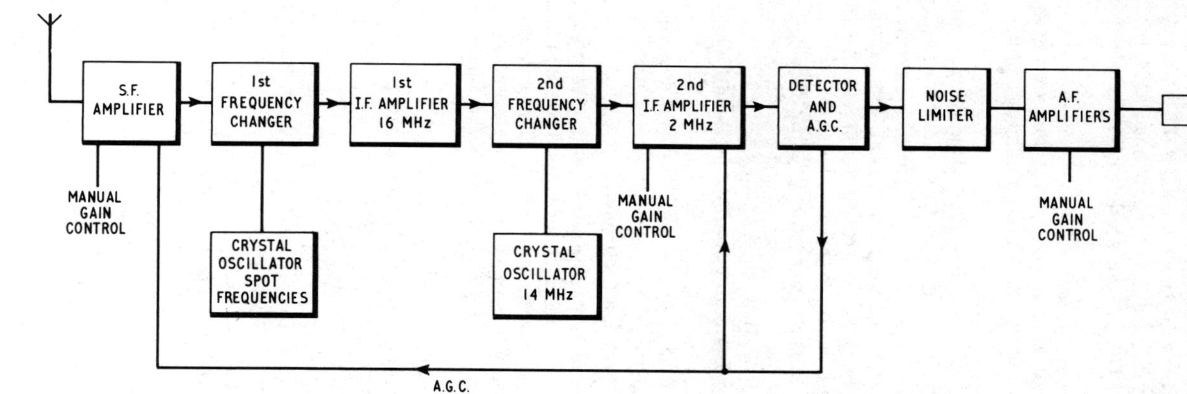

Fig. 10.18 V.H.F. receiver for a.m. signals

Example 10.6

If the first intermediate frequency is 16 MHz and the second is 2 MHz, in order to receive signals on spot frequencies of 97·2 MHz, 98·5 MHz and 99·3 MHz the first crystal oscillator frequencies are 81·2 MHz, 82·5 MHz, and 83·3 MHz. The second oscillator needs one frequency only—14 MHz.

For continuous coverage of frequency, a variable frequency first oscillator is needed. As noted in Chapter 3, oscillator frequency stability is very important.

The first intermediate frequency is high, thus making it easy to remove image channel interference. The second intermediate frequency is low enough to provide adequate adjacent channel rejection, but not so low as to prevent the required bandwidth being obtained. The provision of manual, as well as automatic, gain control in the i.f. section is again a useful means of preventing overloading.

Diode detection, a.g.c., noise limiter and a.f. amplifier circuits follow.

10.9.2. V.H.F. receiver for frequency-modulated signals

Fig. 10.19 shows a block diagram of a single superheterodyne receiver for the reception of f.m. signals. Transistors or valves may be used. The general principles have already been outlined.

For high quality broadcasting reception, the necessary bandwidth approaches one quarter of a megahertz. For simple speech intelligibility, a much narrower bandwidth, a minimum of some 10 kHz, suffices. But the bandwidth must not only accept the signal sidebands but also cater for possible drift in the transmitter frequency and in the receiver oscillator circuit.

A signal frequency amplifier is desirable.

Frequency changing using a variable local oscillator is shown.

There are two or three stages of i.f. amplification—the i.f. is 10·7 MHz.

The limiter ensures that all signals, except the weakest, are presented to the discriminator at the same level. It is very effective in removing impulse-type interference (e.g. that from motor car ignition systems).

Detection takes place in the discriminator circuit, which may be a ratio detector with self limiting for economy, or it may be a Foster-Seeley type with a separate limiter for quality. A good Foster-Seeley arrangement gives less than 1% distortion, the ratio detector about 3%. The detector output is amplified at a.f. in the usual way.

Because the local oscillator is not crystal controlled, the discriminator output is taken from an automatic frequency control circuit for the development of an error voltage should the mid frequency wander. The error voltage is applied to the local oscillator circuit and automatically corrects the frequency in the proper sense to return the mid frequency to its correct value.

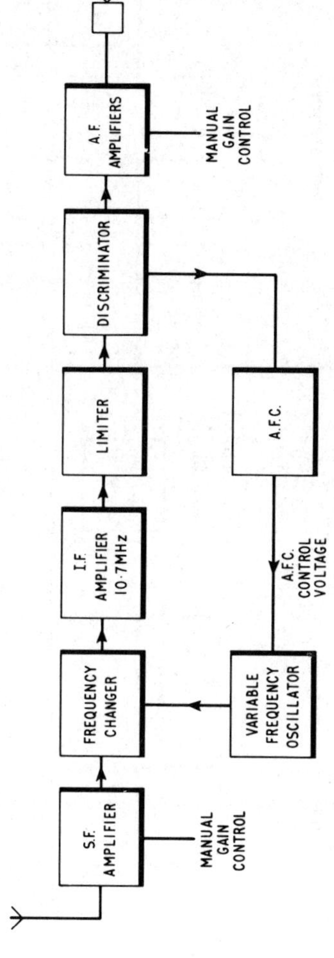

Fig. 10.19 V.H.F. receiver for f.m. signals

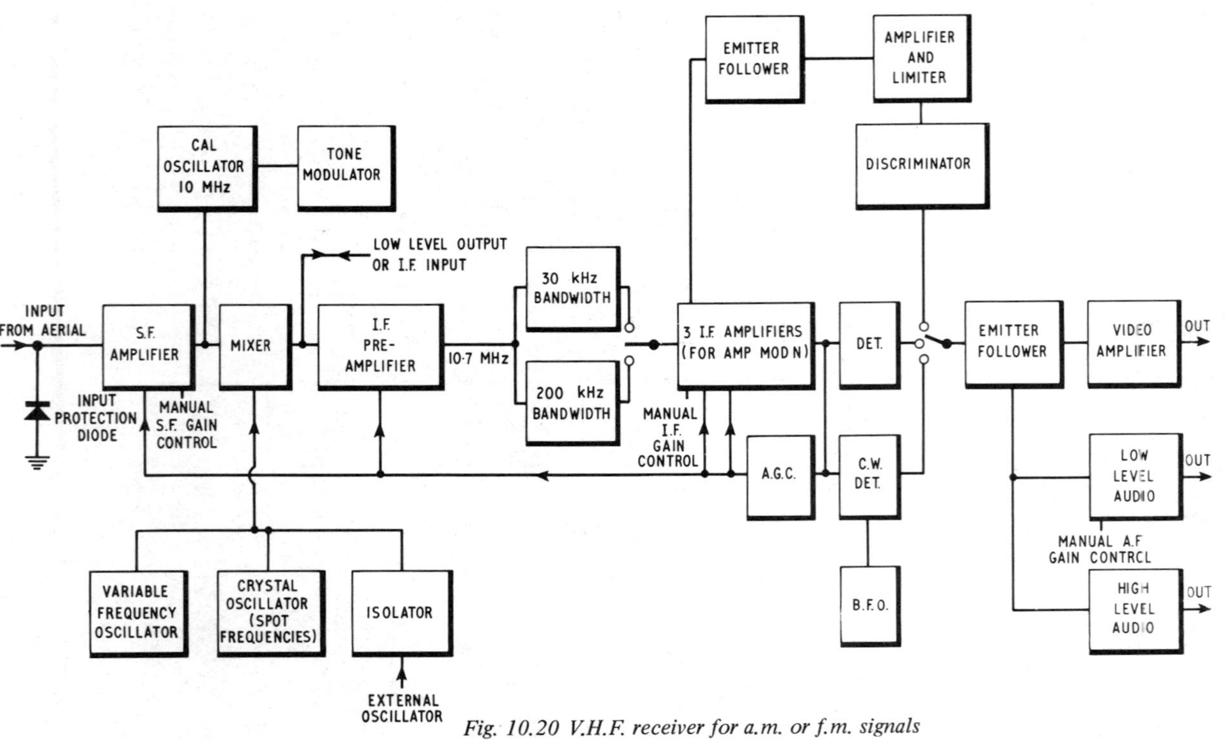

Fig. 10.20 V.H.F. receiver for a.m. or f.m. signals

10.9.3. V.H.F. receiver for amplitude-modulated or frequency-modulated signals

A transistorised v.h.f. receiver, by Eddystone, operates in switched bands covering the range 27-240 MHz and is suitable for the reception of c.w., amplitude-modulated and frequency-modulated signals. Two bandwidths are available and are normally 30 kHz and 200 kHz.

There is a single stage of signal-frequency amplification (Fig. 10.20). This uses a germanium mesa transistor in an earthed-base configuration, and there is manual and automatic gain control. A diode is connected directly across the input to provide some protection when operating close to a high-power transmitter.

In the mixer, which also uses a germanium mesa transistor, the signal may be combined with either the output of a variable frequency oscillator (germanium mesa transistor) or with one of up to eight crystal-controlled channels, or with an externally generated signal.

The intermediate frequency is 10·7 MHz. When the crystal-controlled channels are used for combination with the input signals the crystal frequencies are chosen to be in the range 37–88 MHz and are calculated from the appropriate relationship selected from, in MHz, $(f_s + 10{\cdot}7)$, $\frac{1}{2}(f_s + 10{\cdot}7)$ or $\frac{1}{3}(f_s + 10{\cdot}7)$.

There is a crystal-controlled calibrator with a frequency of 10 MHz. The signal is injected at will into the mixer and provides, by its harmonics, frequency check points at 10 MHz intervals over the whole tuning range.

The main i.f. section is preceded by a pre-amplifier and two i.f. filters, of 30 MHz (crystal) and 200 kHz (*LC* circuit) bandwidth respectively, either of which may be selected. If desired, external input may be made to, or low level output taken from, the input to the pre-amplifier.

The main i.f. amplifier is in two branches, one for a.m. and one for f.m.

For a.m. reception there are three earthed emitter transistor stages using silicon epitaxial planar transistors. Single tuned circuits are employed between the stages. The output to the detector is via a transformer tuned in the secondary. For c.w. there is a separate detector with beat-frequency oscillator. The whole unit is filtered and screened to prevent 10·7 MHz harmonics appearing in the receiver tuning range.

For f.m. reception the first of the earthed-emitter transistor stages referred to in the previous paragraph feeds an emitter follower isolation stage, after this there is amplification and limiting prior to the Foster-Seeley type discriminator circuit. The response is maintained to above 100 kHz.

The appropriate output (from the a.m. stages, or from the c.w. detector, or from the discriminator) is lead to an emitter follower and thence to three different amplifiers from any of which output may be taken:

1. a wideband amplifier;
2. a low power level pair of amplifiers with output at 600 Ω;
3. the four amplifiers, the last of these in push-pull, which constitute the high power level audio circuit, with output at 3 Ω.

Separate manual gain controls are provided for (i) signal frequency, (ii) intermediate frequency, and (iii) audio frequency stages The a.g.c. operates on the signal frequency, intermediate-frequency pre-amplifier and intermediate-frequency stages.

Other circuits, for example muting and metering, are provided but are not shown in our simplified block diagram.

The equipment operates on 12 V d.c., or with any standard a.c. supply.

10.10. V.H.F. TRANSMITTERS

Because of the importance of good frequency stability and narrow tolerance (about a few kilohertz, depending on the type of service), frequency control is by crystal wherever possible.

Apart from some broadcasting transmitters, v.h.f. transmitters are usually of relatively low power–25 W is common, 100–250 W is unusual. This is because:

1. efficient aerials can easily be arranged; and
2. little more than line-of-sight range can usually be achieved at these frequencies however great the transmitter power.

For amplitude-modulation transmitters, using the lower part of the v.h.f. range, operation throughout the transmitter may be at the final radiated frequency, this frequency being generated in the crystal oscillator at the outset. In the higher part of the frequency range, the crystal oscillator may be operated at one half, one third, or smaller fraction of the final frequency. A frequency multiplier is included to raise the frequency to the final desired level.

The radiation of even very low power spurious signals can cause considerable interference with nearby receivers. Such signals are spaced from the carrier by multiples of the crystal frequency. To facilitate their effective attenuation the crystal frequency should be as high as possible.

Corrected phase modulation is often used for f.m. transmissions (see Chapter 8). Considerable frequency multiplication is needed to obtain the final deviation. Hence, either an unduly low initial frequency is needed or, as in the transmitter described in Section 10.10.2, the crystal oscillator may be given some convenient frequency–a few megahertz–and, after some frequency multiplication, a frequency changing process is introduced to bring the mid frequency down to a lower value. This permits the application of any further frequency multiplication necessary to raise the deviation to the required value.

Example 10.7

A crystal frequency of 4 MHz is chosen for a multi-channel f.m. transmitter in which a final deviation of 240 kHz is needed. The maximum possible initial deviation is 1 kHz. The radiated frequency is to be 160 MHz.

To raise the frequency from 4 MHz to 160 MHz necessitates multiplication by 40. To raise the deviation from 1 kHz to 240 kHz necessitates multiplication by 240. Hence the basic problem is to obtain a sixfold increase in the deviation at some stage in the proceedings without any increase in the mid frequency. This may be done as follows:

1. Multiply the frequency by six. We now have a deviation of 6 kHz on a mid frequency of 24 MHz.
2. Separately multiply the crystal frequency by five, thus giving a signal at 20 MHz.
3. Mix the signals derived in stages (1) and (2) and extract the difference frequency. We now have a 4-MHz signal with 6-kHz deviation.
4. Multiply by 40 to obtain the required deviation of 240 kHz on a mid frequency of 160 MHz.

All the usual v.h.f. precautions are needed: short leads, special components, small valves, use of paralleled cathode leads and so on.

Transistors are often used when possible. Valves used in higher power stages are often in push-pull and are generally tetrodes or pentodes. Neutralisation is needed above about 100 MHz. There are special double tetrodes with built-in neutralisation. For high-power f.m. work, earthed-grid triodes are often used.

Transmission line tuning, with $\lambda/2$ or $\lambda/4$ sections of line, is often used at the higher frequencies. Circuit elements may be silver plated and polished to reduce skin effect.

The output circuit and the coupling to the aerial must be carefully designed to prevent radiation of unwanted frequencies. Bandpass filters may be included in the output to the aerial feeder to eliminate harmonics.

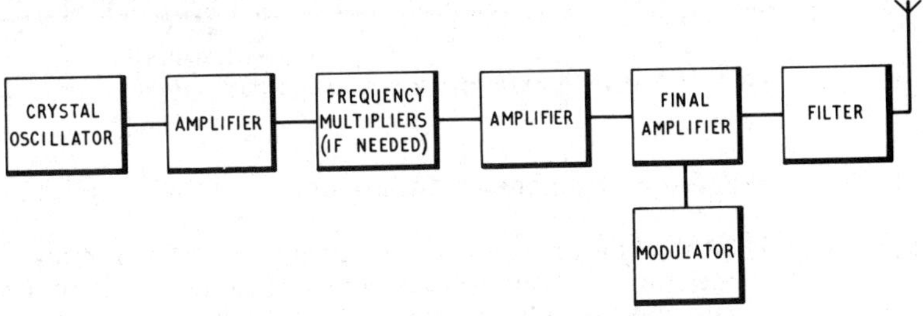

Fig. 10.21 V.H.F. transmitter for amplitude modulation

10.10.1. V.H.F. amplitude-modulated transmitter

Fig. 10.21 shows the block diagram of a simple v.h.f. amplitude-modulated transmitter. It is much the same as the block diagram for a crystal-controlled amplitude-modulated transmitter for other frequencies.

The differences are in the circuit details as already discussed.

Frequency control, where possible, is by crystal oscillator. Frequency multiplication is applied if and as necessary. Frequency stability is important and the tolerance is close.

For single channel working, a bandwidth of some 6 kHz is needed.

If multipliers are used, the gain obtained in the multipliers may be sufficient to enable the final amplifier to be driven without additional amplification stages being provided. Otherwise more amplifiers must be added as shown in the diagram. If valves are used in the multipliers they may be triodes or pentodes. Neutralisation is not needed because the input and output circuits are tuned to different frequencies.

Anode modulation of the final push-pull power amplifier is customary. The usual aerial filter circuit is included.

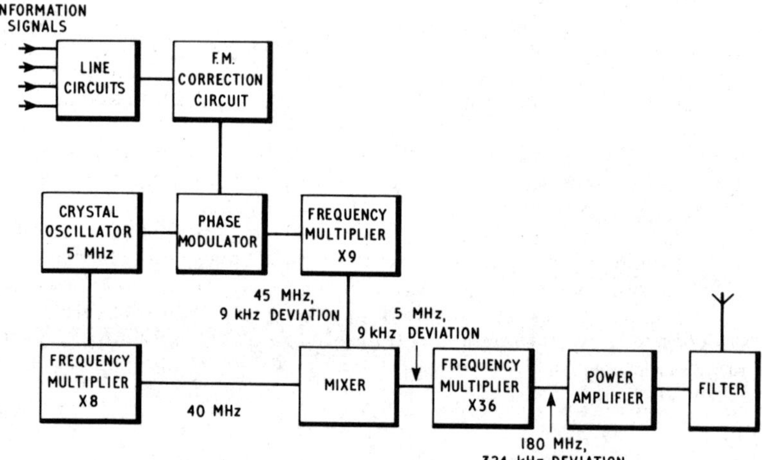

Fig. 10.22 V.H.F. transmitter for frequency modulation

10.10.2. V.H.F. frequency-modulated transmitter

Fig. 10.22 shows the block diagram of a frequency-modulated transmitter with a deviation of about 300 kHz and able to sustain many speech channels. Such circuits are in increasing use in difficult country where there are, for instance, swamps, jungles, rivers, deep valleys and the like, so that the provision of line telephone links is not feasible.

A 5-MHz crystal oscillator determines the mid frequency with good stability and small tolerance. Corrected phase modulation is used and the initial deviation is 1 kHz.

Frequency multiplication by 9 gives a 9-kHz deviation on a 45-MHz carrier. This signal is mixed with one at 40 MHz (eight times the crystal frequency) and the difference frequency selected.

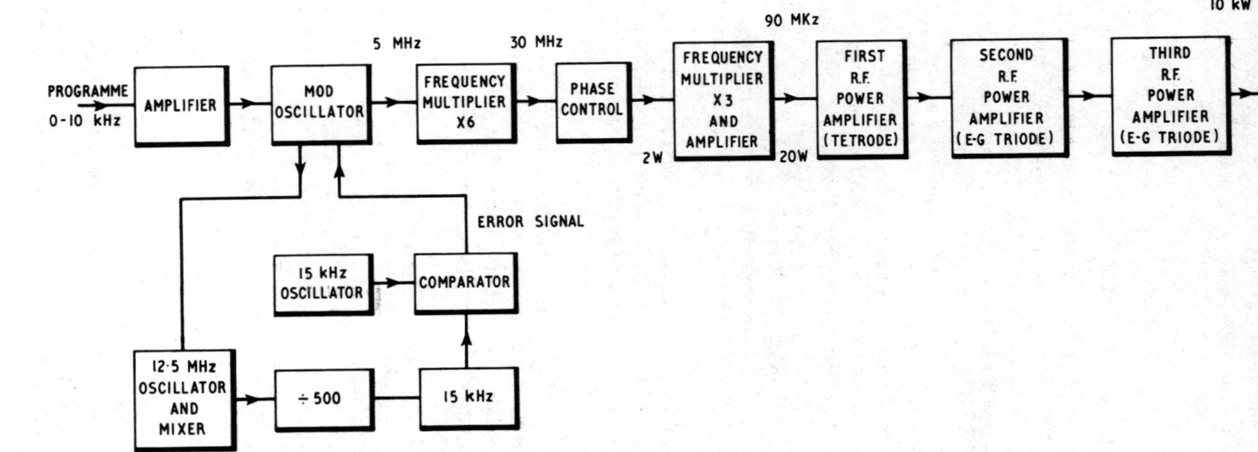

Fig. 10.23 V.H.F. transmitter for f.m. broadcasting

The resulting 5-MHz signal, with a deviation of 9 kHz, is multiplied by 36 to give a final deviation of 324 kHz on a mid frequency of 180 MHz.

Power amplification, using a tetrode or pentode, and aerial filter circuits follow.

10.10.3. V.H.F. frequency-modulated transmitter for broadcasting

The block diagram of Fig. 10.23 shows a transmitter, by Standard Telephones and Cables, designed for unattended v.h.f. broadcasting. It is frequency modulated and special features called for (by the British Broadcasting Corporation) included the need for three independent carriers spaced at 2·2 MHz apart to feed one half of a single wideband aerial, the other half of the aerial to be supplied by three identical transmitters to be kept substantially in phase with their respective opposite numbers (see Chapter 11). These needs were specified to reduce aerial costs (a substantial part of the expenditure on a station of this type), and to provide continuity of service, albeit at reduced power, in the event of failure of part of the equipment or aerial system. These features materially influenced the design.

For broadcasting with an audio-frequency range of some 30 Hz–10 kHz, great frequency stability and deviation linearity are needed. An unmodulated crystal oscillator is, therefore, used as the reference, and the modulation is applied to a free-running oscillator. The centre frequency of the free-running oscillator is compared with that of the reference oscillator and maintained at its correct value by the operation of an error voltage which is developed when the centre frequency departs from the correct value.

The incoming programme material is amplified in the first amplifier and passed to the free-running oscillator circuit, the output of which is thus frequency modulated. The centre frequency of this output is at, say, 5 MHz. Any departure from this value must be corrected automatically. To this end a sample of the signal is mixed with one at exactly 12·5 MHz. The difference frequency (nominally 7·5 MHz) is divided by 500 to give a signal at nominally 15 kHz, substantially devoid of sidebands because of the frequency division. This signal is compared with one known to be at exactly 15 kHz. If the centre frequency of the free-running oscillator is not correct, an error voltage is developed in the comparator. This voltage causes a motor driven tuning capacitor in the free-running oscillator circuit to rotate in the correct direction to reduce the frequency error. When the error is eliminated the error signal disappears and the capacitor drive motor stops.

Because two transmitters each contribute one half of the radiated power, it is necessary to arrange for phase changing in the final output so that the two transmitters shall feed the aerials in phase. A maximum phase change of 180° is necessary in one transmitter. This change is effected automatically or manually. To facilitate the control, it is arranged that it takes place where the frequency is only one third of the final radiated value—here a maximum of 60° of change is needed.

The main 5-MHZ signal from the free-running modulated oscillator is multiplied six times in a frequency multiplier in order to bring the centre frequency up to one third of the radiated value, i.e. to 30 MHz. At this stage the power is about 2 W.

Following the phase change circuit the frequency is tripled in a frequency multiplying circuit. In this circuit the power is raised to 20 W, which is the level needed to drive the first main power amplifier.

There are three stages in the main amplifier and they all work in class C and all use forced air cooled valves. The first stage, with a tetrode, provides a power gain of about 40; the following two, both with earthed-grid triodes, have gains of three and six respectively.

The first two stages have an anode supply of 2·5 kV, and a screen supply of 500 V derived from three-phase full-wave selenium rectifiers. The third stage uses 6 kV for the anodes and xenon rectifiers are used. The filaments are thoriated tungsten and are a.c. heated.

The main power amplifying valves are built into coaxial line tuners, and in the last two stages the grids are directly earthed thus avoiding the need to use grid blocking and neutralising capacitors capable of passing the combined input and output r.f. circulating currents. The design of such capacitors is difficult. The only feedback is that inside the valve, and it is small enough not to need neutralising. The first main r.f. amplifier, the tetrode, is neutralised using an inductive pick-up loop inside the anode cavity.

The valves are self biased and use a combination of grid current and cathode current bias. This is possible because f.m. transmitters may be used with class-C amplifiers with constant grid and anode currents.

10.11. V.H.F. TRANSMITTER/RECEIVERS

Some transmitter/receivers are very small and compact. Even these, in favourable circumstances of terrain and aerials, are capable of maintaining communication over distances of some thirty or forty miles, albeit often with inferior quality. Other equipments are relatively elaborate and sophisticated.

10.11.1. Marine v.h.f. transmitter/receiver

A single unit provides the frequency drive for both the transmitter and the receiver. The unit gives synthesised crystal-controlled outputs at 1/18 of the final frequency. These outputs are made up of the sum of two oscillations, one from each of two groups of oscillators. The first of these has five decade oscillators, the second has five unit oscillators. For any given frequency the appropriate oscillator from each group is chosen by rotation of the *channel selector* control: this connects the

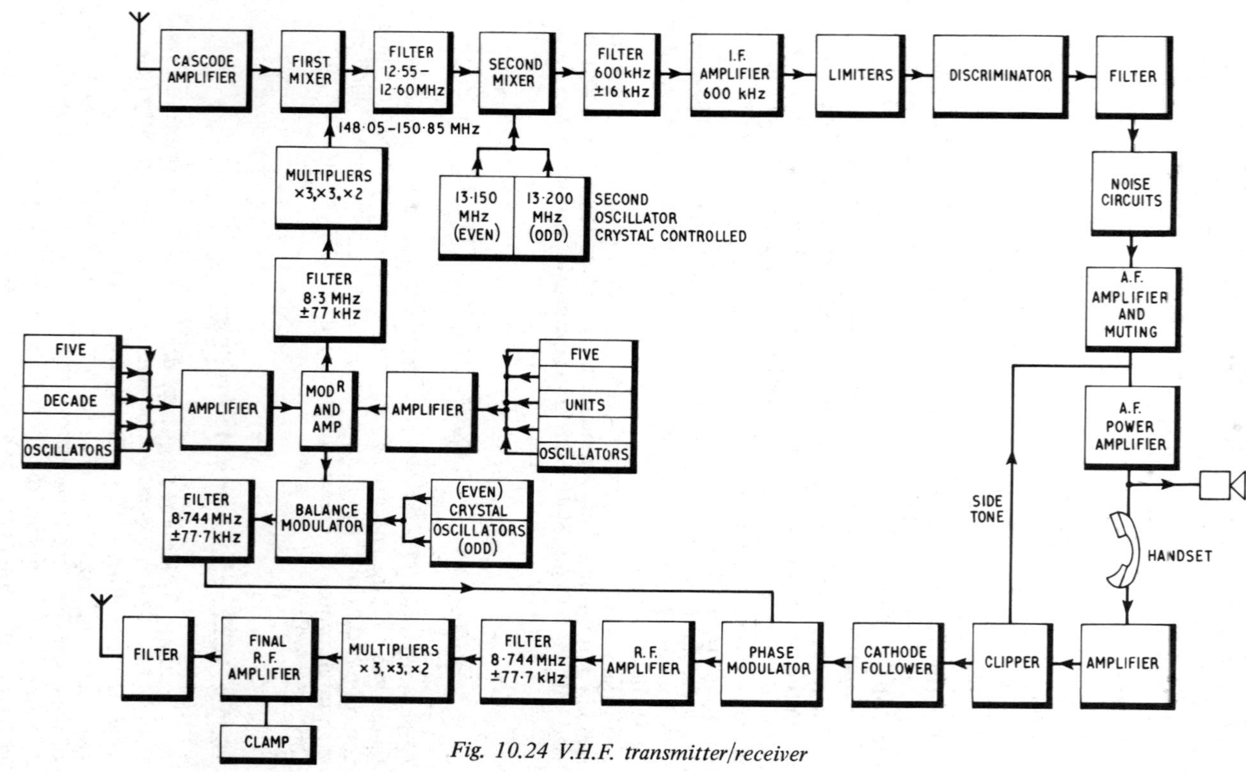

Fig. 10.24 V.H.F. transmitter/receiver

supply voltage to the appropriate units leaving the remainder without power.

The equipment operates at frequencies in the vicinity of 160 MHz. It is suitable for simplex or duplex operation. For simplicity, however, only the duplex line in the receiver will be described. Frequency modulation is employed. The block diagram is shown in Fig. 10.24.

The receiver has a cascode s.f. amplifying stage prior to the first mixer. The oscillator injection frequency is in the range 148·05–150·85 MHz, and it is derived from the synthesiser and frequency multipliers (a total of × 18). The mixer output is taken through a 12·55–12·60 MHz filter and the frequencies within this range continue to the second mixer for mixing with one of the outputs of the second oscillator. These outputs are at 13·15 MHz and 13·20 MHz, and by use of the appropriate frequency the 50-kHz spacing between adjacent channels is effected.

The second intermediate frequency is 600 kHz, and the bandwidth of the amplifier extends to 16 kHz to each side of this value. Limiting, detection in a Foster-Seeley type discriminator, audio-frequency amplification and noise reduction circuits follow. A muting circuit is included with the a.f. amplifier.

For the transmitter the microphone output is amplified and the response shaped to increase the relative sideband power. A side-tone signal is fed to the receiver output.

The a.f. signal phase modulates a signal (of frequency about 8·7 MHz) from the frequency synthesiser. Amplification and filtering in this frequency range follows after which three multipliers (× 2, × 3 and × 3) bring the centre frequency up to the required radiated frequency (156–158·8 MHz).

Final r.f. amplification and harmonic filtering, to reduce second and third harmonics, follows before the signal is fed to the aerial. A clamp circuit prevents the final stage drawing excessive current in the event of failure of the drive circuit.

Some fifty channels are available—all crystal controlled—in the international and private maritime bands in the frequency range 156–162·55 MHz. The full power output is 20 W.

QUESTIONS

1. Draw the block schematic diagram of an independent sideband communications receiver capable of receiving two audio channels on each sideband. Show the frequencies of the local oscillators and the passbands of the filters and amplifiers, and explain the action of the receiver. *(C & G)*

2. Describe with the aid of sketches methods of cooling high power transmitter valves by:

(a) forced air,
(b) forced water, and
(c) vapour.

Tabulate the advantages and disadvantages of the three methods. *(C & G)*

3. What are the advantages of single sideband over double sideband transmitters?

Explain with the aid of diagrams how the level of the carrier relative to the sidebands is controlled in an s.s.b. transmitter. Why is the carrier partially suppressed?

The signal from a conventional double sideband transmitter working at 100% depth of modulation and generating 120 kW of peak power is received by a distant receiver. If both transmitter and receiver are converted to single sideband working, what power must the transmitter generate now to produce the same signal output level and signal-to-noise ratio as before? *(C & G)*

4. Draw the block diagram of a high power short wave transmitter capable of transmitting two independent telephony sidebands with reduced carrier. Explain, concisely, the action of the transmitter.

5. Using the block diagram of a transmitter drive as an illustration, explain in outline how rapid and accurate frequency selection may be obtained using a master oscillator and frequency synthesiser.

6. Draw the block diagram and explain the action of an h.f. telegraph transmitter using frequency shift keying.

7. Explain in principle, with the aid of diagrams, how a telephony transmission may be scrambled in order to provide secrecy of communication.

8. Compare the relative advantages and disadvantages of high-power transmitters which (a) use, and (b) do not use, wideband amplifiers.

9. What special precautions are necessary in designing and manufacturing transmitters for use at v.h.f.?

10. Draw the block diagram and explain the action of a transmitter/receiver for v.h.f. operation.

11

Aerial Systems

11.1. INTRODUCTION

This chapter is mainly concerned with directional aerial systems designed to make the best possible use of transmitter power by confining the radiation to those areas in which it is to be received. Broadly speaking, directional systems can be grouped into those which are called *beam arrays* and those called *travelling wave aerials.* The first group are resonant systems, and each is designed for a particular frequency. The second group may be used over a wide range of frequencies without alteration.

11.2. BEAM ARRAYS, GENERAL PRINCIPLES

Beam arrays consist of a number of resonant elements arranged and spaced so that their radiations reinforce in the direction in which maximum signal strength is required and mutually cancel in the directions in which no signals are required.

The resultant field strength at a distant point is the vector sum of the fields provided by the individual elements. The phase difference between radiations from two elements depends on the spacing of the elements in wavelengths, on the relative phasing of the currents in the two elements, and on the angle which the direction of radiation to the point makes with the aerial array. It is usual and sufficiently accurate to assume that at a distance the directions of radiation from the constituent radiators are parallel.

The directional properties of an aerial are illustrated by polar diagrams. A horizontal polar diagram shows the directional properties of the array in a horizontal plane, while a vertical polar diagram shows directivity in a vertical plane.

Polar diagrams may be drawn to show the directivity of aerials in terms of power or to show relative field strengths as measured in volts per metre or millivolts per metre. Field strength diagrams are the more usual and these are used in the following text. A complete polar diagram is a solid figure, e.g. the complete polar diagram for an isotropic radiator is a sphere, while the polar diagram for a vertical Hertzian dipole is the volume which would be swept out by rotating a figure-eight-shape so as to produce a doughnut shaped volume.

A line of vertical elements correctly and uniformly spaced and energised in phase produce maximum radiation in a direction at right angles to the plane of the aerial and so this is called a *broadside array.* The more elements in the array, the more narrow becomes the beam of the array. The distance between the elements is generally about half a wavelength. If it is much greater, unwanted radiation is directed from the ends of the array, while, if the elements of the array are too closely spaced, mutual inductance between the elements reduces the input impedance too much.

Symmetry indicates that there will be two directions of maximum radiation—one on each side of the array. If it is necessary to direct radiation to one side only, then a system of reflectors is provided on the other side of the array.

An array in which horizontal dipoles are stacked in a vertical plane is also a broadside array but the maximum directivity tends to be in a vertical plane. Such an array may be used to concentrate short-wave broadcast services at a low angle of elevation in order to reflect the radiation from the ionosphere to a very distant receiving area.

When a horizontal line of vertical dipoles is arranged to concentrate radiation in the direction of one end of the array, this arrangement is called an *end-fire array*. With the same number of elements the main beam of radiation is wider using an end-fire array than a broadside array, although there is only one main beam and not two. The end-fire effect is obtained by energising each element with current which lags in phase on the current in the preceding element. If the distance between the elements is d, and the wavelength λ, then the successive phase difference in the currents of adjacent elements must be $2\pi d/\lambda$ rad. This arrangement results in the concentration of radiation in the direction of that end of the array which has the lagging phase of current. The distance between the elements is generally about $\lambda/4$. If it is much greater than $\frac{3}{8}\lambda$, radiation in the reverse direction will occur. The effect of mutual inductance nearly cancels out in this type of array.

So far it has been assumed that the elements, which are usually dipoles, are arranged parallel with one another. But if they are placed

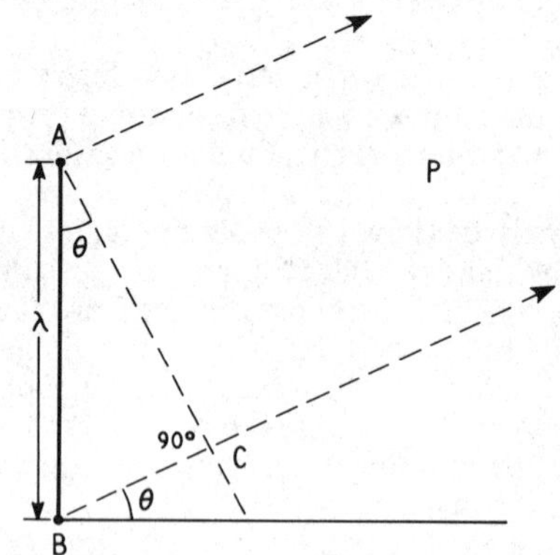

Fig. 11.1 Vertical aerials spaced by one wavelength

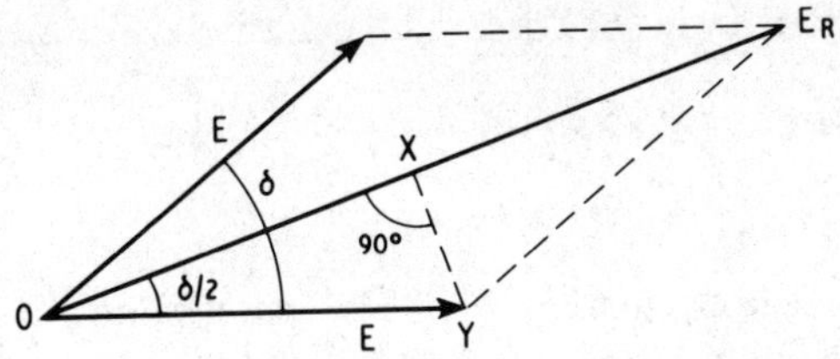

Fig. 11.2 Addition of aerial fields by phasors

in an end to end direction, though with some small spacing in between them, a *collinear array* is formed. Broadside and end-fire effects can be obtained but, since dipoles do not radiate in the direction of their length, a true end-fire effect is not possible. Instead the radiation is concentrated in a more or less narrow cone around the axis of the collinear array.

11.3. EXAMPLES INVOLVING TWO RADIATING ELEMENTS

Now follow six worked examples to show how the polar diagram of field strength may be evolved for two energised aerial elements or for one element and its *image aerial* in the earth.

Example 11.1

Two vertical dipole aerials are spaced at a distance of one wavelength and are energised in phase (Fig. 11.1). Sketch the shape of the polar diagram for a horizontal plane passing through their centre.

The radiations reaching a distant point P from aerial B lag in phase on the radiation from aerial A because of the extra path length BC.

$$BC/AB = \sin\theta$$

$$BC = AB\sin\theta$$

$$= \lambda\sin\theta$$

This is responsible for phase difference of,

$$\lambda\sin\theta/\lambda .\ 2\pi$$

$$= 2\pi\sin\theta \text{ rad}$$

Let $\quad 2\pi\sin\theta \text{ rad} = \delta \text{ rad}$

and E V/m be the field strength at P due to one aerial, then the resultant field strength at P is given by the length of the phasor E_R in the phasor diagram Fig. 11.2.

By construction, angle OXY is a right angle and $OX = E_R/2$. Thus,

$$E_R/2 = E \cos \delta/2 \quad \text{and}$$

$$E_R = 2E \cos \delta/2$$

The resultant field strength E_R has a maximum value if $\cos \delta/2 = \pm 1$ and, therefore, when

$$\delta/2 = 0 \text{ or } \pm \pi$$

That is $(2\pi \sin \theta)/2 = 0$ or $\pm \pi$,

which makes, $\sin \theta = 0$ or ± 1

and the angle $\theta = 0$ or $\pm \pi/2$ or π rad

The resultant field strength has a minimum or zero value if

$$\cos \delta/2 = 0 \quad \text{and therefore,}$$

$$\delta/2 = \pm \pi/2$$

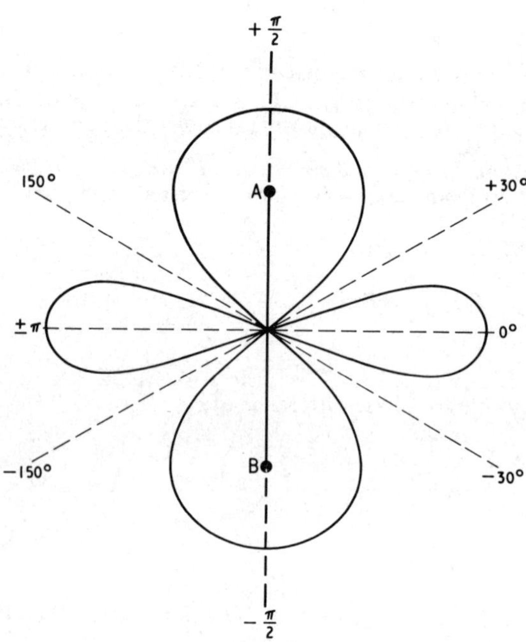

Fig. 11.3 Result for Example 11.1

Therefore $(2\pi \sin \theta)/2 = \pm \pi/2$

and $\quad \sin \theta = \pm 0{\cdot}5$

so that angle $\theta = \pm 30^\circ$ or $\pm 150^\circ$

The anticlockwise direction is called + and the clockwise direction –. The angle θ is measured from a normal to a line joining the two aerial elements. The polar curve of Fig. 11.3 is sketched to show maximum field strength values of $2E$ V/m in the directions determined above and the directions of zero resultant field strength at the appropriate value of the angle θ.

Example 11.2

Two vertical dipoles are spaced by half a wavelength and energised in phase. Sketch the polar diagram for a horizontal plane passing through the centre of the radiating elements.

With an alteration of the aerial element spacing to $\lambda/2$, Fig. 11.1 again applies. But now,

$$BC = \lambda/2 \sin \theta$$

and the phase difference between the radiations received at P from the two elements is,

$$\delta = \pi \sin \theta$$

As in Example 11.1 the amplitude of the resultant field at the distant point P is given by,

$$E_R = 2E \cos \delta/2$$

E_R has a maximum value when

$$\cos \delta/2 = \pm 1$$

which occurs when

$$\delta/2 = 0$$

or when $\quad \pi/2 \sin \theta = 0$

and angle $\quad \theta = 0$ or π

The amplitude is zero if

$$\cos \delta/2 = 0,$$

that is if $$\delta/2 = \pm \pi/2$$

or $$\pi/2 \sin \theta = \pm \pi/2$$

Thus $$\sin \theta = \pm 1$$

and the angle $$\theta = \pm \pi/2 \text{ rad}$$

Fig. 11.4 is the polar diagram drawn from the information deduced above.

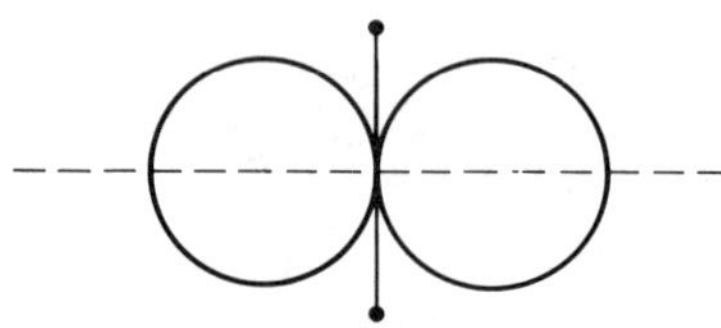

Fig. 11.4 Result for Example 11.2

Example 11.3

Two vertical dipoles are spaced by half a wavelength but energised in antiphase. Sketch the polar diagram of radiation for a horizontal plane passing through their centre. Fig. 11.1 again applies with the distance BC equivalent to $\lambda/2 \sin \theta$. Assuming the radiation from element A to lead by π rad on that from B, the phase difference between the radiated fields received at the distant point P from the two radiators is now:

$$\delta = \pi + \lambda/2 \sin \theta \, . \, 2\pi/\lambda$$

or $$\pi + \pi \sin \theta$$

Thus $$\delta/2 = \pi/2 + \pi/2 \sin \theta$$

The resultant field amplitude is given by,

$$E_R = 2E \cos \delta/2 \text{ V/m}$$

This has a maximum value if

$$\cos \delta/2 = \pm 1$$

or the angle $$\delta/2 = 0 \text{ or } \pi \text{ rad}$$

For this condition,

$$\pi/2 + \pi/2 \sin\theta = 0 \text{ or } \pi$$

therefore, $\pi/2 \sin\theta = \pm \pi/2$

and $\sin\theta = \pm 1$

and angle $\theta = \pm 90^\circ$

The field amplitude is zero if

$$\cos\delta/2 = 0,$$

that is if $\delta/2 = \pi/2$

or $\pi/2 + \pi/2 \sin\theta = \pi/2.$

This gives $\sin\theta = 0$, and the angle θ is, therefore, 0° or 180° for a resultant field strength of zero. The polar diagram which uses this information is Fig. 11.5. It will be noted that the directions of the maximum field strengths have been rotated through 90° from the directions they have in Fig. 11.4 by the reversal of the phase of one of the dipole currents.

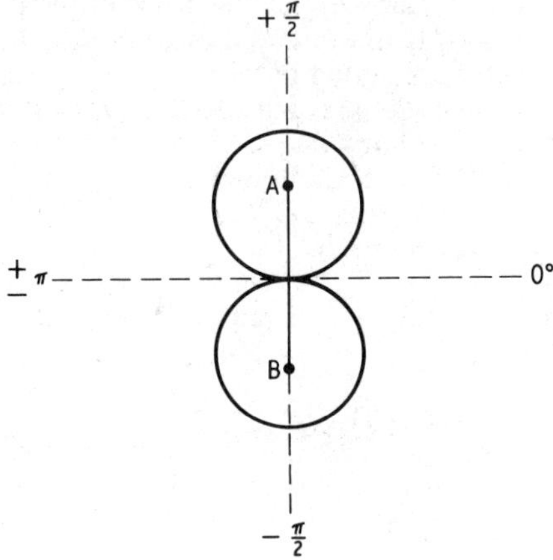

Fig. 11.5 Result for Example 11.3

Example 11.4

Two vertical dipoles separated by a distance d are fed in phase and carry equal currents. Compare the field strengths at a remote point P which lies in a direction at an angle θ to the line of the aerials and is in the

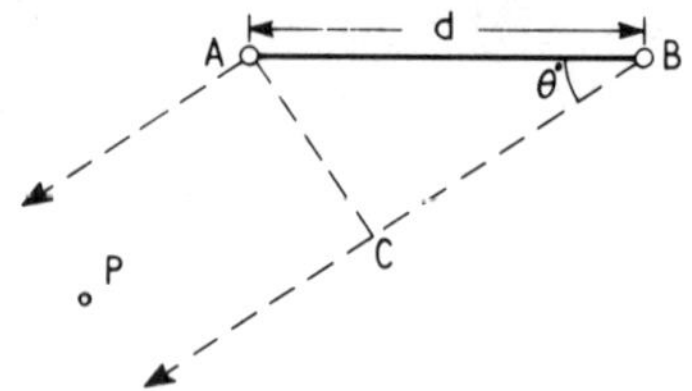

Fig. 11.6 Two vertical aerials spaced d metres

horizontal plane, when the spacing distance d is (a) $\lambda/2$, (b) $\lambda/4$, and the angle θ is 45°.

In Fig. 11.6 CB is the extra distance to be travelled by radiation from element B to reach point P as compared with the distance travelled by radiation from element A.

$$CB = d \cos \theta$$

In the direction of P, therefore, the radiation from element B lags that from A by,

$$CB/\lambda \,.\, 2\pi$$

or $$d \cos \theta/\lambda \,.\, 2\pi \text{ rad}$$

As in the previous three examples, let the phase difference between the two radiations reaching P be represented by δ. In the previous three examples field strengths are stated in terms of r.m.s. values. As an alternative approach the field strengths reaching P from the two elements are now given as instantaneous values:
Let the field at point P due to radiation from A be,

$$e_A = E \sin \omega t \text{ V/m},$$

where ω is the angular frequency of radiation. Then the field reaching P from the element B is,

$$e_B = E \sin (\omega t - \delta) \text{ V/m}$$

where $$\delta = 2\pi \, [(d \cos \theta)/\lambda] \text{ rad}$$

The total field at P is,

$$E \sin \omega t + E \sin (\omega t - \delta) \text{ V/m}.$$

Making use of the trigonometry formula for the addition of the sines of two angles, this total field strength may be written as:

$$2E \cos \delta/2 \sin (\omega t - \delta/_2) \text{ V/m}$$

The amplitude of this alternating quantity is $2E \cos \delta/2$, which is a result which agrees with the phasor diagram addition of Fig. 11.2, with the difference that this working concerns peak field strength and Fig. 11.2 refers to r.m.s. values. Substituting values stated in the problem: (a) $d = \lambda/2$ and $\delta = \pi \cos 45°$; therefore $\delta/2 = \pi/2\sqrt{2}$ or $64°$. The field strength is thus given by, $2E \cos 64°$ or $2E \times 0{\cdot}438$ V/m.

(b) $d = \lambda/4$ and $\delta/2 = \pi/4\sqrt{2}$ or $32°$

and the field strength is now given by, $2E \cos 32°$ or $2E \times 0{\cdot}848$. The ratio of the two field strengths for the two different spacings for the radiated elements is:

$$2E \cos 32°/2E \cos 64° = 1{\cdot}94$$

Hence the reduction of the spacing from $\lambda/2$ to $\lambda/4$ has increased the field strength 1·94 times in the direction of the point P.

Example 11.5

A horizontal dipole is suspended $\lambda/2$ above an earth which can be regarded as perfectly conductive. Sketch the polar diagram for a vertical plane which passes at right angles through the centre of the aerial.

In assessing the resultant radiation in this example, account must be taken of an apparent *image aerial* situated in the earth, the same distance below the surface as the real aerial is above it. As the image aerial is parallel with the actual aerial, it may be imagined that the current is induced in the image aerial by a process of mutual induction and that the current in this effective secondary is opposite in phase to the current in the primary aerial in accordance with Lenz's law. The wave reflected from the earth's surface and which appears to come from the image aerial has a 180° phase change on reflection. The polar diagram is, therefore, that for a plane at right angles to two parallel dipoles, spaced by λ and energised with opposite phase of current. Fig. 11.7 illustrates this.

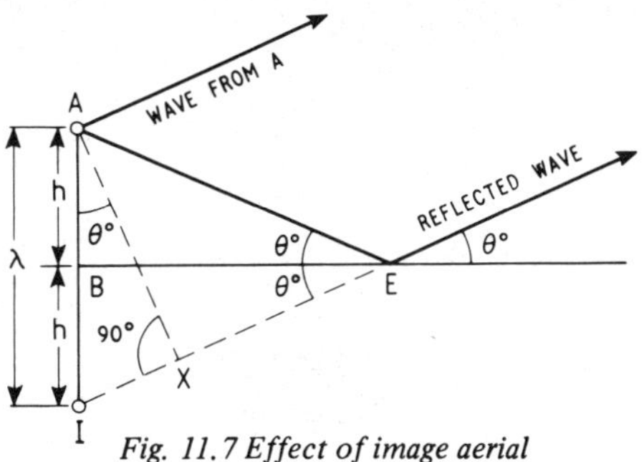

Fig. 11.7 Effect of image aerial

The radiation occurring at an angle of elevation $\theta°$ can now be considered, since from the symmetry of the diagram, IE is equal to AE, IX can be taken as the extra distance of path travel by the radiation from the image aerial I. The distance IX is responsible for a change of $IX/\lambda \,.\, 2\pi$ rad.

But $\qquad IX = 2h \sin \theta.$

Therefore the phase change due to the distance IX is, $2h \sin \theta/\lambda \,.\, 2\pi$ rad. In this example $h = \lambda/2$ and the phase change due to IX may be written as, $2\pi \sin \theta$ rad. To this must be added the phase change caused by the reflection of the wave so that the total phase lag of the radiation from the image aerial is,

$$2\pi \sin \theta + \pi = \pi (2 \sin \theta + 1)$$

As in Example 11.4 , the resultant radiation can be stated as:

$$2E \cos \delta/2 \,.\, \sin (\omega t - \delta/_2)$$

where E is the amplitude of the field due to the aerial alone, δ is the phase difference between the radiations from the aerial and its image aerial and ω is the angular frequency of the wave.

$$\delta/2 = \pi/2 \,(2 \sin \theta + 1)$$

The amplitude of the wave is maximum if $\cos \delta/2 = \pm 1$. This occurs with $\delta/2 = 0$ or π rad.

Thus $\qquad \pi/2 \,(2 \sin \theta + 1) = 0 \text{ or } \pi,$

$$(2 \sin \theta + 1) = 0 \text{ or } 2$$

$$2 \sin \theta = \pm 1$$

$$\sin \theta = \pm 0{\cdot}5 \text{ and,}$$

$$\theta = \pm 30°$$

The amplitude of the wave is zero if $\cos \delta/2 = 0$
This occurs with

$$\delta/2 = \pm \pi/2 \text{ or } 3\pi/2$$

Thus $\qquad \pi/2 \,(2 \sin \theta + 1) = \pm \pi/2 \text{ or } 3\pi/2$

$$2 \sin \theta + 1 = \pm 1 \text{ or } 3$$

$$2 \sin \theta = 0 \text{ or } \pm 2$$

$$\sin \theta = 0 \text{ or } \pm 1$$

$$\theta = 0, \pm 90^\circ \text{ or } 180^\circ$$

Fig. 11.8 is the appropriate polar diagram.

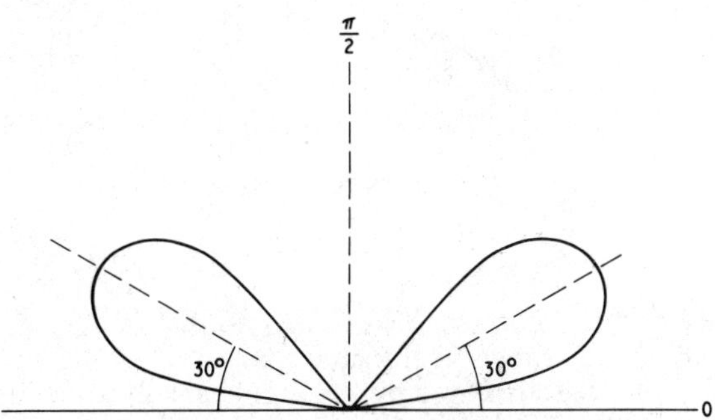

Fig. 11.8 Result for Example 11.5

Example 11.6

A vertical half wavelength dipole is situated one wavelength above the earth. Assuming the earth to be perfectly conductive, deduce an expression from which the polar diagram could be evolved for a vertical plane. Find the strength of field radiated at the following angles to the horizontal relative to the field strength in the direction of maximum radiation: 0°, 30°, 45°, 60°, 90°.

It is assumed that the centre of the dipole is one wavelength above the earth's surface and that there is an image aerial whose centre is one wavelength below the surface of the earth. The phase of the radiation from the dipole and its image are taken to be in phase at the centre of each, and the vertical separation of the two sources of radiation to be two wavelengths. The current in the image aerial may be imagined to be induced by electrostatic induction and for this reason the current in the image aerial is assumed to be in phase with the current in aerial itself. There is no phase reversal of the wave reflection at the ground with a vertically polarised wave as there is when the wave is horizontally polarised.

The polar diagram for a dipole aerial in free space and for a plane parallel with its length is approximately a figure of eight. The field strength in a given direction is proportional to the cosine of the angle

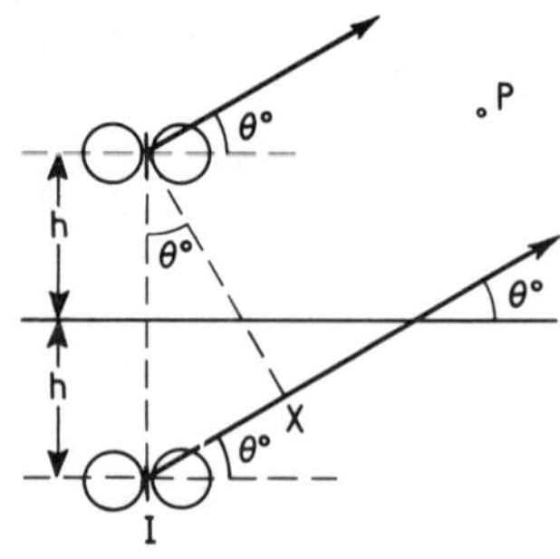

Fig. 11.9 Effect of image of vertical aerial

between the direction chosen and a line at a normal to the length of the aerial. Fig. 11.9 shows the resultant situation.

Let the field strength radiated by the dipole in a direction at 90° to its length have an amplitude of E V/m. In the direction of the distant point P the field radiated by the dipole is of amplitude $E \cos\theta$ V/m, the instantaneous value being (say) $E \cos\theta \sin\omega t$ V/m.

The radiation from the image aerial in the same direction has a similar amplitude if any attenuation on reflection at the earth is assumed negligible, but it lags in phase by an angle proportional to the additional travel distance IX. From Fig. 11.9 it is seen that $IX = 2h \sin\theta$. The phase difference between the two radiations reaching point P is, $2h \sin\theta/\lambda \,.\, 2\pi$ rad. As in previous examples, let the phase difference between the two components of the total field be δ rad.

As $\quad h = \lambda,$

$$\delta = 2\pi \,.\, 2 \sin\theta \text{ rad}$$

The field radiated by the image aerial in the direction of point P is,

$$E \cos\theta \sin(\omega t - \delta) \text{ V/m}$$

and the resultant field is therefore,

$$E \cos\theta \sin\omega t + E \cos\theta \sin(\omega t - \delta) \text{ V/m}$$

This may be written as,

$$2E \cos\theta \cos\delta/2 \,.\, \sin(\omega t - \delta/2) \text{ V/m}$$

where $\quad \delta/2 = 2\pi \sin\theta$ rad

The amplitude of this resultant field is then,

$$2E \cos\theta \,.\, \cos 2\pi \,.\, \sin\theta \text{ V/m}$$

and the biggest value which this can have is $2E$ V/m.

The expression from which the polar diagram may be evolved is, $\cos\theta \cos 2\pi \sin\theta$. The value of this factor for a number of selected values of is shown in Table 11.1 under the column headed F. If the

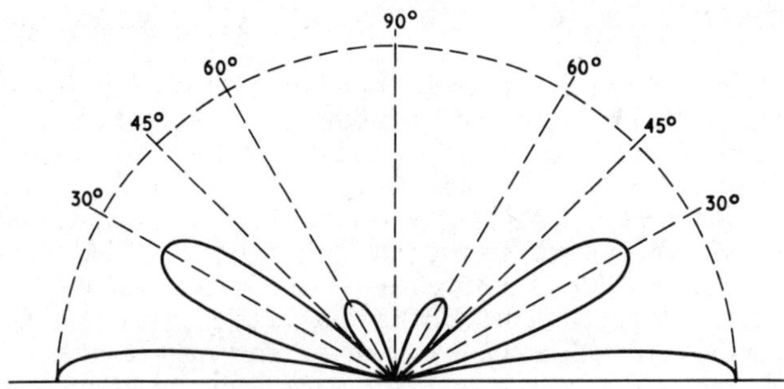

Fig. 11.10 Polar diagram result for Example 11.6

factor F is calculated for a range of angles from 0 to 180°, the polar diagram may be constructed and is found to be as in Fig. 11.10. The practical polar diagram is modified a little by the attenuation of the reflected wave by the earth, especially at low angles of radiation. Indeed most attempts to predict aerial polar diagrams with precision fail on account of the unknown extent to which reflections on the actual aerial site will modify the theoretical diagram. The negative sign for the values of F where θ is 30° and 45° indicates an opposite phase of radiation in those directions to that at equal distances for θ equal to 0° or 60°.

Table 11.1

θ	Sin θ	360 Sin θ	Cos (360 Sin θ)	Cos θ	F
0	0	0	1	1	1
30	0·5	180	−1	0·866	−0·866
45	0·707	254	−0·2756	0·707	−0·196
60	0·866	316	0·7193	0·5	0·3596
90	1·0	360	1·0	0	0

It should be noted that the figure-of-eight polar diagram assumed for a dipole radiator in the above working is only precise for a Hertzian dipole. For a halfwave dipole the function which shows the exact variation of field strength with the angle θ of Fig. 11.9 is,

$$\cos\left(\frac{\pi}{2}\cos\theta\right)\Big/\sin\theta.$$

11.4. FINDING THE ARRAY FACTOR FOR A LINE OF n UNIFORMLY SPACED VERTICAL DIPOLE AERIALS

The *array factor* of an arrangement of radiators is the factor by which the field strength due to a single omni-directional element of the array must be multiplied in order to calculate the total field strength of the array at a given distance. The field strength at a given distance due to one element of the array is considerably less than the same element would produce if the whole power of the transmitter were confined to that one element. If the available transmitter power is P watts, then with this power divided equally between n elements the power in each element is P/n watts. The field strength at a given point due to a single element in the array is then $E'/\sqrt{n}$ V/m where E' V/m is the field strength at the same point if all the transmitter power is fed into the single element. The array factor is a function of the number of elements in the array and of the angle which the direction of radiation considered makes with a line at a normal to the line of the elements. If the radiating elements are not energised in phase, then the array factor also depends on the phase difference between the currents in adjacent elements of the array. The presence of a reflecting curtain of conductors or a reflecting earth will also alter the array factor.

In Fig. 11.11 n vertical dipoles are assumed although to save space only five of them are shown. The dipoles are spaced at regular intervals in a straight line by d metres. In assessing the polar diagram for a horizontal plane, earth reflections need not be considered. No reflectors are present and the mutual coupling between the dipoles is assumed to be negligible. The working also assumes that the point P at which the field strength is calculated, is far enough away from the array for the signals reaching P to be considered as travelling along parallel paths. Let ϕ rad be the phase difference between the currents in successive elements. If the current in dipole C leads on the current in B by ϕ rad, then the signal reaching the distant point P from C leads the signal reaching P from B because:

1. The current leads by ϕ rad.
2. The radiation from B travels an additional distance x metres where $x = d \sin\theta$ metres.

The total phase difference between the signals reaching P from the two aerials is thus:

$$\phi + 2\pi d \sin\theta/\lambda \text{ rad} \quad \text{say, } \delta \text{ rad}$$

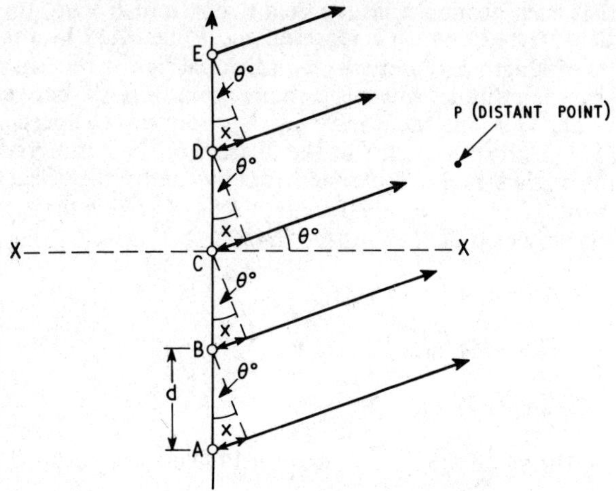

Fig. 11.11 A line of n vertical aerials

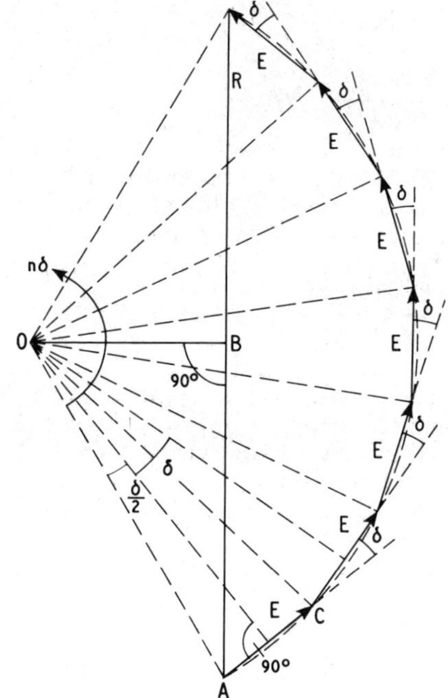

Fig. 11.12 Addition by phasors of the fields of n uniformly spaced vertical aerials

With n aerials each producing at P a field strength of E V/m, there is a resultant field strength which is represented in Fig. 11.12 by the phasor R. The phasors of length E represent the individual field strengths produced at point P by each dipole. Angle δ is the phase difference between consecutive phasors. The same angle δ exists between the perpendiculars drawn to O from the mid points of the phasors E. By symmetry it is seen that the angle δ is also subtended at O by the length of each phasor. The angle subtended at O by the resultant R is $n\delta$. OB is the perpendicular bisector of R. Thus $R = 2AB$.

Also $$AB = OA \sin . n\delta/2 \tag{11.1}$$

But $$E/2 = OA \sin \delta/2$$

and thus, $$OA = E/2 \sin \delta/2 \tag{11.2}$$

Substituting the value of OA of Equation 11.2 into Equation 11.1 gives,

$$AB = \frac{E \sin n\delta/2}{2 \sin \delta/2}$$

Since $$R = 2\,AB,$$

$$R = \frac{E \sin n\delta/2}{\sin \delta/2} \text{ V/m.}$$

If the field strength E is written in terms of the field strength which the one dipole gives at the same point if it has the whole transmitter power ($E = E'/\sqrt{n}$), of the array is:

$$R = \frac{E' \sin n\delta/2}{\sqrt{n} \sin \delta/2} \text{ V/m.}$$

The factor by which E' is multiplied is then the voltage gain factor of the array in terms of the field strength which would be produced by the same transmitter feeding all its power to one element similar to those used in the array. If R is written as $E'F$,

where $$F = \frac{\sin n\delta/2}{\sqrt{n} \sin \delta/2}$$

then the power gain is 20 log F dB.

The following examples illustrate the application of these results.

Example 11.7

Eight vertical dipoles uniformly spaced by half a wavelength in a horizontal line are energised in phase. Deduce an expression for the array factor and sketch the polar diagram for the horizontal plane.

With reference to Fig. 11.11, the phase difference between radiations from successive aerials, calculated for a direction at an angle θ to the normal to the line of the array is equal to:

$$\lambda/2 \sin\theta \, . \, 2\pi/\lambda \text{ rad.}$$

Hence $$\delta = \pi \sin\theta \text{ rad}$$

and $$\delta/2 = \pi/2 \sin\theta \text{ rad.}$$

From Section 11.4, the array factor is

$$\frac{\sin n\delta/2}{\sin \delta/2} \quad \text{where } n = 8.$$

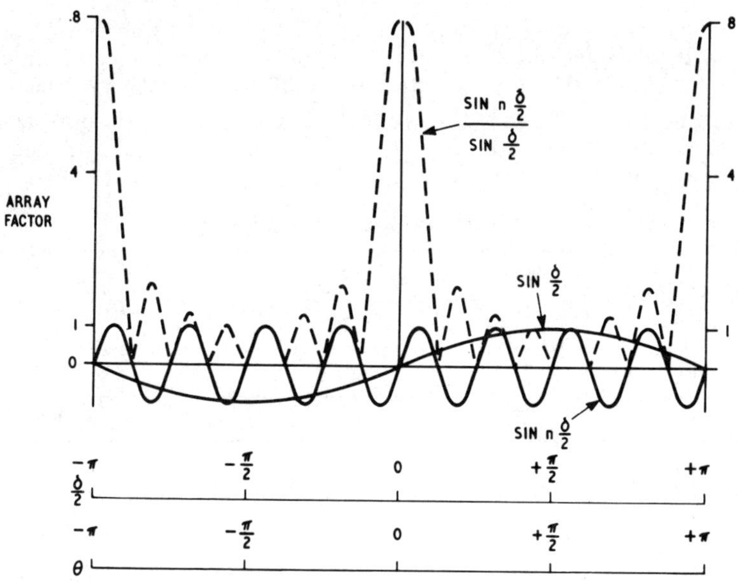

Fig. 11.13 Graphs relating array factor to angle θ

In the graphs of Fig 11.13, $\sin \delta/2$ is plotted for values of $\delta/2$ between $\pm\pi$. The graph of $\sin n\delta/2$ is shown completing eight cycles for the single period of $\sin \delta/2$. The third graph shows how the ratio $\sin n\delta/2$ to $\sin \delta/2$

changes and its ordinates are obtained by dividing the ordinates of the graph of sin $n\delta/2$ by those of the graph of sin $\delta/2$, regardless of sign. It is to be noted that the array factor is zero if sin $n\delta/2 = 0$ but sin $\delta/2$ is not at the same time zero. But as sin $\delta/2$ approaches zero, sin $\delta/2$ also approaches $\delta/2$. Likewise sin $n\delta/2$ approaches $n\delta/2$ as $n\delta/2$ approaches zero. As both these angles tend to zero, their quotient tends to the value n. From the graph of the array factor the following information is available:

1. The polar diagram has two major lobes whose maxima lie in the directions given bv $\theta = 0$ and $\theta = \pi$.
2. The array factor for the directions of maximum radiation is 8 so that the field strength in these directions is eight times the field due to one of the dipoles.
3. The angular width of each major lobe is from the scale of the graph seen to be, from $\delta/2 = -180/8$ to $+ 180/8$ degrees or δ from $-45°$ to $+45°$. This gives

$$\sin\theta = \delta/\pi = \pm 0{\cdot}25 \quad \text{and}$$

$$\theta = -14°\ 30' \text{ to } + 14°\ 30'$$

which is an angular width of 29°.
4. Between the major lobes there are six minor ones having amplitudes as indicated by the ordinate scale of the graph for the array factor.

The tabulation in Table 11.2 sets out the values of $\delta/2$, δ, and θ for which in one quadrant there are nulls. The nulls for all quadrants are marked on the polar diagram Fig. 11.14.

Table 11.2

$\delta/2$	$\pi/8$	$\pi/4$	$3\pi/8$	$\pi/2$
δ	$\pi/4$	$\pi/2$	$3\pi/4$	π
Sin $\theta = \delta/\pi$	0·25	0·5	0·75	1·0
θ	14°30′	30°	48°36′	90°

Example 11.8

Six vertical half wave dipoles are arranged in a horizontal straight line and energised so that successive dipole currents lag by $\pi/2$ on the currents in the adjacent aerials. The radiators are regularly spaced by $\lambda/4$. Deduce an expression for the array factor and sketch the polar diagram for the horizontal plane.

With reference to Fig. 11.11, it is now assumed that the distance d represents $\lambda/4$ and that the current in A lags that in B by $\pi/2$, that the

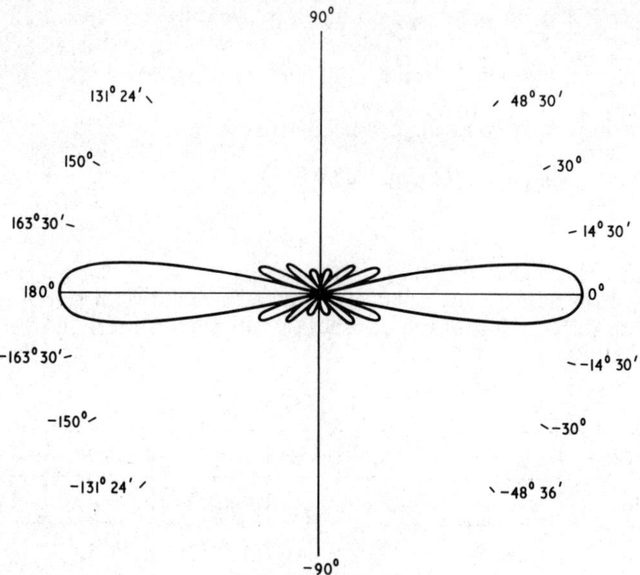

Fig. 11.14 Polar diagram result for Example 11.7

current in B lags that in C by $\pi/2$, and so on. The phase difference between signals from successive aerials radiated in a direction at angle θ to the normal to the line of aerials is equal to:

$$\delta = \pi/2 + \lambda/4 \,.\, 2\pi/\lambda \,.\, \sin\theta$$

and
$$\delta/2 = \pi/4 + \pi/4 \,.\, \sin\theta$$

Using the result of Section 11.4, the array factor is

$$\frac{\sin n\delta/2}{\sin \delta/2}$$

where $n = 6$. Fig. 11.15 shows the graphs of $\sin \delta/2$, $\sin 6\delta/2$ and

$$\frac{\sin 6\delta/2}{\sin \delta/2}$$

for $\delta/2$ between $\pm\,\pi$.

Half the polar diagram shape will be given by the graph of the array factor for the range within which θ varies from $+\,\pi/2$ to $-\,\pi/2$.

If $\theta = +\pi/2$, $\sin\theta = +1$ and $\delta/2 = \pi/2$

If $\theta = -\pi/2$, $\sin\theta = -1$ and $\delta/2 = 0$.

The values of $\delta/2$ and angle θ calculated from,

$$\sin\theta = 4/\pi\,(\delta/2 - \pi/4)$$

for the null directions and lobe maxima are set out in Table 11.3. Fig. 11.16 is the appropriate polar diagram and this shows the array to be an end-fire array with four minor lobes of radiation in addition to the one major lobe.

Table 11.3

$\delta/2$	0	$\pi/6$	$\pi/4$	$\pi/3$	$5\pi/12$	$\pi/2$
Sin θ	−1	−0·3	0	+ 0·3	0·6	1·0
θ	$-\pi/2$	$-19°26'$	0	$+19°26'$	$41°50'$	$+\pi/2$

11.5. THE SLOT AERIAL

A slot formed by cutting a narrow rectangular strip out of a plane sheet of material of high conductivity may be used as a radiator. The resonance frequency of the slot is almost equal to that of the rectangular strip of metal which has been removed. The slot may be fed from a waveguide, a coaxial feeder or from a balanced transmission line connected to the mid points of the opposite long sides of the slot. When the slot is energised at the correct frequency, current alternates round the metal sheet and an alternating electric field is sustained across the slot. Fig. 11.17 indicates paths of current flow while Fig. 11.18 shows a plane at 90° to the plane of the slot containing lines of electric flux.

A vertical slot radiates a wave which is horizontally polarised, while the radiation from a horizontal slot would be vertically polarised. It may be said that the slot dipole is analogous to the conductor dipole but with space and metal interchanged and with magnetic and electric field planes interchanged.

The area of the metal sheet in which a slot aerial is cut, should be large compared with the dimensions of the slot. Nine or ten wavelengths from edge to edge is a satisfactory size. The width of the slot is not critical, though a wider slot gives a wider bandwidth. The resonance frequency impedance of a slot is theoretically 471 Ω but in practice it is about 350 to 360 Ω resistive. When fed with a signal frequency which is less than the resonance frequency of the slot, the slot has inductive

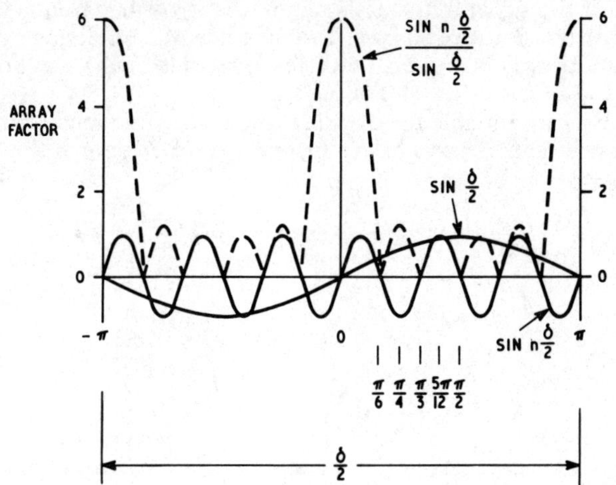

Fig. 11.15 Graphs relating array factor and angle δ/2

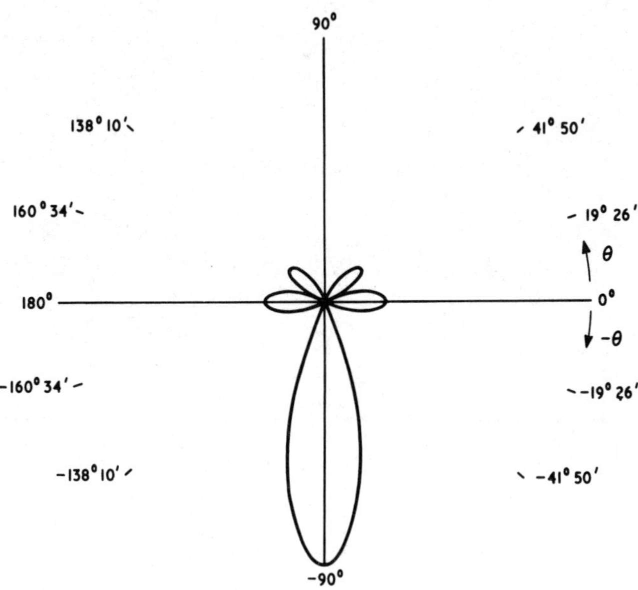

Fig. 11.16 Polar diagram result for Example 11.8

reactance. If the signal frequency is above the resonance frequency, the reactance of the slot is capacitive. The presence of a reflecting cavity behind the slot makes the aerial radiate to one side of the sheet only. The reflector also increases the slot impedance.

Slot aerials are suitable for the v.h.f. and higher frequency bands since they are small radiators and have a rigid structure. Although slot aerials

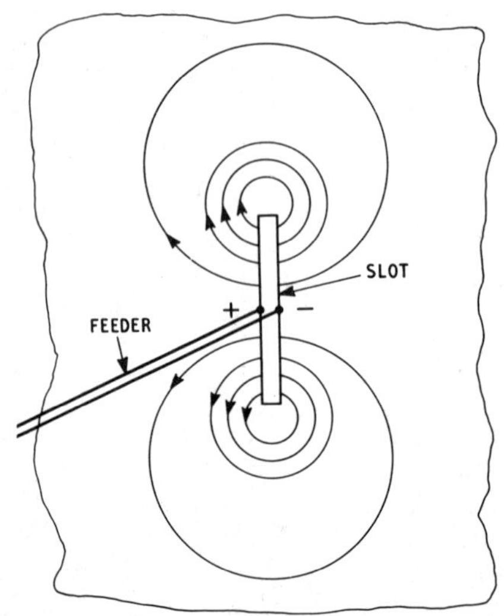

Fig. 11.17 Circulation of current round a slot

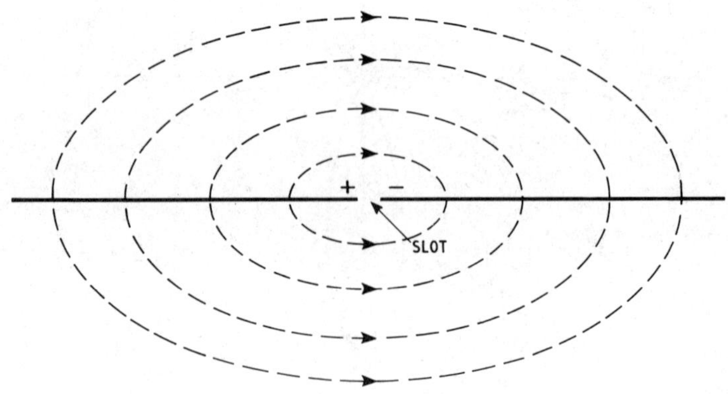

Fig. 11.18 Electric flux on opposite sides of slot

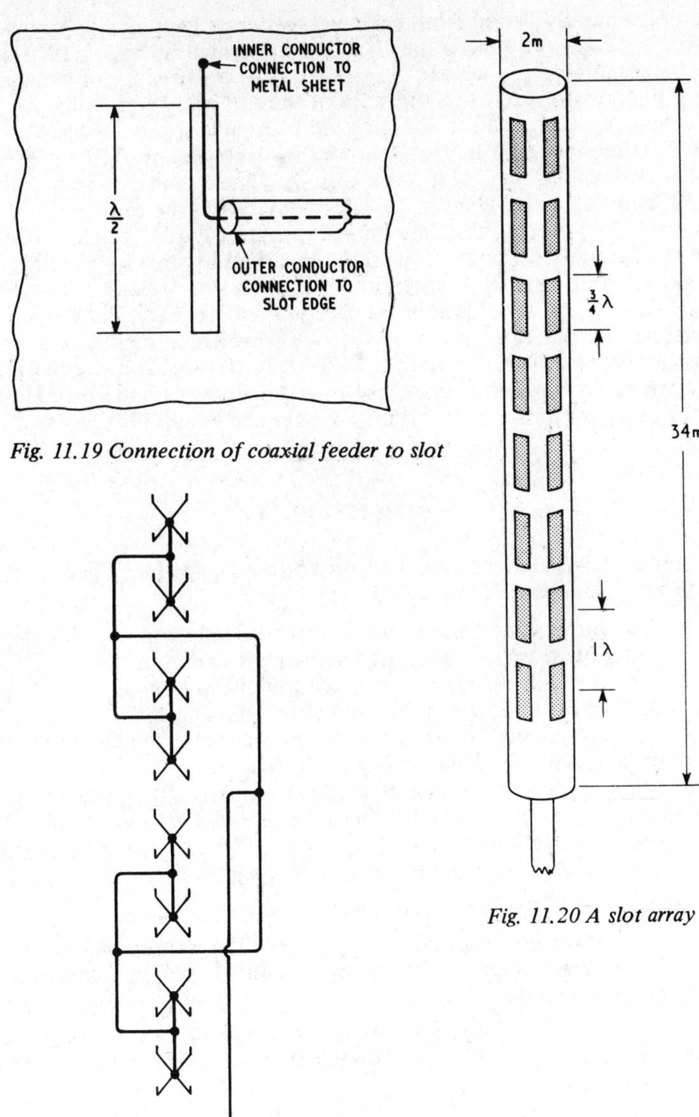

Fig. 11.19 Connection of coaxial feeder to slot

Fig. 11.20 A slot array

Fig. 11.21 Branching feeders for slot array

can conveniently be fed from balanced feeders they may also be matched to coaxial feeders by using the arrangement shown in Fig. 11.19. The centre conductor of the coaxial feeder acts as the primary of a magnetic coupling transformer while the sides of the slot are the secondary.

Slot arrays are used for radiating v.h.f. broadcast programmes by the B.B.C. (Chapter 10). Fig. 11.20 shows the arrangement of 32 slots in eight tiers around the surface of a cylinder. The cylinder is 34 m tall and 2 m in diameter. The vertical spacing between the centres of adjacent slots is approximately one wavelength. Fig. 11.21 shows how concentric feeders branch at the points marked by dots to carry the services of three different programmes to each slot radiator. The total bandwidth of the aerial is approximately 4·5 MHz. Each slot has a horizontal conductive bar spaced 0·1 of a wavelength behind it to prevent inward radiation. It is the object of the array to provide a circular polar diagram in the horizontal plane but to avoid skyward radiation. The gain of the array over a single slot dipole is between 8 and 9 dB.

11.6. RHOMBIC AERIALS

Rhombic aerials are so named because of their shape (see Fig 11.22). Their advantages are as follows:

1. They are untuned and provide a broad-band aerial suitable for rapid switching from one working frequency to another.
2. They are usually relatively simple and cheap to erect, and those for shortwave working require only a low mast height.
3. The vertical angle of radiation is low and therefore rhombics are suitable for long distance F-layer propagation.
4. Rhombic aerials have unidirectional properties of radiation and reception with a gain which may be between 10 and 15 dB compared with a single vertical dipole. The longer the arms of the rhombic aerial the greater its gain is likely to be.

Disadvantages of rhombic aerials are:

1. There is a considerable loss of power in the terminating resistance.
2. These aerials require a large area of land. How much, depends on the wavelengths involved.

Such aerials may be designed to operate at frequencies as low as 3 MHz or as high as 200 MHz. The side lengths of the rhombus may be between 2 and 6 wavelengths.

The rhombic aerial is a development of the terminated long-wire aerial. A single long conductor fed by a transmitter at one end and terminated at the other end by an absorbing characteristic resistance, has a travelling wave current whose amplitude decreases exponentially towards the termination. The longer the wire, the greater is the total amount of radiation and the less power is absorbed in the termination. To be effective, such an aerial must be several wavelengths long. Its considerable

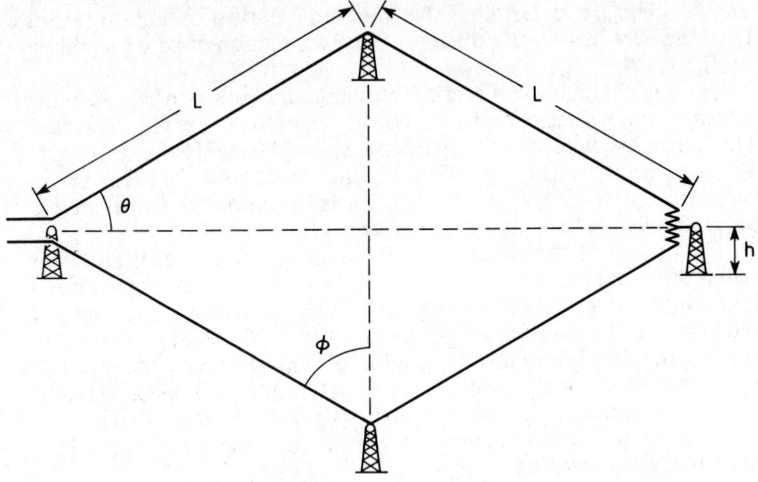

Fig. 11.22 The rhombic aerial

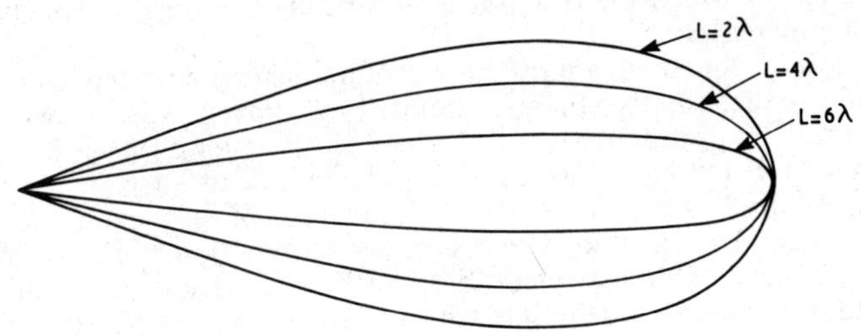

Fig. 11.23 Showing increase of directivity with side length

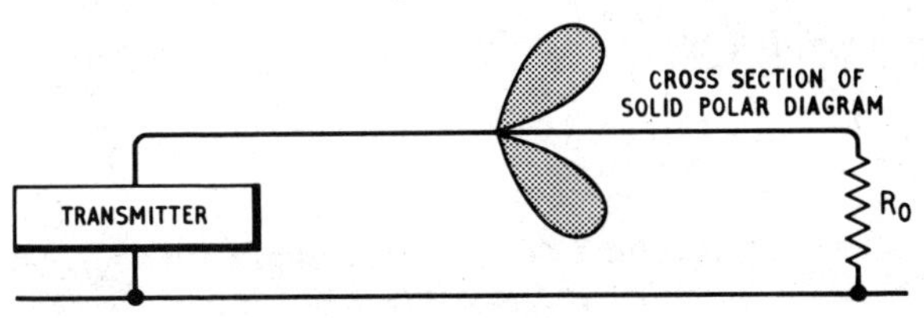

Fig. 11.24 A long wire aerial

length, therefore, determines that this aerial must be a horizontal wire. The long wire has a polar diagram the cross section of which is shown in Fig. 11.24.

The angle which the direction of maximum radiation makes with the length of the wire is determined by the length of the wire in wavelengths. The longer the wire, the smaller is the angle—being 30° for a 2λ wire and 17° for a 6λ wire. The long wire can be considered to be made up of a very large number of short collinear radiators whose phase of current lags progressively towards the terminating resistance.

The rhombic has four such long wires arranged so that their several radiations add up in the same forward direction. The angle θ in Fig. 11.22 is chiefly determined by the angle which the direction of maximum radiation makes with the length of the wire. It, therefore, depends on the length of each arm in wavelengths. The height above the ground is important as this is the chief factor in determining the angle which the direction of maximum radiation makes with the horizontal. The relationship between the angle of elevation and the aerial height is approximately given by

$$\sin \delta = \lambda/4h,$$

where λ and h are expressed in the same units. If the frequency is reduced below that for which the rhombic is specifically designed, the angle of elevation increases.

As lower frequencies are used for communicating with receiving points nearer to the transmitter than for points further away, the changed angle of elevation is appropriate to the geometry of F-layer propagation (Chapter 1). Rhombic aerials operate well over a 2 to 1 frequency range and can be used over a wider ratio of frequencies if some reduction of efficiency can be tolerated. The dissipative termination marked as R_o in Fig. 11.22, has a value between 600 and 800 Ω. It may be constructed from an iron alloy twin wire line with an attenuation coefficient of 10 dB. Dimensions of two types of rhombic aerial in use for h.f. transmission at the Post Office station at Rugby are listed below:

Type 1	*Type 2*
Height h = 23m	h = 46m
Leg length L = 95m	L = 164m
angle 2ϕ = 140°	2ϕ = 135°

These dimensions represent a rhombic side length of about 4λ in the 12-MHz band and about 8λ in the 25-MHz band for the first type and about 2λ side length in the 4-MHz band with 4λ in the 8-MHz band for the second type of aerial.

The use of multiple wire rhombics gives a gain of 0·5 to 1·5 dB over a single wire aerial with similar dimensions. Fig. 11.23 shows how both directivity and gain increase if the side length of the rhombic is made a greater number of wavelengths long.

11.7. AERIAL FEEDERS

An m.f. transmitter situated immediately below its aerial can be connected directly to the aerial by a single-wire feeder which forms a part of the tuned-aerial circuit. A coupling capacitor and tuning indicator provide for matching and tuning as explained in Chapter 7.

A spaced twin-wire feeder may be used to feed an h.f. transmitter which is near to its aerial. The feeder is tuned and consists of spaced twin wires cut to an integral number of quarter wavelengths. Standing waves form along the feeder. Coupling to the transmitter is via a series-tuned circuit if the voltage at the feeder input is low (the feeder in this

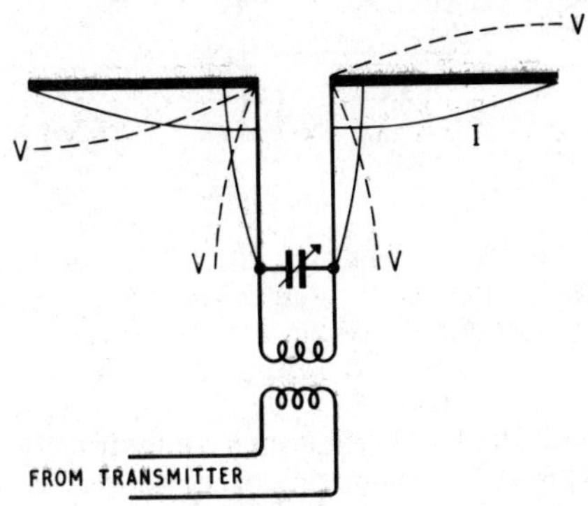

Fig. 11.25 Standing waves on a λ/4 feeder

instance is an even number of quarter wavelengths long), or by a parallel-tuned circuit if the voltage at the feeder input is high. Figs 11.25 and 11.26 illustrate these two alternatives. The standing waves of current which form on the feeders and the aerials are shown by full line curves. The dotted curves show the distribution of voltage standing waves. In Fig. 11.25 the feeder has a high input impedance but the feeder in Fig. 11.26 has a low input impedance. Such arrangements have only limited applications. The transmitter must be of relatively low power and the frequencies which can be used are only those which are harmonically related to the fundamental frequency of the feeder system.

Longer feeders carrying r.f. power from a transmitter to its aerial are matched at the transmitter to the optimum load impedance required by the power amplifier valves and at the aerial to the input impedance of the aerial. Accurate matching prevents reflection of travelling waves at the

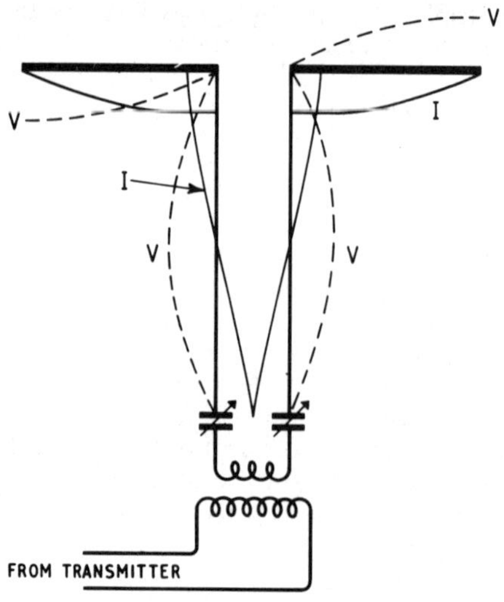

Fig. 11.26 Standing waves on a λ/2 feeders

termination of the feeder and so avoids the formation of standing waves. If standing waves exist, the copper loss is heavy at the current antinodes, while heavy leakage loss and possibly even a flashover occurs through the insulation between the lines at voltage antinodes.

Feeders may be of the twin-wire type, supported by poles and insulators which hold the two conductors symmetrically disposed to earth and about 12 in apart. Such feeders are said to be balanced. Each conductor has the same impedance to earth and the r.f. potential of one feeder is equal to but opposite in phase to that of the other. A small amount of radiation occurs from a transmitter feeder of this type and pick up of r.f. power from adjacent transmitter feeders is possible. Radiation and pick up can be reduced by using a four-wire feeder with the conductors arranged so that a cross section would show them at the corners of a square. Diagonally opposite conductors are connected in parallel. Such a four-wire feeder is suitable for an m.f. transmitter.

Almost complete freedom from feeder radiation or pick up is provided by coaxial feeders. These, however, are expensive and cannot be used for high-power transmitters because the conductor spacing is relatively small and flashover could easily occur. A single coaxial feeder is unbalanced since the outer conductor is earthed or has a very low impedance to earth while the inner conductor is insulated from earth. Two coaxial conductors can be used as a balanced pair in which the inner conductors provide the go and return paths for aerial current. Both outer conductors of this balanced system are earthed.

11.8. MATCHING

Methods of matching a transmission line to an aerial may be listed as follows:

1. Direct matching.
2. Delta matching.
3. Use of a quarter-wave, open-wire line transformer.
4. Use of stubs.
5. Use of a quarter-wave matching line.
6. Tapered matching line.
7. R.f. coil matching transformer.

A direct match is usually only practicable for a receiving dipole connected to a coaxial cable of characteristic impedance about 70 Ω. The input impedance of a centre connection to a dipole is nominally 73 Ω. Although the direct connection of the unbalanced feeder line to the dipole may result in slight pick up of noise or signal voltages by the screen, the interference is small and tolerable in conditions of reasonable signal strength.

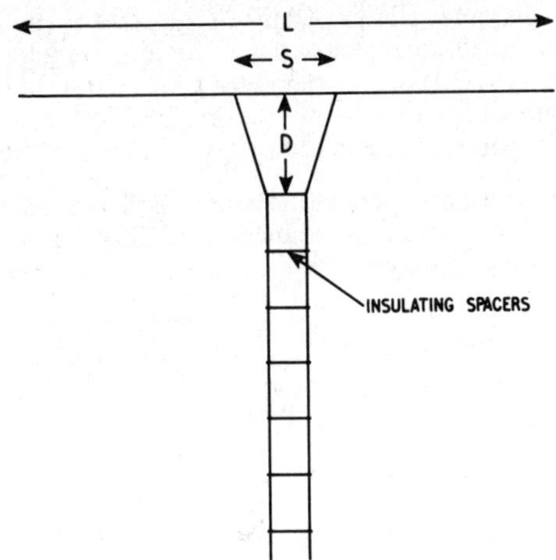

Fig. 11.27 Delta match

A simple method of matching a twin-wire feeder to a horizontal dipole is by the delta match illustrated in Fig. 11.27. This is a type of auto transformer match in which, for a receiving aerial, the whole aerial is the primary and the tapped section is the secondary. The feeder is tapped on to the aerial at positions where the ratio of the r.m.s. voltage to the r.m.s. current gives an impedance value which matches the characteristic impedance of the feeder. The exact positioning of the tapping points may be found by moving them outwards symmetrically from the centre until the voltage standing wave ratio on the feeder becomes as near unity as possible. Approximate dimensions for a $\lambda/2$ dipole for a frequency f MHz can be found by using the empirical formulae:

Length L in metres = $150 \times 0{\cdot}95/f$ m.
Spacing of tapping points, $S = 45/f$ m.
Clearance between aerial and feeder, $D = 37{\cdot}5/f$ m.

These dimensions are for a typical twin-wire feeder impedance of 600 Ω.

A $\lambda/4$ transmission line which is terminated by an impedance Z_L, has an input impedance equal to Z_o^2/Z_L where Z_o is the characteristic impedance of the line. (The explanation of this is to be found in most standard works on transmission line theory.) By designing a quarter-wave line having a suitable characteristic impedance, it is possible to arrange that the input impedance to this line will match the transmission line proper when the aerial of impedance Z_L is connected at its output terminals.

Example 11.9

An open-wire line of resistive characteristic impedance 600 Ω is to be matched by a quarter-wave line to an aerial of resistive input impedance 70 Ω. Find the characteristic impedance required for the $\lambda/4$ line and the centre to centre conductor spacing if the matching line is constructed of two parallel lengths of tubing of diameter 1·27 cm. (Given that the characteristic impedance of a twin-wire line is: $276 \log_{10} D/r$ Ω where D is the conductor spacing and r is the radius of each conductor.)

The $\lambda/4$ line must have a characteristic impedance such that when it is terminated by the aerial impedance of 70 Ω, it has an input impedance of 600 Ω. It may then be connected to the 600 Ω line without mismatch. Thus:

$$600 = Z_o^2/70$$

Therefore $$Z_o = \sqrt{(70 \times 600)}$$

$$= 205\ \Omega$$

Also, $$205 = 276 \log_{10} D/r$$

and $$\log_{10} D/r = 0{\cdot}743$$

$\therefore$ $$D/r = 5{\cdot}543$$

$\therefore$ $$D = 5{\cdot}543\, r$$

$$= 5{\cdot}543 \times 0{\cdot}635 \text{ cm}$$

$$= 3{\cdot}5 \text{ cm},$$

spacing between the centres of the conductors.

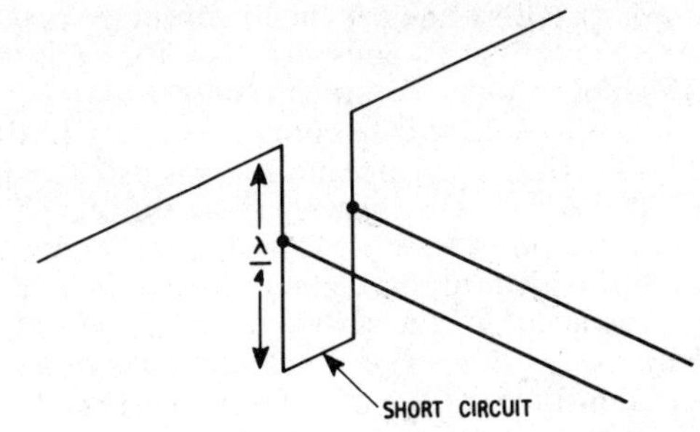

Fig. 11.28 A quarter-wave matching line

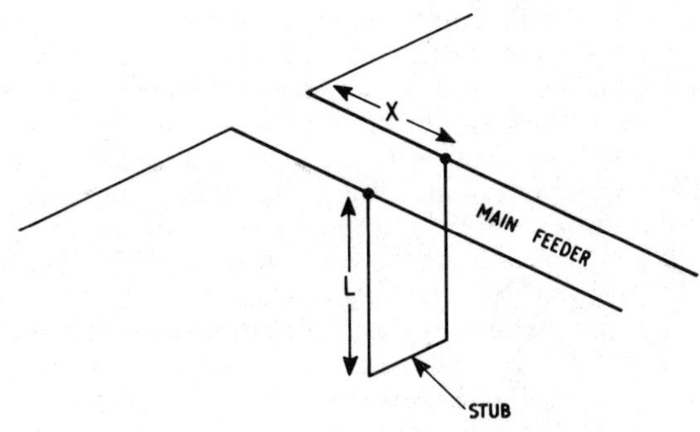

Fig. 11.29 A pendant stub

The λ/4 transformer match is suitable when the ratio of feeder to aerial impedance is of the order of 5 or more. A disadvantage of this system is that matching is only achieved at the one frequency for which the matching line is λ/4. A broader band of operation is obtained if the impedance transformation is carried out in two steps using two λ/4 lines of different spacings. For example, the load of 70 Ω in Example 11.9 can be matched to a transmission line of 600 Ω by using a λ/4 line next to the aerial, of characteristic impedance 140 Ω, and a second λ/4 line between the first and the transmission line, of characteristic impedance 410 Ω. The two-stage match reduces the amplitude of any standing waves which form for frequencies which differ from the λ/4 lines' resonance frequency.

An alternative way of using a quarter-wave line for matching is shown in Fig. 11.28. The quarter-wave line is open circuit at the lower end if the aerial terminal impedance is low but is short circuited if the aerial input impedance is high. In either arrangement, standing waves of voltage and current form on the quarter-wave line. The feeder is tapped on to the matching line at a point where the ratio of the r.m.s. voltage to the r.m.s. current represents an impedance equal to that of the feeder line.

The quarter-wave matching line has the disadvantage that its impedance is reactive except at its ends and, therefore, fails to provide an exact match to a line impedance which is purely resistive. An improvement is the non-resonant *stub* connected to the feeder line, as shown in Fig. 11.29. The impedance along a mismatched line varies along its length. The distance X is chosen so that the real part of the line admittance, at the point where the stub is to be connected, is equal to the conductance provided by the characteristic resistance of the transmission line. The length of the stub line is adjusted so that it presents a susceptance to the feeder line equal in value, though opposite in sign, to that of the line admittance at the point of connection. If, for example, the characteristic resistive impedance of the feeder is 600 Ω then the distance X is such that without the stub the line has an admittance at this distance from the aerial of $(1{\cdot}67 \pm \mathrm{j}B)$ mS. The stub is then connected and adjusted in length so that it presents a susceptance of $(\mp \mathrm{j}B)$ mS. Standing waves exist on the section of the feeder X so that this length should be kept as short as possible. The nearest point to the aerial which presents the required conductance value is, therefore, chosen for the connection of the stub. An open-circuited stub leads to a minimum distance X if the aerial impedance is less than the characteristic impedance of the line, but a short-circuited stub is better in this respect if the aerial impedance is bigger than that of the transmission line impedance. As a first step in carrying out the matching process the voltage standing wave ratio for the aerial and feeder, without the stub connected, is measured. This is done by measuring the line voltages at points of antinode and nodes and then dividing the larger voltage by the smaller. Charts such as that of Fig. 11.30 are available from which the distances X and L can be found in terms of the wavelength of the transmission frequency. Final adjustment of the stub length is trial and error in which a standing wave ratio of unity is sought.

Any method of matching which makes use of resonant sections of line or tuned stub provides effective matching over a small band of frequencies only. Matching for a broader band of frequencies is provided by a tapered transmission line section. For matching a balanced load to a balanced line the tapered line is an open twin wire, the spacing of which decreases from one end to the other. The end of the matching line with the wider spacing is connected to the higher of the two impedances which are to be matched. The inductance per unit length of the tapered line decreases and the capacitance per unit length increases towards the end of the line with the smaller spacing. For a line with an exponential taper,

$$L_2/L_1 = C_1/C_2 = \epsilon^{4\pi l/\lambda_c}$$

where l = line length in metres,

L_2 = line inductance in H/m at the end with wider spacing,

L_1 = line inductance in H/m at the end with narrow spacing,

C_1 = line capacitance in F/m at the narrower end,

C_2 = line capacitance in F/m at the wider end,

λ_c = the cut-off wavelength of the line. This wavelength corresponds to the cut-off frequency of a low-pass filter whose elements are L_1 and C_1.

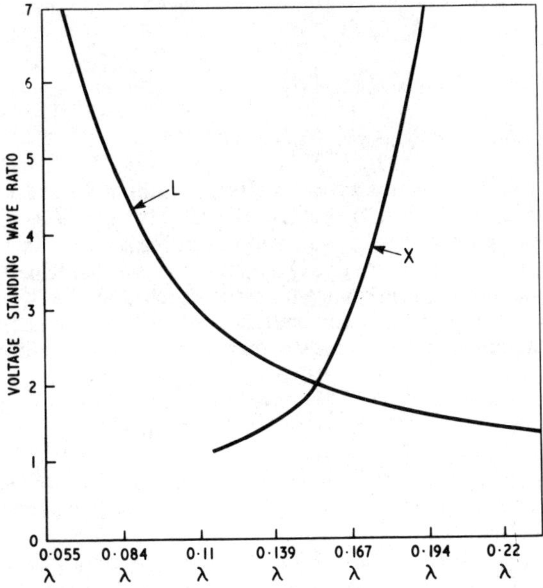

Fig. 11.30 Graphs for finding position and length of shorted matching stubs in terms of wavelength

The ratio of the impedances matched by the tapered line is approximately $\sqrt{(Z_{o2}/Z_{o1})}$ where, Z_{o2} is the characteristic impedance in terms of the line constants at the wider end and Z_{o1} is the characteristic impedance in terms of the line constants at the narrower end of the tapered line. A ratio of matched impedances is typically 2 to 1.

At medium and low frequencies, stubs and $\lambda/4$ lines occupy too much space and an r.f. coil transformer matching system is used. Fig.

11.31 shows a matching circuit suitable for matching a balanced open-wire line to an aerial which presents an unbalanced load. The variable capacitor C_v tunes the primary to resonance at the carrier frequency. The dynamic impedance of C_v in parallel with the load transformer primary is arranged to match the charateristic impedance of the balanced line. The e.m.f.s in L_3 and L_4 are phased relative to the primary e.m.f.s, as indicated by the arrows, so that these e.m.f.s act as generators in parallel as far as the aerial is concerned. On the primary side of the transformer, however, L_1 and L_2 are coils whose e.m.f.s of self induction act in series aiding. Nevertheless, L_1 and L_2 have equal capacitances to earth so that the transmission line is connected across a balanced termination. C_k is the coupling capacitor. Fig. 11.32 represents this as a capacitive reactance in parallel with an aerial of resistive impedance R_{ae}, and also as an equivalent series arrangement of X_s and R_s.

$$X_p = 1/j\omega C_k$$

$$X_s = X_p \cdot R_{ae}^2 \Big/ \left(R_{ae}^2 + X_p^2\right)$$

$$R_s = R_{ae} \cdot X_p^2 \Big/ \left(R_{ae}^2 + X_p^2\right)$$

The reactance X_s resonates with the effective inductance reactance of the secondary circuit of the transformer. The transformer ratio transforms the series resistance value R_s to an equivalent resistance in series with the primary of the transformer. The Q value of the parallel-tuned primary with the reflected load resistance present is such that the dynamic impedance of the parallel circuit matches the line impedance. Choice of a suitable capacitor makes a transformer of small turns ratio practicable.

11.9. THE YAGI AERIAL

The Yagi array, named after its Japanese inventor, may be used either for transmission or reception. It is widely familiar as a receiving aerial for television signals.

The elements of the array are: a dipole aerial, a reflector and a number of directors (typically four to eight in number). Fig. 11.33 represents such an array. Assuming it were designed for the reception of a v.h.f. radio signal of carrier frequency 94 MHz the appropriate dimensions of the elements are:

Length of reflector R	1.63 m (0.51λ)
Length of dipole A	1.52 m (0.475λ)
Length of Director D_1	1.43 m (0.45λ)
Length of Director D_2	1.32 m (0.41λ)
Reflector spacing X	0.84 m (0.253λ)
Director spacing Y	0.485 m (0.152λ)

The elements should be aligned within a plane parallel with the plane of polarisation of the signal to be received. The reflector is placed so that the reradiated signals due to the currents induced in it by the signal will reach the dipole in phase with the signal fields coming to the dipole directly from the

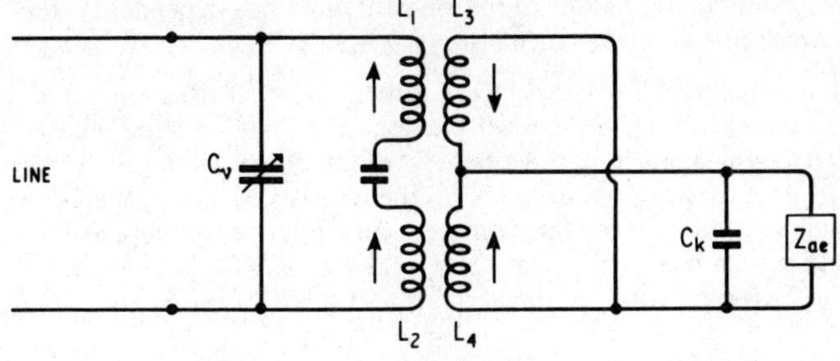

Fig. 11.31 Matching balanced line to unbalanced aerial

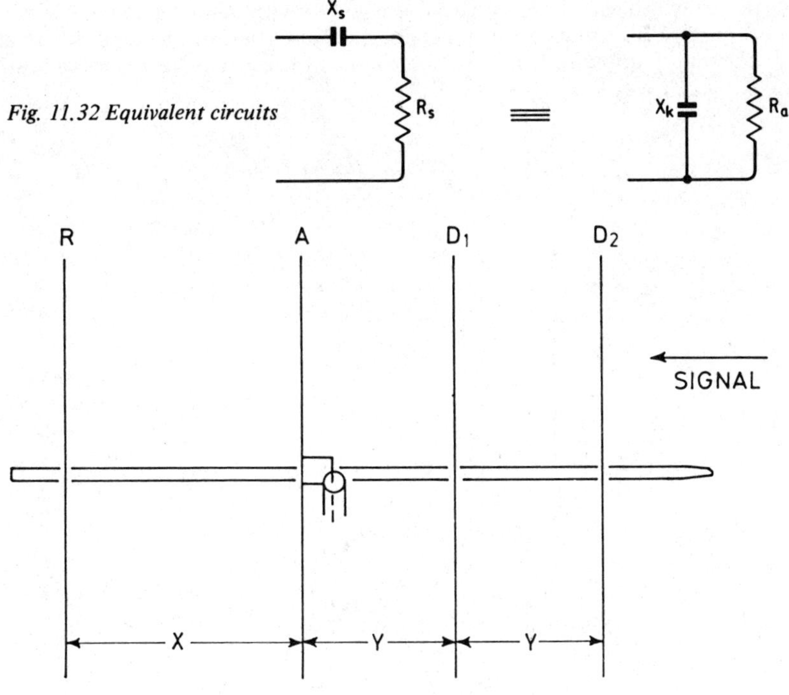

Fig. 11.32 Equivalent circuits

Fig. 11.33 A Yagi aerial

transmitter. The directors similarly are placed so that their reradiation reinforces the directly received signals at the dipole. Practical results show that the dimensions in terms of signal wavelength (shown in brackets) are optimum. Successive directors decrease in length by about 8 % of the previous element. The phase of the reradiated signal reaching the dipole from a parasitic element depends on:

1. The spacing between the elements and the dipole, and

2. The impedance angle of the element, an angle dependent on its length, thickness and resistance.

Signals which reach the array from directions other than along the line joining the directors to the dipole, reach the dipole wrongly phased relative to the indirect signals reaching the dipole from the directors and the reflector.

The gain of the aerial depends on the number of directors used. Five elements give a gain of approximately 7 dB relative to a dipole, ten elements may give a gain of 10 dB relative to a dipole.

The 3 dB bandwidth of the aerial is about 6% of the frequency for which the array is designed.

A large number of parasitic elements arranged near to a dipole aerial reduce the normal aerial impedance of 73 ohms to a lower value making it difficult to match to the television or v.h.f. receiver feeder. For this reason the dipole in the array is often a folded dipole (see Fig. 11.34). In this aerial the mutual coupling between the two parallel arms increases the impedance at the aerial terminals fourfold so that direct matching to the coaxial feeder is again possible.

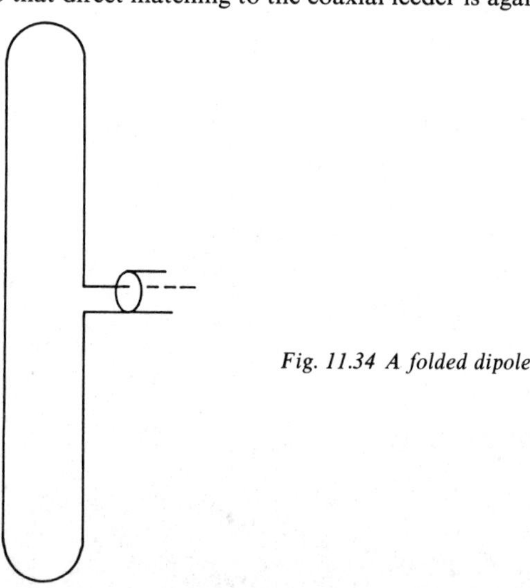

Fig. 11.34 A folded dipole

11.10. THE LOG PERIODIC DIPOLE

The *log periodic dipole* aerial is useful for radiation or reception in the h.f., v.h.f., or u.h.f. bands. Like the rhombic aerial it has a broad bandwidth of operation but it has the advantage of requiring less ground space than the rhombic.

The elements of a log periodic dipole array are connected to a feeder as in Fig. 11.35. The lengths of consecutive elements are reduced by a constant scaling factor whose value is likely to be within the range 0.85 to 0.95. Let this factor be represented by S. Then the geometry of the array is such that:

$$\frac{l_2}{l_1} = \frac{l_3}{l_2} = \frac{l_4}{l_3} = \frac{l_5}{l_4} = S.$$

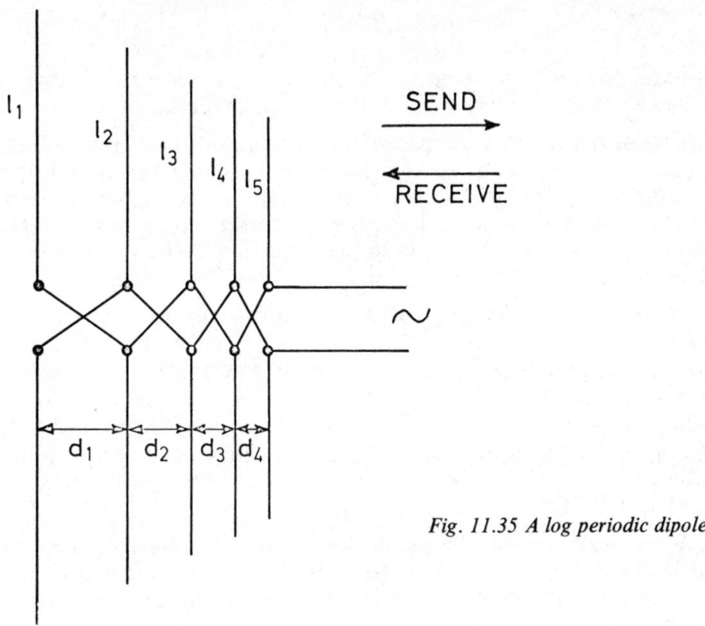

Fig. 11.35 A log periodic dipole

The element spacings have the same ratio to one another so that:

$$\frac{d_2}{d_1} = \frac{d_3}{d_2} = \frac{d_4}{d_3} = S.$$

As the resonant frequency of each element is inversely proportional to its length, it follows that:

$$\frac{f_2}{f_1} = \frac{f_3}{f_2} = \frac{f_4}{f_3} = \frac{1}{S}$$

where f_1, f_2, f_3, f_4 and f_5 are the resonant frequencies of the elements. Consequently, $(\log f_2 - \log f_1) = (\log f_3 - \log f_2)$, so that the logarithm of the resonant frequency increases by a constant increment from one element to the next, which is the basis for the name of the array.

Each dipole responds to those frequencies near to its own resonant frequency but collectively the elements may respond to a bandwidth of 3:1 ratio.

The spacing of the elements and the transposed connection of alternate dipoles to the feeder causes the main lobe of the polar diagram to be in the direction towards the shorter elements. A horizontal array has a half-power beam width in the horizontal plane of some 60 degrees. There are few minor lobes in the polar diagram and the aerial has a good front-to-back ratio. The gain relative to an isotropic radiator is some 10 or 12 dB. A typical input impedance is 300 ohms.

The log periodic dipole array is useful for shore-to-ship communication, ground-to-air communication or for communicating with a number of fixed points in the broad direction of forward radiation.

QUESTIONS

1. What do you understand by the terms: broadside array, end-fire array, collinear array? Describe briefly the essential features of these arrays.

2. Two vertical aerials are spaced at a distance of one wavelength and energised equally in phase. At distance of d km in a direction in the horizontal plane at right angles to the plane containing the aerials, they produce a field strength of 10 mV/m. What field strength will these aerials produce at d km in a direction in the horizontal plane which makes 45° to the plane containing the aerials?

3. Two vertical aerials are positioned with one due north of the other by a quarter of a wavelength at the radiation frequency. Both aerials have the same amplitude of current but the more northerly aerial current lags by 90° on the current in the other aerial. How many decibels down is the field strength radiated on a bearing of 030° relative to the field radiated in a northerly direction, and how many decibels down on the field radiated in a northerly direction is the field radiated on the reciprocal bearing to 030°.

4. A horizontal dipole is suspended a quarter of a wavelength above an earth which can be regarded as being perfectly conductive. Find the angle of elevation above the horizontal at which maximum radiated field strength can be expected.

5. Ten vertical aerials are energised equally in phase. They are uniformly spaced at half wavelength intervals along a straight horizontal line. Find the angular width of the main lobe of radiation in the horizontal plane.

6. Describe briefly the constructional features of a rhombic aerial and state the particular advantages of this type of aerial.

7. Explain the necessity for matching an aerial feeder line to the aerial input impedance at the one end and to the transmitter optimum load impedance at the other. Name and illustrate by sketches seven different methods of matching a feeder to an aerial.

8. Find the characteristic impedance of a quarter-wave transmission line suitable for matching an open wire line of characteristic impedance 600 Ω to a load of 200 Ω.

9. What advantage has a log periodic aerial over a rhombic aerial? State a possible application of a log periodic array. A log periodic array has a scaling factor of 0.8 and consists of five dipole elements. The shortest dipole is resonant at 75 MHz; find the resonant frequency of the longest element. If the spacing between the shortest element and the next one to it is 1.25 m, what spacing should exist between the longest element and the one adjacent to it?

10. Give a brief description of a slot aerial and say what disadvantages it has as a radiator of v.h.f. signals.

12

Receiver and Transmitter Tests and Measurements

Measuring instruments used for receiver and transmitter tests and measurements include signal generators, electronic voltmeters, power meters and multi-range moving-coil instruments. In addition, certain auxiliary pieces of equipment, such as dummy aerials, are required. Most of this apparatus has been covered in the previous volumes but the following additional comments may now be made.

Signal generators for use in receiver tests should be accurate to better than 1% and should be directly calibrated. The output voltage should be precisely known and variable from 1 μV upwards. It is convenient if the output can be varied using a control which is calibrated in decibels. Alternatively an extra calibrated attenuator may be used.

Standard modulation frequencies which should be available are 400 Hz and, less frequently used, 1 000 Hz. It is often useful to be able to vary the modulation between 50 Hz and a few thousand hertz. The standard modulation depth is 30% but it is convenient, additionally, to be able to vary the modulation. The output impedance should be low; it is often 50 Ω, sometimes 10 Ω.

The signal generator should be well screened so that signals can not be coupled into the equipment under test except through the proper connection. Likewise, output from the generator should not be able to reach the equipment being tested through common mains supply leads or in any other extraneous way.

During any tests on receivers the impedance at the receiver input terminals should be about the same as the impedance of the aerial system with which the receiver is likely to be used. The input impedance of communication receivers is commonly 75 Ω. This may substantially match the signal generator impedance. For 10-Ω signal generators (Fig. 12.1(a)) a 65-Ω resistor in series with the output suffices to provide a reasonable match. For a balanced input receiver the signal generator resistance and the additional input resistance should be balanced, i.e. divided between the two input lines.

For broadcast-type receivers an 'average' aerial is simulated by use of the dummy aerial circuit shown in Fig. 12.1(b). This, over the range of

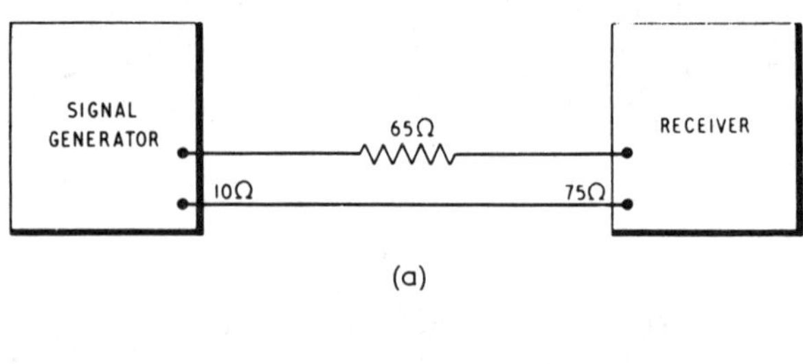

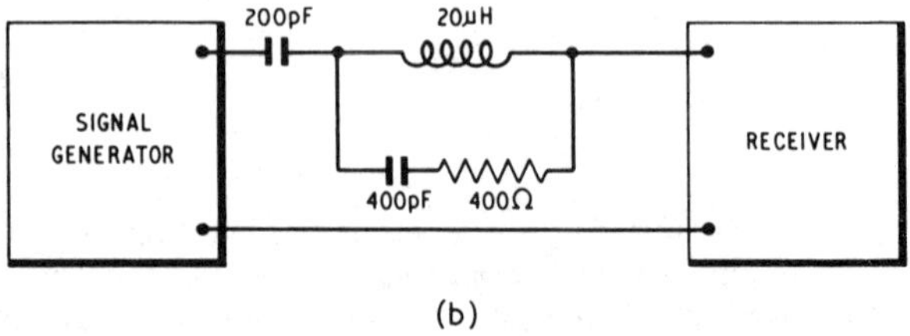

Fig. 12.1 Receiver testing input circuits

frequencies from below 1 MHz to above 20 MHz, roughly equates to the average aerial impedance likely to be encountered. Bearing in mind the wide variety of aerial types in use, it is clear that a large degree of approximation is involved.

For a sound receiver the output should properly be measured and evaluated at a loudspeaker or telephone headset. To do this is very complicated and methods are not standardised. In practice, therefore, the output is measured either as power, in a power meter, or as a voltage. Output power should be measured in a non-inductive resistance of ohmic value equal to the effective speech coil impedance at 400 Hz.

Standard output power for loudspeaker sets is usually 50 mW, but where the output power is great the standard may be 500 mW. For headphone receivers the figure may be 0·1 mW: if the output is to be taken to a landline or, if the headphone operation is to take place in a noisy situation, the higher standard of 1 mW may be adopted.

12.1. RECEIVER MEASUREMENTS

12.1.1. Sensitivity

For equipments of relatively low gain, such as broadcast receivers, the sensitivity is determined by applying from the signal generator, through the dummy aerial, sufficient input modulated 30% at 400 Hz to give standard output from the receiver. If this level brings the a.g.c. into operation, this should be disconnected without reducing the sensitivity.

The manual s.f. and a.f. gain controls should be at maximum. The sensitivity may be defined as the input voltage. Note that the lower the figure the better the sensitivity.

For receivers of high gain, and this includes communication-type receivers, noise must be considered, because with all controls at maximum a sensitive receiver may deliver 50 mW of noise without any signal input at all. This would, on the face of it, set the sensitivity at zero volts for standard output—a quite meaningless figure. In these circumstances sensitivity may be determined by finding the input signal, modulated 30% at 400 Hz, which is required to give an output signal-to-noise ratio of at least a certain minimum value. The ratios usually specified are 15 dB for good speech intelligibility, and 40 dB for entertainment programmes. For the standard output value of 50 mW the signal-to-noise ratio of 15 dB is obtained when the noise output is 1·6 mW. The measuring procedure is, therefore, as follows.

The signal generator input is adjusted so that the a.g.c. comes into operation, and the a.f. gain is set at a value which gives 50-mW output. Then, with the a.g.c. off, an input of a level approximately equal to the sensitivity is applied and the s.f. and i.f. gain controls adjusted for 50-mW output. The modulation is switched off, but the signal generator carrier left at unchanged level, and the noise measured. If the noise level happens to be 1·6 mW, the value of the signal generator input is the sensitivity. If, as is probable, the noise level is not 1·6 mW, the signal generator output and the s.f. and i.f. gain controls are adjusted until the two required output values are obtained, i.e. until the output with the signal generator modulation on is 50 mW, and the output with the modulation off (but with the carrier on) is 1·6 mW. The settings of the s.f., i.f., and a.f. gain controls should be noted as they are used for other tests.

Measurements of sensitivity are usually made at three or four frequencies in each range. With communication-type receivers the sensitivity variation between one part of a range and another is not considerable.

For narrow bandwidth c.w. receivers, which give little power output with a 400 Hz-modulated signal, special methods need to be applied—the output, resulting from a c.w. input, normally being measured as a voltage at the detector stage.

Representative sensitivity values for good communication-type receivers are as follows, in each instance for 50-mW output:

1. For m.f./h.f. amplitude-modulated reception: better than 3 μV for 15 dB signal-to-noise ratio.
2. For v.h.f. amplitude-modulated reception: better than 5 μV for 15 dB signal-to-noise ratio.
3. For u.h.f. amplitude- or frequency-modulated reception: better than 10 μV for 15 dB signal-to-noise ratio.

It should be noted that when the noise level is low, hum may render the results of dubious value. In such circumstances a filter, cutting off below about 200 Hz, should be used to eliminate the hum.

12.1.2. Noise factor

Noise factor has been referred to in Section 2.1.4. For communication receivers, values range between about 3 dB and 15 dB (the lower the value the better). Values of some 3–5 dB may be obtained at the lower frequencies (say up to about 30 MHz), and the higher figures in the v.h.f. and u.h.f. ranges.

To measure the noise factor the detector must be supplied with a carrier of the proper frequency and of level sufficient to give linear operation. The a.g.c. should be off and the gain controls set to the positions determined during the sensitivity tests.

A noise generator, consisting basically of a temperature-limited diode (i.e. one operating with an anode voltage high enough to draw off all the electrons emitted by the cathode), is connected to the receiver input. First, however, with the noise generator delivering no output, the receiver output noise (P_0, say) is measured. The noise generator is then switched on and adjusted to increase the noise output to three times the value due to the receiver alone (i.e. to $3P_0$). If the noise generator consists of a temperature-limited diode shunted by a resistance R the noise factor is $0{\cdot}02IR$, where I is the mean diode anode current in milliamperes.

Alternatively, to avoid instrument errors due to waveform, an attenuator may be introduced between the receiver and the output power meter. With the attenuator set at zero, and the noise generator off, the receiver internal noise output (P_0) is measured. The attenuator is then set to provide 4·8-dB attenuation, and the noise generator switched on and adjusted to bring the noise power registered on the output meter to the same value as before (i.e. to P_0). The noise factor is again given by $0{\cdot}02IR$.

If adjustment of the noise generator fails to increase the noise level three times, it should be adjusted to give an increase to twice the internal noise value. The noise factor is now given by $0{\cdot}01IR$. Alternatively, if the attenuator is used, it should be set, for the second reading, to provide 3 dB attenuation.

Like sensitivity, the noise factor values are generally determined at three or four frequency values in each frequency range. However, there is generally not much variation between one end of a range and the other.

12.1.3. Selectivity

For c.w. transmissions a considerable degree of selection may be effected in the audio-frequency stages, particularly by judicious use of the beat-frequency oscillator to yield a well defined note from the wanted station and a near inaudible one from a closely adjacent interfering station. Apart from this possibility, the selectivity (as distinct from the elimination of second channel interference) in a superheterodyne receiver is mainly derived from the intermediate-frequency amplifier(s).

To measure the selectivity, therefore, we may concentrate on the i.f. stages.

In fact, the operation of a receiver when tuned to a wanted station is influenced by the presence of unwanted signals. In order to reduce the complexities to a comprehensible, if undue, simplicity, however, it is not uncommon to use one signal only in selectivity measurements, and to ignore the complications of interaction between wanted and unwanted signals. This is not altogether unreasonable because the magnitude of possible additional troubles which may arise due to interaction of signals depends on the results obtained in the absence of an interfering signal and can be deduced with reasonable accuracy by an experienced interpreter of test results from the single signal values.

With the a.g.c. off the selectivity is measured by applying an unmodulated signal to the frequency changer input. A number of readings is taken over the required range of frequencies—perhaps 20 kHz to each side of the intermediate frequency. For each frequency the input is adjusted to give the same voltage output at the receiver detector and the selectivity curve is plotted as a relative response in decibels versus frequency difference from the middle of the band.

Communication-type receivers often have a number of different selectivities available, and a curve should be plotted for each.

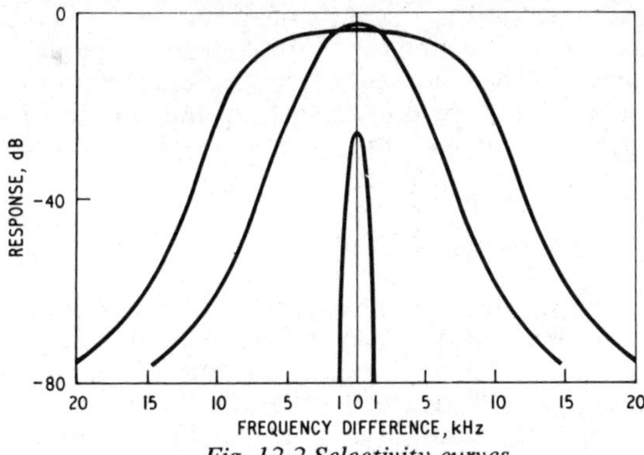

Fig. 12.2 Selectivity curves

For receivers which are not very selective a modulated input may be used and the input level adjusted, and noted, for constant output power at the different frequencies of input. The relative output power in decibels may then be plotted against frequency difference from the middle of the band. Typical curves are shown in Fig. 12.2 for three different bandwidths.

Other representative figures are as follows:

1. For h.f./m.f. and 6-dB bandwidth: 6 kHz for a.m., 3 kHz for s.s.b., 1·3 kHz for c.w., 50 Hz for very narrow band (crystal filter).
2. For v.h.f. and 6-dB bandwidth: 30 kHz for a.m./c.w., 80 kHz for f.m. (narrow bandwidth), 300 kHz for f.m. (wide bandwidth).
3. 40-dB bandwidth: 200 kHz for a.m./c.w., 320 kHz for f.m. (narrow bandwidth), 700 kHz for f.m. (wide bandwidth).

12.1.4. Image channel rejection ratio

The signal-frequency amplifiers may not contribute much to the adjacent channel selectivity of a receiver but they are required to reject image-frequency signals.

The performance of a receiver in respect of image channel rejection is specified as the ratio, expressed in decibels, of the wanted signal to the unwanted image channel transmission. The measurement is effected by first determining the sensitivity of the receiver in the usual way. The signal generator carrier frequency is then altered to the image frequency and, with all other controls unaltered, the sensitivity is again determined. The ratio of the two sensitivities so measured, expressed in decibels, is the required image rejection ratio. Three or four values are usually obtained in each wave range.

A typical value for the ratio at m.f. is better than 75 dB. At v.h.f. and u.h.f. smaller values are obtained, perhaps 20 dB at the top of the tuning range and 40 dB at the lower frequencies.

If desired, adjacent channel selectivity may be determined in a similar way.

If two signal generators are used, one tuned to the wanted frequency is first used alone and the output noted. The second generator is then switched on and tuned to the image channel, leaving the first generator untouched (i.e. switched on and tuned to the wanted frequency). The output from the image channel generator is increased until the output is doubled. The ratio, expressed in decibels, of the outputs of the two generators is again the image channel rejection ratio.

12.1.5. Automatic gain control

To measure the effectiveness of the a.g.c. system of a receiver a graph of output power versus input signal voltage is needed. The signal generator, often adjusted to about the mid-frequency of one of the frequency ranges and modulated 30% at 400 Hz, is connected to the receiver input through an appropriate resistor or dummy aerial in the usual way. The a.g.c. is switched on and the s.f. and i.f. gain controls set to maximum. The a.f. gain control is adjusted so as to prevent overloading when the maximum signal input is applied. If there is a choice, an average bandwidth setting, say 5 kHz, is used. If hum is troublesome to the extent of unduly limiting the drop in noise power with increasing input, the 200 Hz high-pass filter is inserted.

The input signal is increased in steps from 1 μV to the maximum desired level, and the output power recorded for each value. The noise output power (that is the output power with the modulation switched off) is also recorded for each input voltage. The readings are repeated at one or two (middle) frequencies in other frequency ranges. Fig. 12.3 shows one typical set of results (the noise level falls with increase of signal because of the reducing gain caused by a.g.c. action).

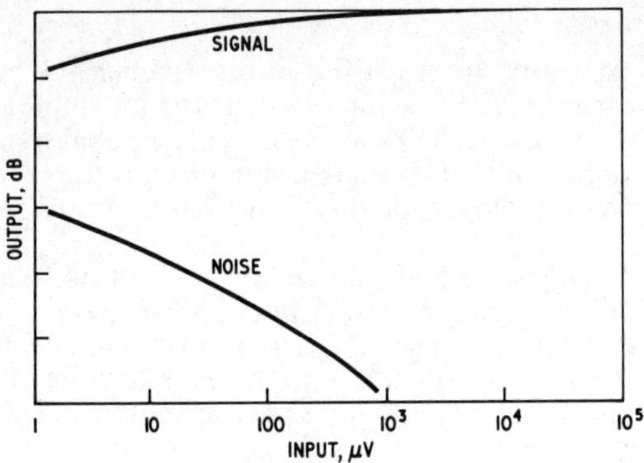

Fig. 12.3 Automatic gain control

12.2. RECEIVER ALIGNMENT

By *alignment* of a receiver (or of a transmitter) is meant the adjustment of pre-set controls of the tuned circuits, using signal generator and indicating instrument (e.g. electronic voltmeter or output meter) to obtain optimum, or specified, response over the appropriate tuning range.

12.2.1. Intermediate-frequency amplifier

The i.f. tuned circuits may need to be *stagger tuned,* that is tuned to slightly different frequencies in order to give a wide bandwidth, or they may all be tuned to the same frequency. In either case the required frequencies, for a given receiver, are known.

The signal generator is tuned to the frequency appropriate to the circuit which is being tuned. The generator output level is always adjusted to the minimum necessary for useful readings to be obtained at the receiver output. The output may be measured in one of a variety of ways as follows. For the first method the input signal is unmodulated and the receiver a.g.c. switched on, for the other methods the input is modulated and the receiver a.g.c. is switched off.

1. By the use, if fitted in the receiver, of a tuning meter.
2. By the use of an electronic voltmeter to measure the output voltage from the detector.
3. By the use of an a.c. voltmeter to read the voltage at the output of the last amplifier.
4. By the use of a power output meter—substituted for the loudspeaker.

The procedure is as follows:

The signal generator output at the correct frequency is applied through a small value capacitance to the input of the last i.f. amplifier. The secondary and afterwards the primary of the associated transformer are tuned for maximum output. If there is not a transformer, but only a single tuned circuit at this stage, then this circuit is tuned for maximum output.

The process is repeated, reducing the signal generator output as necessary, with the signal generator input applied to the next to the last i.f. stage to enable the secondary and primary circuits of the associated i.f. transformer (or the single *LC* circuit) to be tuned for maximum output at the appropriate signal generator frequency or frequencies, and so on.

For the first i.f. tuned circuit the signal generator is applied to the frequency changer input with the local oscillator stopped (e.g. with the output short circuited). If the signal generator is likely to be considerably damped by the preceding tuned circuit, which is resonant at a frequency quite different from the intermediate frequency, then it may be necessary either to raise the signal generator output level considerably, or to temporarily disconnect the preceding circuit.

For receivers with crystal filters the filter should be out of circuit while the foregoing procedure is carried out. When the alignment is completed, the filter is switched in and the signal generator frequency readjusted for maximum output. For this the signal generator output should be unmodulated, because the crystal selectivity cuts the side frequencies, and the tuning meter used. If a tuning meter is not available, an unmodulated input is used and the beat-frequency oscillator switched on to provide a modulated output. The i.f. tuners are then readjusted for optimum output.

12.2.2. Oscillator

For the oscillator the procedure may be as follows:

1. Switch the receiver to the desired range and tune to a specific frequency near to the low-frequency end of the range.
2. Tune the signal generator to the same frequency and connect the signal generator output to the receiver input in the usual way.
3. Adjust the oscillator inductor for maximum output.
4. Tune the receiver to a specific frequency near to the high-frequency end of the range.
5. Tune the signal generator to the same frequency.
6. Adjust the oscillator trimmer for maximum receiver output.
7. Return to the frequency selected in (1) above and make any necessary slight readjustments of the oscillator inductor.
8. Return to the frequency selected in (4) and make any necessary slight readjustment of the oscillator trimmer.
9. This to-and-fro adjustment must be continued until no further readjustment is found to be necessary at either end of the range.

12.2.3. Signal-frequency amplifier

With the signal generator and the receiver tuned towards the high-frequency end of the range, first the mixer trimmer and then the signal-frequency circuit trimmers are adjusted for maximum output from the receiver. All being well, the low-frequency end of the band will now be found to be properly tuned without further adjustment, but this must be checked by varying the trimmers (always bearing in mind that tracking is not continuously perfect). If the low-frequency end alignment is found to be wrong, the oscillator circuit tuning should be checked and if necessary the process re-started from this point.

12.3. USE OF CATHODE RAY OSCILLOSCOPE FOR THE DISPLAY OF RECEIVER AND TRANSMITTER PERFORMANCE

A cathode ray oscilloscope may be used as an electronic voltmeter by applying the voltage to be measured to the *Y* input with the time base switched off. The display on the tube face then consists simply of a vertical line. The length of the line is proportional to the input voltage. Using a graticule—a transparent screen with engraved equidistant lines—the voltage magnitude may be closely estimated, given an initial reliable calibration. Similarly, with the time base switched on, the peak-to-peak value of an alternating waveform may be determined.

Alternatively a calibrated decade attenuator may be used in series with the vertical deflection input terminals. Unknown voltages are determined by noting the amount of attenuation necessary to bring the deflection to a standard value, the voltage corresponding to which is precisely known. This is a particularly convenient method when conducting receiver gain tests.

12.3.1. Preliminary checks

Before using an oscilloscope, as for any other test apparatus, its performance must be checked. The linearity of the time bases may be verified by viewing a trace with marker pips at evenly spaced time intervals: when necessary the calibration may also be checked. For less accurate indication of departure from linearity a known waveshape (e.g. 50-Hz supply mains) may be viewed.

The linearity, and if necessary the calibration, of the vertical deflection should be checked against a suitable voltmeter of known accuracy. When the linearity or calibration is suspect, one or two known amplitudes of trace may be used as standard and all signals reduced to this level by use of a calibrated variable attenuator.

A useful check of the *X* and *Y* circuits of the oscilloscope is to apply a signal to both *X* and *Y* inputs (Fig. 12.5 (a)). If the resulting trace is a straight line it may be assumed that the circuits introduce no distortion or phase shift. Imperfections in the oscilloscope circuits result in the trace either showing departure from linearity or, because of phase shift between X and Y voltages, becoming elliptical.

The input capacitance of an oscilloscope may be unduly high for the tests which are to be carried out. For instance, an oscilloscope connected across a tuned circuit may appreciably detune the circuit and so alter the performance. To reduce the input capacitance an arrangement similar to that shown in Fig. 12.4 may be used.

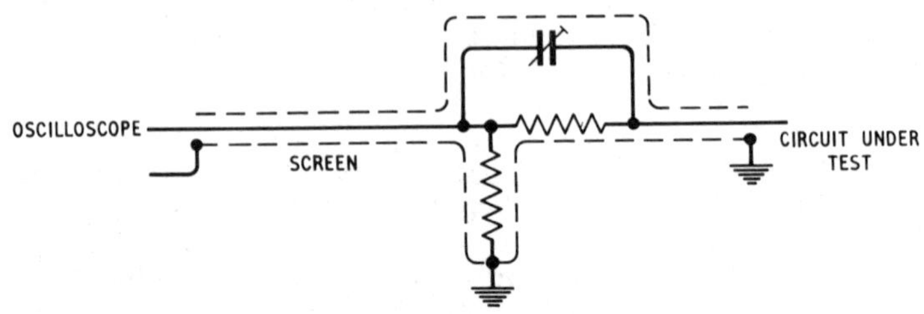

Fig. 12.4 Reducing the effective input capacitance of a c.r.o.

The signal input to the oscilloscope is reduced in the same ratio as the capacitance. To simplify the operation, particularly in conjunction with decade attenuators, it may be convenient to employ a reduction ratio of 10.

If the oscilloscope is to be used along with other test equipment, for example with a signal generator, to determine some aspect of receiver performance, say linearity, it is essential to check that the test equipment itself is entirely satisfactory, perhaps by viewing the appropriate trace on the oscilloscope screen before including the apparatus to be tested.

12.3.2. Amplifier distortion

Having checked the linearity of the oscilloscope horizontal and vertical deflection circuits, and taking care that they are not overloaded, the oscilloscope may be used to investigate amplifier stages for distortion by comparing the traces of (a) a signal applied at the amplifier input, and (b) the output signal from the amplifier. This may most conveniently be done by using a double beam oscilloscope. If the traces are superimposed, an even closer comparison may be made.

Applying the output of an oscillator to both the X- and Y-deflection circuits (Fig. 12.5(a)) results in a straight line trace on the screen. The angle which the line makes with the horizontal depends (for electrostatic deflection) on the relative voltages applied to the X- and Y-deflection systems. Similar considerations apply for electromagnetic deflection systems.

Interposing, as shown in Fig. 12.5(b), the amplifier stages under test (operating at normal power and terminated in the proper load) in

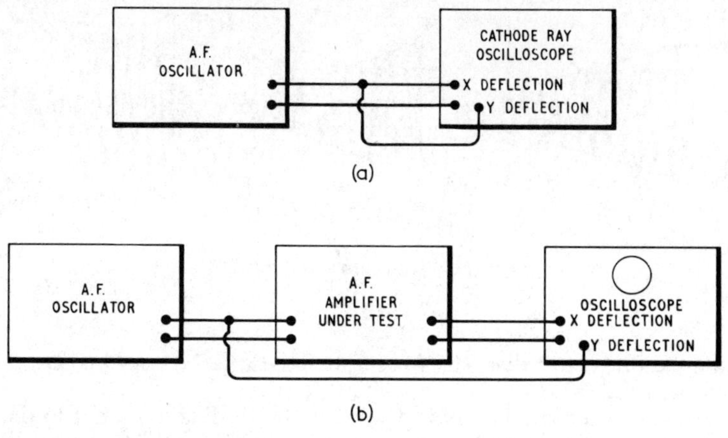

Fig. 12.5 Checking the linearity of (a) apparatus and (b) an amplifier under test

the leads to one pair of oscilloscope deflection terminals, and with suitable attenuation to equalise approximately the voltages applied to the horizontal and vertical deflection systems, the oscilloscope trace may be compared with the original. If the linearity of the trace is substantially unchanged, there is no undue distortion in the amplifier. The extent to which the trace linearity is degraded is a measure of the distortion introduced by the amplifier. If the trace becomes elliptical, it indicates phase shift between the X- and Y-deflection voltages.

12.3.3. Modulation tests

Applying the modulated carrier signal of a transmitter to the Y-deflection input, and the modulating waveform signal to the X-deflection input of a cathode ray oscilloscope we obtain a trapezium-shaped trace, filled with r.f. as shown in Fig. 12.6. The pattern is independent of the modulating waveform. If there is no phase change between the modulator and the carrier, the sloping sides of the trapezium are straight. If there is a phase change, the lines form an ellipse. The percentage modulation is given by:

$$100 \times \frac{V_1 - V_2}{V_1 + V_2}$$

A flattened top to the trapezium indicates that over this range increases of modulation do not result in any increase in r.f. output. For 100% modulation the trapezium becomes a triangle (Fig. 12.6(b)), $V_2 = 0$. For

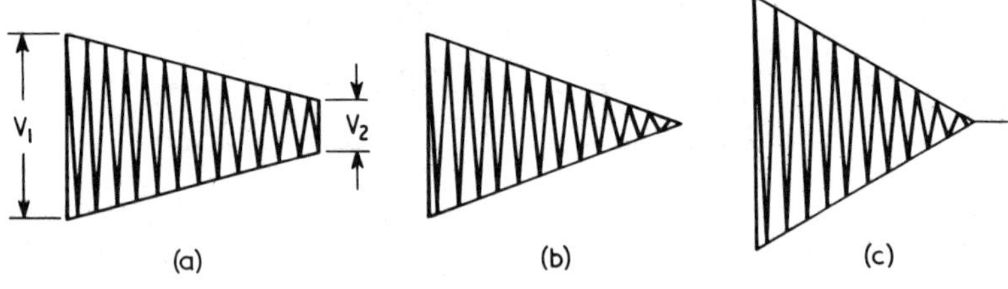

Fig. 12.6 Transmitter modulation

modulation greater than 100% (over modulation) the trace becomes that shown in Fig. 12.6(c).

To detect second-harmonic distortion the modulation input to the oscilloscope may be reduced to zero—to leave a single vertical line on the screen—and then steadily increased. If the trapezium expands unequally to each side of the vertical line, the indication is that there is second-harmonic distortion.

The presence of third-harmonic distortion is usually shown by a line of increasing brilliance across the trace.

12.3.4. Use of frequency-modulated oscillator in conjunction with an oscilloscope

A *frequency-modulated oscillator* (often called a *sweep oscillator,* or a *wobbulator*) is a signal generator the output of which may be swept linearly over a known and controlled band of frequencies. The amplitude of the output remains constant at a pre-determined value over the swept frequency range.

The use of such a frequency-modulated generator in conjunction with a cathode ray oscilloscope, the timebase of which is synchronised with the change of frequency of the oscillator, enables rapid investigation to be made of frequency/response characteristics.

Fig. 12.7 shows a block diagram of an arrangement for investigating the response curve of an i.f. amplifier. The a.g.c. circuit should be rendered inoperative.

For all the tests using an oscilloscope, a resistor of value about 0·1 MΩ should normally be included in the lead to the oscilloscope as shown in the figure.

The oscillator frequency, centred on the intermediate frequency (which we shall for the time being assume to be 470 kHz), sweeps linearly over the desired range, from say, 455 kHz to 485 kHz, thirty or so times per second. The sweep occurs during the time intervals t_0 to t_1, t_2 to t_3, etc as shown in the figure. At the end of each sweep the frequency is rapidly returned to the value 455 kHz (this occurs during the time intervals t_1 to

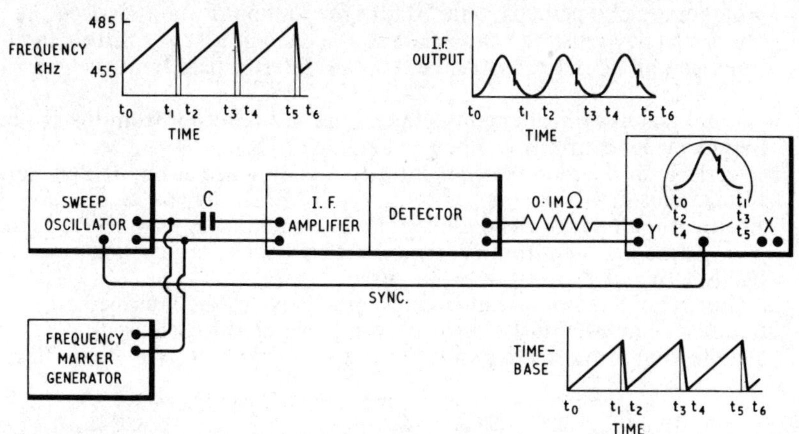

Fig 12.7 Use of sweep oscillator, frequency marker and oscilloscope to determine circuit response curve

t_2, t_3 to t_4, etc. in the figure). The time base sweep is in synchronism with the oscillator frequency sweep. In fact the same voltage may be used both for the time-base and to determine the frequency change of the oscillator.

The output of the i.f. amplifier, after detection, varies with the input frequency in accordance with the overall frequency response curve of the amplifier tuned circuits. The trace drawn on the cathode ray tube face is thus the response curve of the amplifier.

It is useful, and for good accuracy essential, to provide frequency marker points on the curve so that the exact centre frequency and bandwidth may be known. The marker signals may be provided by the application of known frequencies from a signal generator. The point of connection may be the same terminal as that at which the sweep oscillator signal is injected. The small deflection of the trace which is shown on the diagrams is the marker signal. Alternatively, marker signals may be applied as narrow pulses to the c.r.t. grid or cathode to produce a brief black out or a brightening of the trace at known frequencies.

The sweep oscillator and marker signals are applied to the circuit under test through a small capacitor (*C* in the figure).

With the response curve clearly visible on the c.r.t. screen the effect of varying the tuning at any part of the amplifier can immediately be seen and the circuits, therefore, rapidly aligned. For amplifiers which are badly off tune, the sweep frequency generator is applied initially to the last stage so that the associated tuned circuit(s) may be properly trimmed. The generator output is then transferred to the preceding stage and the corresponding circuits trimmed, and so on.

For frequency-modulation receivers the required bandwidth of the sweep oscillator needs to be about 0·25 MHz for entertainment programmes and perhaps some 50 kHz for communication services. The frequency must be appropriate to the band in use–for British and American broadcast receivers the standard intermediate frequency is 10·7 MHz.

For receivers which employ limiters the signal output from the sweep frequency generator should be great enough to bring the limiter effectively into action. Since rectification takes place at the limiter the voltage thus developed (across *AB* in Fig. 12.8(a)) may be used as the *Y*-deflection input to the oscilloscope. The circuits are trimmed so as to give the maximum output consistent with obtaining the desired bandwidth and shape of response curve.

For ratio-detector circuits the response curve of the intermediate-frequency amplifier is determined by connecting the oscilloscope *Y*-deflection terminals to point *A* in Fig. 12.8(b) and to the chassis line.

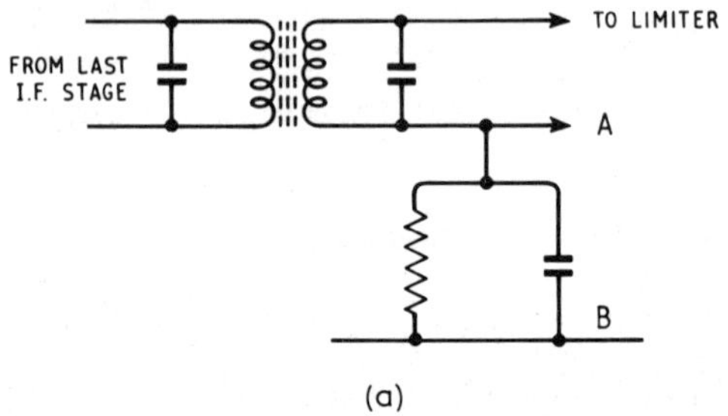

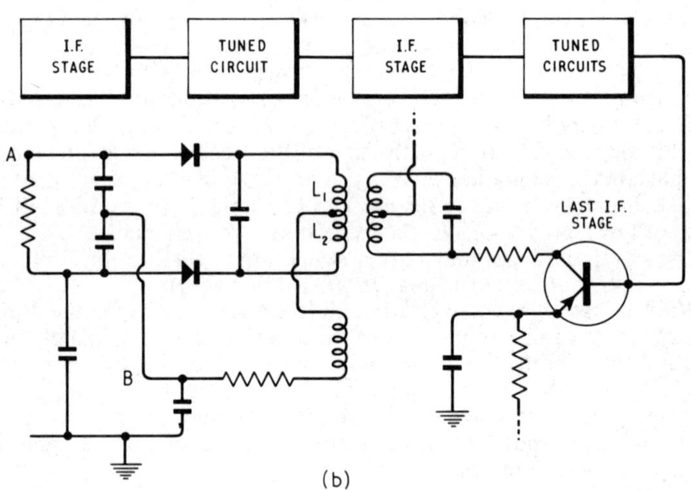

Fig. 12.8 F.M. receiver: alignment of i.f. stages

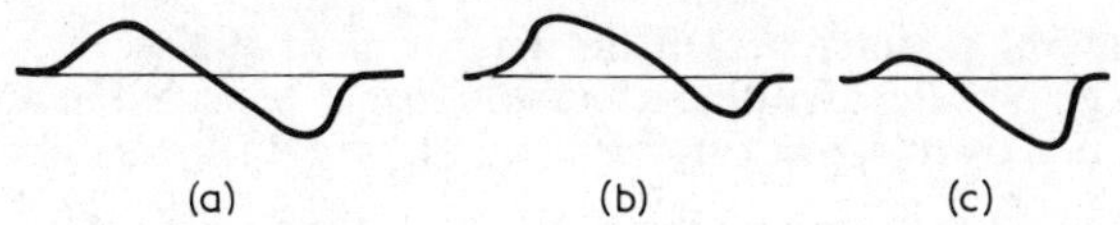

Fig. 12.9 F.M. receiver: ratio detector S *curves*

The sweep oscillator and the frequency marker generator are connected as usual through a small capacitor to the input of the i.f. stage concerned. The previously described procedure is then carried out.

The *S*-curve of the detector is checked by connecting the oscilloscope *Y*-deflection terminals between *B* in the figure and the chassis line. The sweep oscillator and the frequency marker generator are connected between the i.f. amplifier input and the chassis line. L_1 and L_2 are trimmed to give a symmetrical *S* curve like that of Fig. 12.9(a) (and not like that of Fig. 12.9(b) or Fig. 12.9(c)) centred on the intermediate frequency with the straight part occupying the correct bandwidth.

QUESTIONS

1. Why does the procedure adopted for measuring the sensitivity of low-gain receivers differ from that applied for a high-gain receiver?

Describe in outline how the sensitivity of a communication-type receiver could be determined (a) at normal bandwidth, (b) at narrow bandwidth.

2. Explain clearly what is meant by the term *noise factor* when used in specifying the performance of a radio receiver. State typical values of noise factor for communication receivers for the h.f., v.h.f. and u.h.f. ranges. Does the noise factor value usually vary much within a given range?

3. What do you understand by the term *selectivity* as applied to a radio receiver?

Explain briefly what are the selectivity needs in the different sections of a radio receiver.

Give some typical figures which indicate the measure of selectivity to be expected for receivers used for different services in several frequency bands.

4. Define the term *image channel rejection ratio.*

Why is it important that a receiver should perform well in terms of image channel rejection?

Describe how the image channel rejection ratio of a receiver may be found using (a) one signal generator, and (b) two signal generators.

5. (a) Explain why the presence of hum may be troublesome in carrying out receiver sensitivity and a.g.c. measurements. What may be done to avoid the trouble?

(b) How is a quantitative measure of the adjacent channel selectivity of a receiver defined?

6. How may the effectiveness of the a.g.c. system of a receiver be illustrated graphically? Outline a suitable test to provide the necessary information. Sketch typical graphs of results for different types of a.g.c.

7. (a) Describe a method by which the oscillator stage of a radio receiver may be aligned.
(b) Describe concisely a test to determine the selectivity of a radio receiver.

8. Describe how to check the linearity of (a) the time base, and (b) the vertical deflection of a cathode ray oscilloscope.
How could a cathode ray oscilloscope be used to investigate the amount of distortion introduced in an amplifier?

9. Explain concisely the function of a frequency-modulated oscillator used in conjunction with a cathode ray oscilloscope for test purposes.
Describe with the aid of sketches how the pair of instruments may be used to determine the response curves of stages of an i.f. amplifier.

10. (a) Describe how the noise factor of an item of equipment may be determined.
(b) Explain how a sweep oscillator and cathode ray oscilloscope may be used to check the response curve of (a) the i.f. stages, and (b) the ratio detector, of an f.m. receiver.

ANSWERS

Chapter 1:

(3) 11·7 MHz;
(4) 14·14 MHz;
(7) 79 m.

Chapter 2:

(5) 100: 1, 3 (or 5 dB);
(7) 17·6 MHz, 19·2 MHz.

Chapter 5:

(1) 78%;
(2) (a) 1·5 A, (b) 1 100 V, −31·4 V, (c) 324 W, (d) 64·8%;
(4) 7.85 kΩ, 204 mW, 314 mW, 111 mW;
(5) 40 kΩ, 5 W, 22·5%;
(6) 62·5 W, 95·4 W, 65·5%;
(7) 69·5 pF;
(8) 10·1, 314 Ω;
(10) 2 pF.

Chapter 6:

(1) 41·67 W;
(3) (a) 0·6, (b) 470 mA, (c) 0·32 kW, (d) 1·74 kW.

Chapter 7:

(2) 16 kΩ, 4 kΩ;
(3) 50·5 Ω or 3·5 Ω;
(7) 600 W, 525 W, 75 W;
(8) 43·25;
(10) 30 μH, 2·45 kΩ.

Chapter 8:

(1) (a) Doubled, (b) halved, (c) doubled, (d) doubled;
(2) 93 015 kHz, 92 985 kHz, 30 kHz;
(3) 2 kHz, 71·06 W;
(4) 40 kHz;
(10) 105·6 MHz, 15 kHz.

Chapter 9:

(2) (a) 424 mV, 424 mV, zero,
(b) 491 mV, 344 mV, −73·5 mV,
(c) 344 mV, 491 mV, 73·5 mV;
(5) 144 Ω;
(8) 4 dB;
(9) 32 dB.

Chapter 10:

(3) 15 kW (average) for same signal output
7·5 kW (average) for same signal-to-noise ratio.

Chapter 11:

(2) 12 mV/m;
(3) −0·1 dB, −19·6 dB;
(4) 90°;
(5) 23°;
(8) 346 Ω.

Index